AF834728

Johnson

Practical Statistical Methods for Injection Molding

Bradley G. Johnson

Practical Statistical Methods for Injection Molding

Using Excel and Minitab

HANSER

The author: *Bradley G. Johnson, Penn State Behrend*

Print-ISBN: 978-1-56990-561-6
E-Book-ISBN: 978-1-56990-966-9

All information, procedures, and illustrations contained in this work have been compiled to the best of our knowledge and is believed to be true and accurate at the time of going to press. Nevertheless, errors and omissions are possible. Neither the authors, editors, nor publisher assume any responsibility for possible consequences of such errors or omissions. The information contained in this work is not associated with any obligation or guarantee of any kind. The authors, editors, and publisher accept no responsibility and do not assume any liability, consequential or otherwise, arising in any way from the use of this information – or any part thereof. Neither do the authors, editors, and publisher guarantee that the described processes, etc., are free of third party intellectual property rights. The reproduction of common names, trade names, product names, etc., in this work, even without special identification, does not justify the assumption that such names are to be considered free in the sense of trademark and brand protection legislation and may therefore be used by anyone.

The final determination of the suitability of any information for the use contemplated for a given application remains the sole responsibility of the user.

Bibliographic information of the German National Library:
The German National Library lists this publication in the German National Bibliography; detailed bibliographic data are available on the Internet at http://dnb.d-nb.de.

This work is protected by copyright.
All rights, including those of translation, reprint, and reproduction of the work, or parts thereof, are reserved. No part of this work may be reproduced in any form (photocopy, microfilm, or any other process) or processed, duplicated, transmitted, or distributed using electronic systems, even for the purpose of teaching – with the exception of the special cases mentioned in §§ 53, 54 UrhG (German Copyright Law) – without the written consent of the publisher.
No part of the work may be used for the purposes of text and data mining without the written consent of the publisher, in accordance with § 44b UrhG (German Copyright Law).

© 2026 Carl Hanser Verlag GmbH & Co. KG, Munich
Vilshofener Straße 10 | 81679 Munich | info@hanser.de
www.hanserpublications.com
www.hanser-fachbuch.de
Editor: Dr. Mark Smith
Production Management: Cornelia Speckmaier
Cover concept: Marc Müller-Bremer, www.rebranding.de, Munich
Cover design: Max Kostopoulos
Cover picture: © Max Kostopoulos
Typesetting: le-tex publishing services GmbH, Leipzig

Preface

My undergraduate and graduate education in chemical engineering provided only a cursory introduction to statistics. Shortly after entering the workforce in the mid-1980s, a QC technician at the plating operation where I held my first engineering position showed me how to construct control charts (Chapter 3) by hand. A few years later, while working at an injection molding operation, I attended workshops on statistical process control (SPC) (Chapter 3) and process capability (Chapter 4). As I developed injection molding processes and molds, I learned firsthand, often the hard way, the importance of obtaining accurate dimensional and physical property measurements. It was during this period that I was introduced to gage R&R studies (Chapter 5) and their role in understanding measurement systems.

After approximately three years at the injection molding operation, I attended a week-long workshop entitled Understanding Industrial Experimentation, taught by Don Wheeler. That workshop sparked my interest in design of experiments (DOE) (Chapters 7–9) and the practical application of statistical methods in manufacturing. When I returned, I enthusiastically began applying DOE techniques to better understand injection molding processes and, later, several post-molding operations. Not everyone on the production team shared my enthusiasm, and many of the lessons learned from those experiences helped shape the perspectives presented in this book.

Not long after implementing DOEs, I began investigating process monitoring techniques (Chapter 10) and invited Rod Groleau of RJG to visit our facility as we evaluated cavity pressure monitoring. Unbeknownst to me, his visit coincided with a quality meeting being held at our division headquarters in the same building. When the quality manager learned that Rod was onsite, he invited him to speak briefly to the group. That brief presentation generated significant interest in process monitoring and ultimately helped secure the funding needed to implement cavity pressure monitoring in our operation.

As discussed in Chapter 11, I learned injection molding on open-loop machines that relied on manual adjustment of pressure and velocity valves. I later transitioned to closed-loop control machines and implemented primarily D2 processes during my years in industry. The remaining processes discussed in Chapter 11 were ones that I later studied and taught after joining the Plastics Engineering Technology faculty at Penn State Behrend in 1994. Interestingly, the technical presentation I delivered during my faculty interview at Behrend was a modified version of a presentation I had previously given to an OEM describing a DOE that improved the sealing performance of an electrical connector.

I began using Excel for engineering calculations shortly after learning about DOE techniques. At the time, our quality department used specialized software for most statistical analyses, so my use of Excel remained focused on calculations and data organization. When I joined Penn State Behrend, Excel became one of my primary teaching tools for statistical methods.

My introduction to Minitab came in the early 1990s while taking a statistics course as part of an MBA program. That course also provided my first formal exposure to hypothesis testing (Chapter 6) and regression analysis, a topic that plays an important role in the DOE chapters. I later incorporated Minitab into my courses at Behrend, where it has remained an important instructional tool.

This book is a compilation of material developed over more than four decades of industrial practice, teaching, consulting, and workshop instruction. Much of the content originated from my undergraduate courses in Plastics Processing & Statistical Methods and Advanced Injection Molding, as well as from my industrial workshop, DOE for Injection Molding.

The goal of this book is not simply to explain statistical techniques, but to demonstrate how they can be applied effectively to understand, improve, and control manufacturing processes. Whether you are a student learning these concepts for the first time or an experienced practitioner looking for a reference and refresher, I hope the material provides both practical guidance and useful insight. Many former students have told me that they still keep these notes within easy reach and refer to them regularly. My hope is that this book will serve the same purpose for you.

Bradley G. Johnson
Summer 2026

Contents

1 Brief History and General Concepts

This book is about the use of design of experiments (DOE), statistical quality/process control (SQC/SPC), and other statistical methods that are used in the manufacture of plastic products, primarily injection molded parts. The main goal in using these methods is to make the manufacturer more competitive by increasing quality and reducing costs. It is important for the reader to understand that the techniques that will be introduced in this book are not the solutions to manufacturing problems; they are only tools and, as such, are only as good as the way in which they are applied. If a carpenter has the world's best hammer, it really doesn't make him/her a better carpenter if he/she doesn't know how to swing it and hit the nail. The intent of this book is not only to introduce the tools; it is also to help the reader "hit the nail" when using them.

1.1 What is Quality?

Quality has more than one definition. Two of the ways in which quality is generally defined are described here.

Having features which help meet a customer's need or a perceived need

If a product has better or more features than another product, it is usually described as being a higher quality product. Look at the pairs of items shown in Figure 1.1. Which are the higher quality products? Design features are added to products to make them more attractive to the customer and, hence, increase increase customer satisfaction. However, the reason that some customers still buy the lower quality items shown in Figure 1.1 is that they cost less. In the case of features, higher quality almost always costs more.

Figure 1.1 Similar products with different feature levels

Freedom from deficiencies

Although some of the tools introduced in this book will be useful to test out product features, the main definition of quality in a manufacturing environment is freedom from deficiencies. Any unacceptable product that a manufacturer makes is going to increase costs. Using this definition, there is a cost involved whenever a manufacturer makes poor quality product and, therefore, higher quality costs less. The following are the four categories that make up the cost of poor quality and lists some of the types of things included in each category:

1. ***Prevention Costs:***
   a. Quality planning & engineering
   b. New-product review
   c. Product/process design
   d. Process control
   e. Quality data acquisition & analysis
   f. Supplier quality evaluation
   g. Training
2. ***Appraisal Costs:***
   a. Incoming inspection and testing
   b. In-process inspection and testing
   c. Final inspection and testing
   d. Product quality audits

   e. Maintaining accuracy of testing equipment
   f. Inspection and testing of materials and services
   g. Evaluation of stock
3. ***Internal Failures Costs:***
   a. Scrap
   b. Rework
   c. Failure analysis
   d. Scrap and rework supplies
   e. One hundred percent sorting inspection
   f. Reinspection and retesting
   g. Downtime
   h. Yield loss
   i. Downgrading (off-spec'ing)
4. ***External Failure Costs:***
   a. Warranty charges
   b. Complaint adjustment
   c. Returned product/material
   d. Liability costs
   e. Lost business

For most manufacturing situations, the only one of the four categories of costs that is generally acceptable is the prevention costs. As the name implies, these are the costs that are needed to minimize the number of poor-quality products that are made. Unless the quantity to be produced is very small, the investment in prevention generally is worth the cost because of the reduction in the costs from the other three categories. In other words, the more often a bad product is made, the more often the product needs to be inspected and tested, the more often the product will have to be sorted, reworked, or scrapped, and the more often a bad product will get to the customer.

A couple of other definitions of quality that go along with the freedom from deficiency definition are 1) what has been called the modern definition of quality: Quality is inversely proportional to variability, and 2) what has been called the definition of "world-class quality": On-target with minimum variance.

If every product that was produced was exactly the same as every other product and they were made exactly as the designer intended them to be, then internal and external failure costs should disappear... assuming the designer was perfect. This, of course, is impossible (God doesn't even make snowflakes the same), so the most that can be hoped for is the minimization of variability.

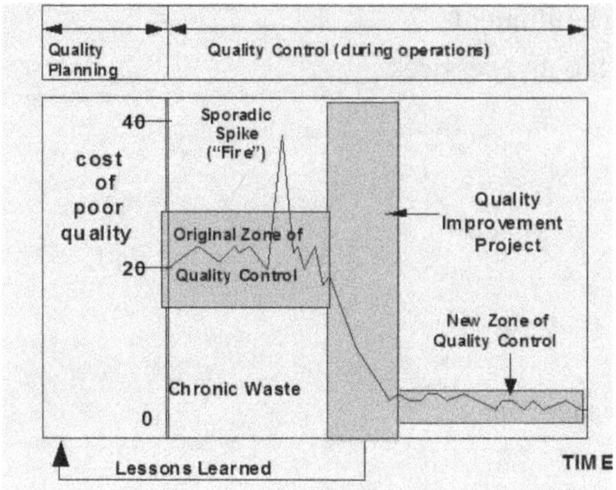

Figure 1.2
Juran's Quality Trilogy

Because of this inherent variability, once a product is released to the operating personnel and full production is begun, it cannot be expected that the cost of poor quality will be zero. Joseph Juran's Quality Trilogy, Figure 1.2, illustrates this trilogy. The first process in the trilogy is quality planning. These are the upfront design and engineering decisions that are made as a product moves from "art to part". Some things that are included here are material specification, product design, tool design, equipment design and set-up, and process design and set-up. Often, if designers do not have a good understanding of the processes and materials that they are designing products for, the initial cost of poor quality is high. For injection mold design, it is not uncommon for things like the gate location, cavity balance, cooling layout, ejection layout, and venting to be suboptimal and cause problems during production. It is also surprising how many processors do not have a standard methodology when setting up their processes. Chapter 11 goes through a recommended procedure for starting up a new injection process. In Figure 1.2, about 20% of the product made was deficient in some way. The best that the operating personnel could be expected to do is to maintain that 20% using the second process in the trilogy, quality control, and by putting out the "fires" that occur during operations. SPC is one of the major tools used for quality control (much more on this later). Examples of "fires" would be equipment malfunctions, tooling problems, and material contamination. To lower the cost of poor quality, the third process in the trilogy, quality improvement, is needed. The goal of quality improvement is to decrease the variability in the products. DOEs and some of the other methods discussed in this book are often used for quality improvement projects. Note that feedback on the results, lessons learned, from quality improvement projects should be passed back to those doing the quality planning so that the same problems are not repeated for the next product.

1.2 Overview & Timeline for SPC and DOE

This section will go through some of the history behind the use of statistical process control (SPC) and design of experiments (DOE). We will get into the technical details in later chapters, but a basic understanding of what each is and when each is primarily used will be helpful here. It will also be helpful to gain an understanding of when and why these techniques have been used in the past, to better understand how to apply them in the future.

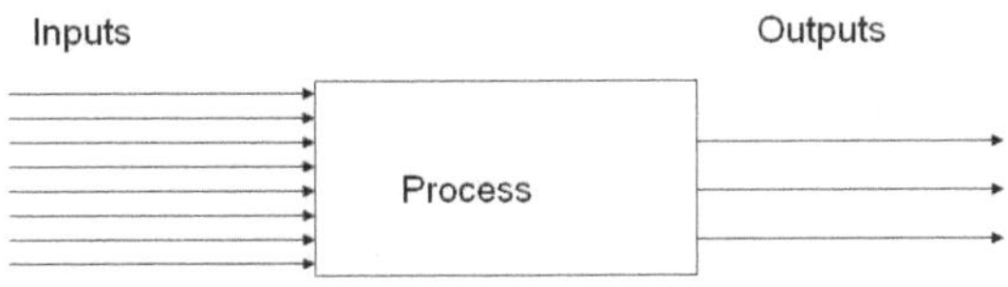

Figure 1.3
Process inputs and outputs

For any process that is used to make products, there are inputs to the process and outputs from the process (see Figure 1.3). The inputs are also called factors or process variables. Examples are temperatures, pressures, speeds, positions, material properties, and tooling variations. Outputs are the things that can be measured on the products that come out of the process. Examples are dimensions, physical properties, and aesthetic properties. As can be imagined, variation in the inputs will contribute to variation in the outputs. There are two primary methods used to study process and product variation: 1) informed observation and 2) directed experimentation.

Informed observation involves monitoring the process outputs and/or inputs to see what happens over time as various, unplanned events happen to the process. By studying the moon and the sun over time, eventually you will learn what conditions are necessary for an eclipse and be able to predict when it is going to happen. This is the method used with **SPC**. This technique increases the probability of catching an event that causes product problems before it does too much damage or before it does any damage at all. The product outputs of the process are what are typically monitored, but the process outputs can also be watched to see if their variation can be correlated with the outputs. Process outputs are dependent variables, such as cavity pressure, injection pressure, and screw position that vary with the combination of the inputs to the process. For example, the cushion (screw position) can change when the material changes, if the equipment wears, or if various input set points are changed.

Directed experimentation does not wait for the information to come; it goes out and gets it. Ben Franklin did not extract electricity from lightning by just watching an electric storm. He purposely interfered with the environment by flying a kite, with key attached, to make something happen. A **DOE** purposely makes large changes to a process. Systematic changes are made to the inputs or factors so that the effect on the outputs or responses can be observed.

The study of process variation really goes back to the beginning of mass production. One of the first attempts at mass production was made by Eli Whitney. In the late 1790s, he was contracted by the US government to produce 10,000 muskets. Up until that time, just about everything was made by a craftsman and one assembly at a time was done. The fifty components of the musket would be hand-fitted (filed, bent, and whittled) until everything fit together nicely. Whitney had stated that low-cost, high-quality muskets could be made by making interchangeable parts. Whitney did understand the benefits of interchangeable parts, but he didn't completely understand, and never did conquer, the inherent variability that is associated with manufacturing. Only 14 out of the first 700 of his muskets were successfully assembled and much customization and reworking was required to produce the 10,000 muskets.

About a hundred years after Whitney's experiment, two other important developments occurred that would later prove important to the development of both SPC and DOE methods. In 1893, a French mathematician by the name of Jacques Hadamard published a classic proof about complex matrices. The gist of the proof was that a small subset of a large matrix of numbers could be used to obtain most of the solution for the large matrix. This information was a basic step needed for many of today's DOEs, but would not be applied to DOEs for several decades. Around the same time, early 1900s, Frederick Taylor was refining his method of scientific management. He systematically studied every step of the process to optimize the movements and the time that it took to complete the tasks. His documentation of processes was probably the first attempt at a quality control system and was an attempt to improve throughput and quality by minimizing variation in the process.

World War I (1914–1918) provided an important "lesson learned" for US manufacturing. This was America's first experience of supplying vital materials to remote locations in large volumes. Because of the number of products, mass production was utilized, and 100% inspection was impractical. As a result, many defective products were discovered on the battlefield which, undoubtedly, led to the loss of American lives.

The 1920s brought two more important events. The first was Walter Shewart's application of the bell curve to manufacturing. Shewart, a physicist at Bell labs, was working on developing a standard-sized headset for the US military. To do this, he measured a large sample of head widths. Figure 1.4 shows the distribution of head widths that he found. He had seen this same distribution when studying Brownian motion and thought that it was more than a coincidence that head sizes and Brownian particles shared a common distribution. Upon measuring many manufactured objects, he determined that the bell-shaped curve was almost always what was found. Shewart used this discovery to develop Average and Range ($\overline{x}$, R) charts to monitor production processes and, thus, became the father of SPC.

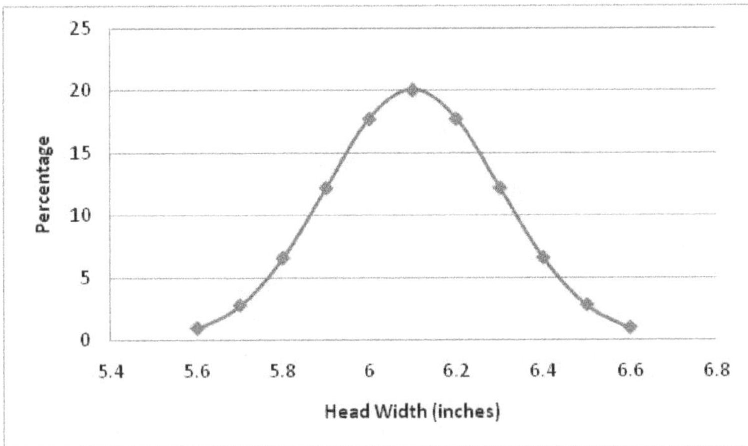

Figure 1.4 Shewart's head measurements (the bell curve)

The second important happening in the 1920s was Sir Ronald A. Fisher's use of the matrices developed by Hadamard to study agricultural experiments. Fisher was a mathematician and geneticist in England. The output from his agricultural experiments was the yield or how much crop he could get from a plot of land of given size. The inputs that he studied were many: seed varieties, watering conditions, soil type, plow type, fertilizers, pesticides, etc. Different blocks of land were prepared and cared for differently. The techniques that Fisher developed to study the data generated by these DOEs are still commonly used today.

The war materials sent overseas during World War II (1939–1945) were much more reliable than those sent during World War I. This was, in large part, due to the mandate by the war department that SPC be used in the making of all war materials. This was the first widespread implementation of SPC and it proved quite successful. Edward Deming, a colleague of Shewart's, was one of the trainers commissioned to work with US manufacturers during the war. It should be noted that DOEs were used in the war effort also. Plackett & Burman, two British scientists, used them in the development of proximity fuses for bombs.

DOEs continued to be used and developed, mainly in England, after World War II. George Box, one of the most influential statisticians of the 20th century, used DOEs to optimize processes at Imperial Chemical Industries (ICI) in the 1950s. With chemical processes, input factors such as time, temperature, pressure, concentration, flow rate, and agitation could easily be manipulated to optimize output. Box moved to the US in 1960 and created the Department of Statistics at the University of Wisconsin-Madison. The following quote attributed to him is a good indication of how important he believed DOEs are as a tool for quality and productivity improvement:

> *"If we could begin by just getting engineers to run a simple design, this will usually whet their appetite for more. For there are hundreds of thousands of engineers in this country, and even if the 2^3 was the only design they ever used, and even if the only method of analysis that was employed was to eyeball the data, this alone could have an enormous impact on the experimental efficiency, the rate of innovation, and the competitive position of this country."*

What a 2^3 design is will be discussed later in this book, but suffice it to say here that it is a fairly simple DOE to carry out and analyze. This quote also shows that Box understood that statistics are a tool to help the engineer analyze data and most of the time sophisticated analysis isn't necessary; "eyeballing" the data will usually give the answers required, once one has a trained "eye".

A very important development after World War II took place in Japan. They had been devastated by the war. They wanted to rebuild their manufacturing base, with the goal of becoming an economic power since they could no longer be a military power. Japan began to seriously focus on improving the quality of their products. At that time, "made in Japan" was synonymous with low quality, "cheap" products. Japanese teams went abroad to learn how other companies managed quality. This author has heard numerous accounts of how these teams came into US manufacturing plants during the 1950s and 1960s, with their cameras, and "stole" US technology. In fact, at this time, the US companies were having no problem selling everything they made and really had no fear of losing business to the Japanese. Because of this, the focus of the US manufacturer was on increasing production, not quality.

Practitioners of quality management were also invited to Japan to speak on the topic. Two notable ones, mentioned earlier, were Edward Deming and Joseph Juran. The US manufacturers looked at the quality methods that they were forced to implement during the war as a burden that got in the way of maximizing production, so they were not interested in learning more from these experts. The Japanese, on the other hand, bought into SPC, DOE, and the accompanying quality philosophy and used the methods to improve on what they had learned from the US and other countries.

The next several decades saw Japanese manufacturing thrive. Several well-known Japanese Quality gurus, such as Ishikawa and Taguchi, from that era are familiar to people all over the world. Taguchi is the one who coined the definition of World Class Quality in 1960: "On target with minimum variance". One example of the impact of minimizing variation about the target dimension occurred in the automatic transmission business of a major US automobile company. Ford had contracted a Japanese supplier, Mazda, to make a certain portion of their front-wheel-drive automatic transmissions, with the balance of production made at a US plant in Batavia, Ohio. Both sites were making transmissions to the same set of print specifications and the transmissions were being installed only on American cars. Mazda's version, as warranty records showed, had a substantially lower claim rate than the Batavia version. Ford investigated this phenomenon and found that Mazda's transmissions were made much more consistently than the ones they made. In fact, when they first started measuring them, they thought something was wrong with their measurement system because the first few parts were all identical. On some critical control valve components (valves, valve bores and springs) which make a transmission shift automatically, Mazda was using only 27% of the allowed tolerance range, while Batavia was using 70%. Ford thought its plant was doing well and, by US standards at the time, it was. It became quite apparent, however, that the Mazda plant was superior. Not only were all the parts made to print, as were the US-made parts, they were also more nearly like one another.

In the early 1980s, Ford invited Dr. Taguchi to speak and do seminars for their top executives. Taguchi used both SPC and DOE concepts in his training and, although late to the table, many US companies began to adopt the "new", "Japanese" quality philosophy. By this time, both Deming and Juran were in their 80s, but both continued to help move the Quality movement forward almost up to the time of their deaths. Deming worked into his 90s and this author remembers seeing the news of Dr. Juran giving a speech on his 100th birthday in 2004.

Another important event that occurred in the 1980s was Motorola's development of Six Sigma. Working for a Motorola supplier at the time, this author can tell you that the original definition meant that you had to do a very good job of making products that were "on-target with minimum variance" like Mazda had done with the Ford transmissions. The calculations will be discussed later in the book, but the definition is basically that a process cannot produce more than 3.4 defects per one million opportunities to be deemed six sigma capable. Six Sigma has come to have a much broader meaning over the last two decades and now describes a data-driven methodology of continuously improving processes and products. SPC and DOE are key components of the Six Sigma approach.

Today, driving out variation from processes and products continues to be a major focus of manufacturing companies around the world. The increasing use of automation magnifies the need to make more consistent parts, as robots and automation are not always as adaptable as human beings when doing assembly.

2
Basic Statistics

This chapter will be the foundation for the statistical methods that are discussed elsewhere in this book. Statistics are the language of data and it is important to understand the basic concepts that will be presented in this chapter before the next few chapters can be approached.

2.1 The Bell Curve

As mentioned in the previous chapter, Shewart started applying the bell curve to manufacturing in the 1920s. Figure 2.1 again shows this distribution. It was also mentioned earlier that most things vary in the manner shown by this graph.

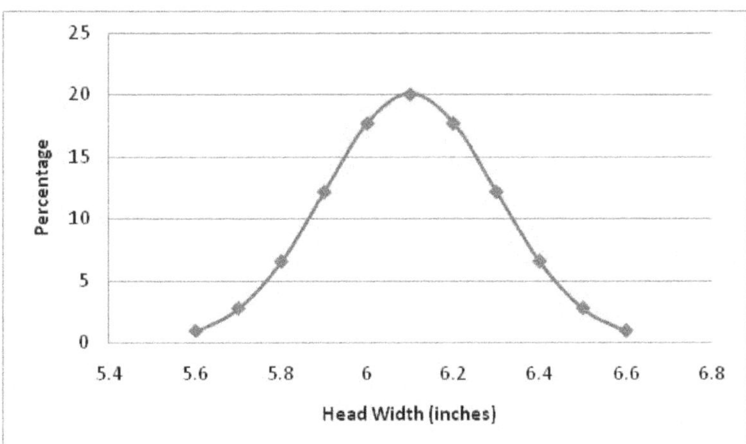

Figure 2.1 Shewart's head measurements (the bell curve)

The equation or mathematical model used for the normal curve is:

$$f(x) = \frac{1}{\sqrt{2\pi\sigma_x^2}} \exp\left[\frac{-(x-\mu_x)^2}{2\sigma_x^2}\right] \tag{2.1}$$

For Shewart's head measurement example, it can be seen in Equation 2.1 that to calculate $f(x)$ percentage, for each x, two parameters are needed; mean, μ_x, and standard deviation, σ_x. If 6.1 inches is used for the mean and 0.2 inches is used for the standard deviation, the bell curve shown in Figure 2.1 is the approximate result. The **mean** is the **location parameter**, as it indicates where to find the data on the number line (always somewhere between negative ∞ and ∞). The **standard deviation** is the **dispersion parameter**, as it indicates the spread of the data. It should also be pointed out that the standard deviation squared, σ_x^2, is called the **variance**. Note that the bell curve is normally considered to be about six standard deviations (6σ) wide, but it is infinitely wide (try putting a very large number in for x in the above equation). Figure 2.2 shows the percentage of a set of data that is theoretically found (by taking the integral of Equation 2.1) within certain distances from the mean. For this example, the mean is zero and the standard deviation is one. As can be seen, a 6σ width (or $\pm 3\sigma$) would theoretically include 99.7% of the data.

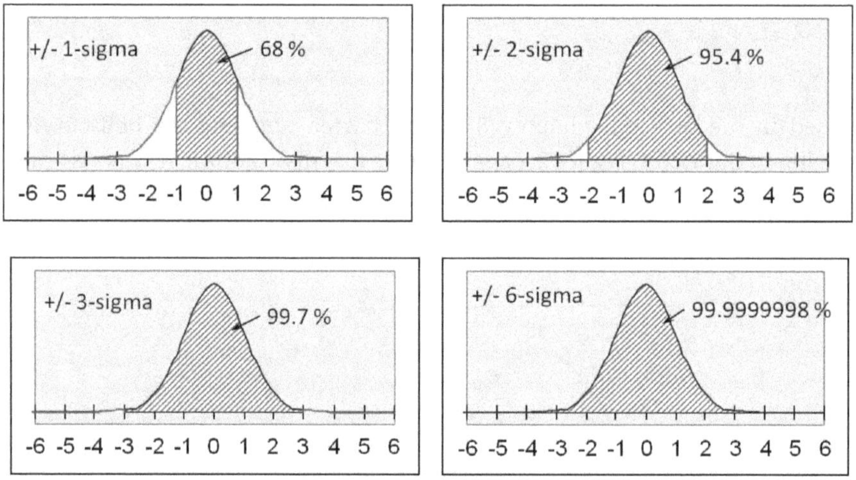

Figure 2.2 Standard normal-curve data distribution

So that it is not necessary to integrate Equation 2.1 every time one wants to estimate the percentage of data points (part dimensions, physical properties, etc.) between two values, above a certain value, or below a certain value, Table 12.3 in the Appendix (Chapter 12) has been provided. Because there are infinite combinations of μ_x

and σ_x, it is necessary to calculate a standardized normal variable, Z, before Table 12.3 can be used. The standard normal curve has a mean of 0 and a standard deviation of 1. When integrated from negative ∞ to ∞, the area under the curve is 1. Without a standard variable, a new table would be required for each combination of μ_x and σ_x. Z, also called the standard normal score, describes how far a data point is from the mean in terms of the number of standard deviations. The transformational formula for Z is:

$$Z = \frac{X - \mu_X}{\sigma_X} \tag{2.2}$$

The following example will illustrate how to use Table 12.3.

2.1.1 Example 2.1: Using the Standard Normal Distribution Table

The diameter of a hole in a large plastic panel is normally distributed, with a mean of 575 mm and a standard deviation of 5 mm. The bell curve shows this distribution (Figure 2.3).

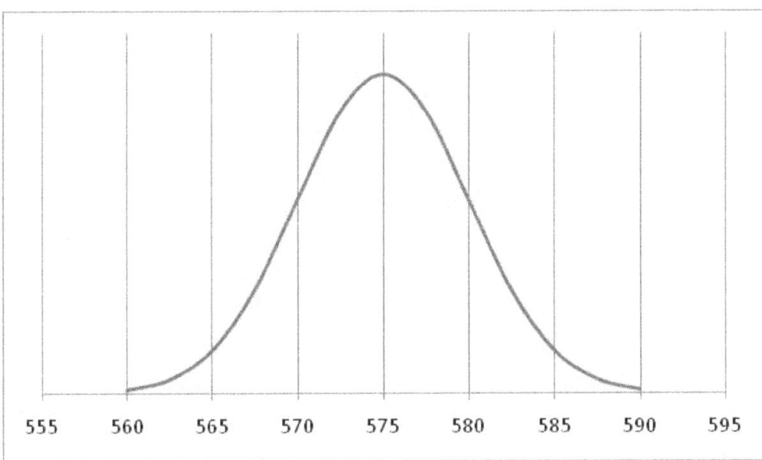

Figure 2.3 Bell curve for Example 2.1 data

1. What is the probability that a randomly selected panel has a hole with a diameter:
   a. Between 575 and 579 mm?

      This question is asking what percentage of the data is in the shaded region shown in Figure 2.4.

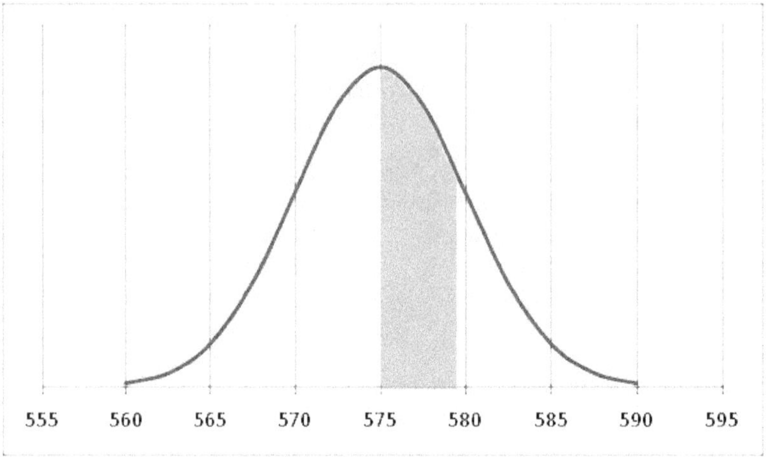

Figure 2.4 Bell curve shaded between 2 numbers

Instead of integrating Equation 2.1 to find this area, we will convert the *x*-values to *Z*-values (Figure 2.5) and use Table 12.3. Using Equation 2.2:

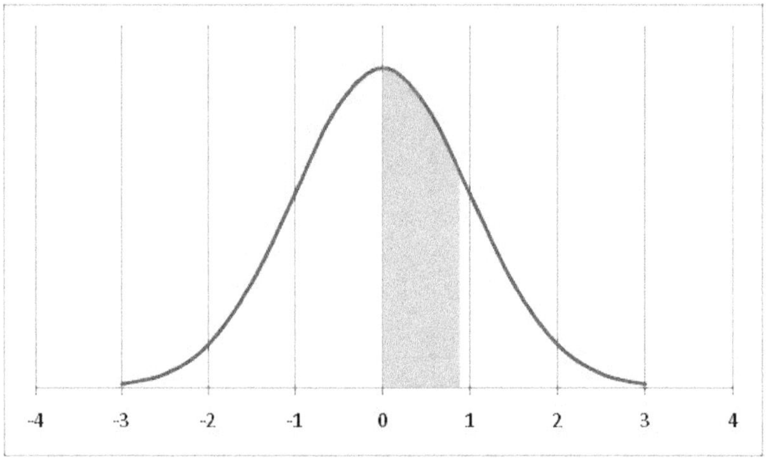

Figure 2.5 Normal distribution shaded between the mean and a positive *Z*-value

For 575: $Z = \frac{X-\mu_X}{\sigma_X} = \frac{575\text{ mm}-575\text{ mm}}{5\text{ mm}} = 0$, since 575 is the mean

For 579: $Z = \frac{X-\mu_X}{\sigma_X} = \frac{579\text{ mm}-575\text{ mm}}{5\text{ mm}} = 0.8$

From Table 12.3, $P(575 < x < 579) = P(0 < Z < 0.8) = 0.2881$ or 28.81%

b. Between 572 and 580 mm (Figure 2.6)?

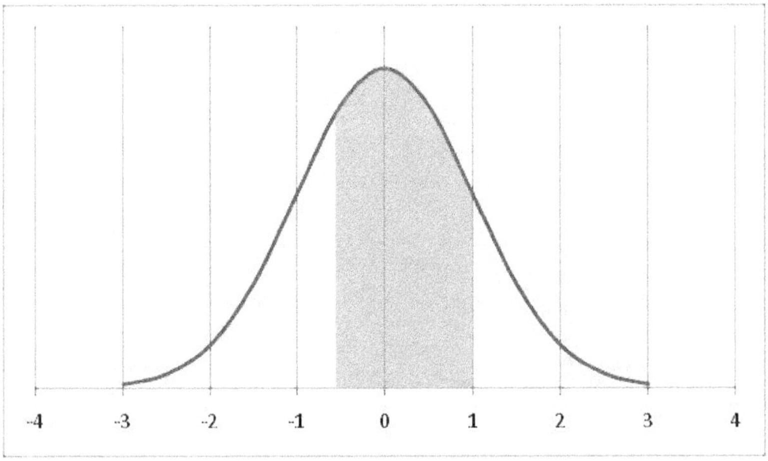

Figure 2.6 Normal distribution shaded between a negative and a positive Z-value

For 572: $Z = \frac{X-\mu_X}{\sigma_X} = \frac{572\text{ mm}-575\text{ mm}}{5\text{ mm}} = -0.6$

For 580: $Z = \frac{X-\mu_X}{\sigma_X} = \frac{580\text{ mm}-575\text{ mm}}{5\text{ mm}} = 1.0$

$P(572 < x < 580) = P(-0.6 < Z < 1.0) = P(-0.6 < Z < 0) + P(0 < Z < 1.0) = P(0 < Z < 0.6) + P(0 < Z < 1.0) = 0.2257 + 0.3413 = 0.5670$ or 56.70%

Note that since Table 12.3 only gives values for positive Z scores, the symmetry of the normal distribution allows finding areas for negative Z scores (Figure 2.7).

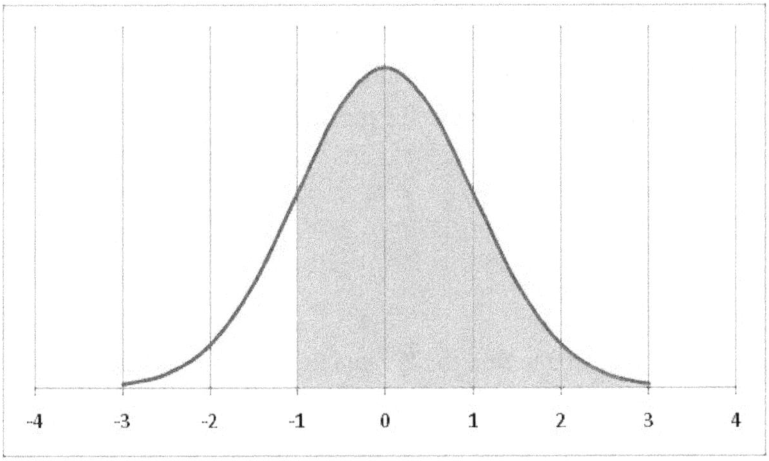

Figure 2.7 Normal distribution above a negative Z-value

c. Above 570 mm (Figure 2.8)?

For 570: $Z = \frac{X-\mu_X}{\sigma_X} = \frac{570\ \text{mm}-575\ \text{mm}}{5\ \text{mm}} = -1.0$

$P(x > 570) = P(Z > -1.0) = P(-1.0 < Z < 0) + P(0 < Z < \infty) = P(0 < Z < 1.0) + P(0 < Z < \infty)$
$= 0.3413 + 0.5 = 0.8413$ or 84.13%

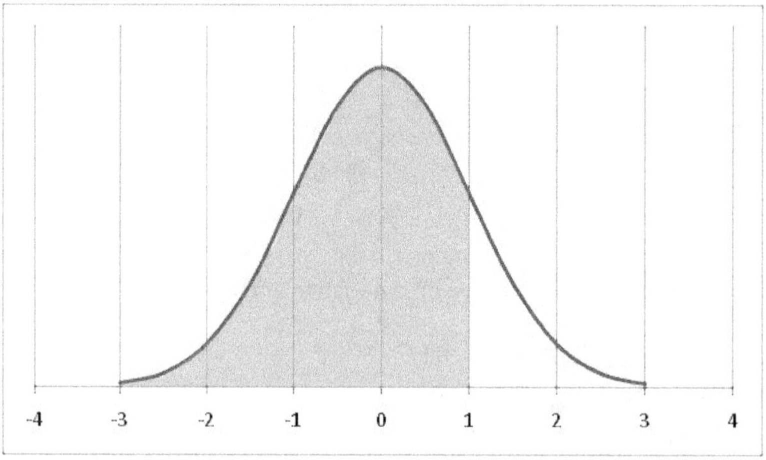

Figure 2.8 Normal distribution shaded below a positive *Z*-value

d. Below 580 mm?

For 580: $Z = \frac{X-\mu_X}{\sigma_X} = \frac{580\ \text{mm}-575\ \text{mm}}{5\ \text{mm}} = 1.0$

$P(x < 580) = P(Z < 1.0) = P(\infty < Z < 0) + P(0 < Z < 1.0) = 0.5 + 0.3413 = 0.8413$ or 84.13%

2. 75% of the holes will have diameters above what value?

It is known that 50% of the values are above the mean, therefore, a value of *Z* must be found such that:

25% or $0.25 = P(? < Z < 0) = P(0 < Z < ?)$

Looking at Table 12.3:

$P(0 < Z < 0.67) = 0.2486$

$P(0 < Z < 0.68) = 0.2517$

Linear interpolation can be used to find the *Z*-value for a 0.25 area (Figure 2.9).

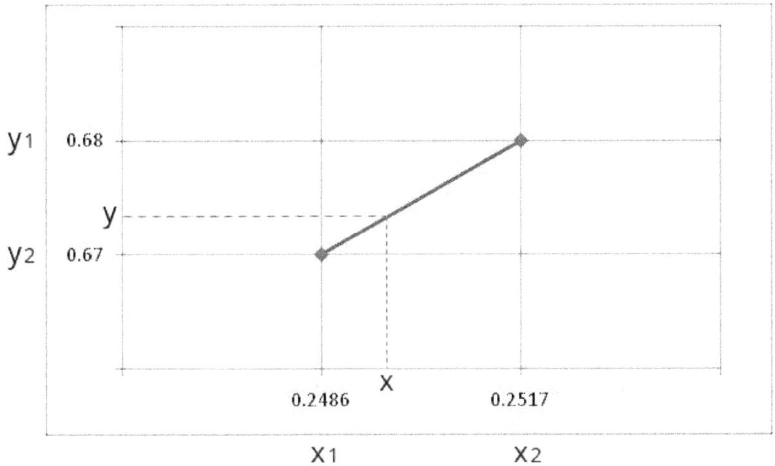

Figure 2.9 Linear interpolation for *Z*-value

$$\frac{y - y1}{y2 - y1} = \frac{x - x1}{x2 - x1}$$

$$y = y1 + \frac{y2 - y1}{x2 - x1}(x - x1) = 0.67 + \frac{0.68 - 0.67}{0.2517 - 0.2486}(0.25 - 0.2486) = 0.6745$$

So, $P(-0.6745 < Z < 0) = 0.25$.

Using Equation 2.2:

$$-0.6745 = \frac{X - 575\text{ mm}}{5\text{ mm}}$$

$$X = 571.6\text{ mm}$$

3. 75% of the holes will have diameters between what two values (37.5% on each side of the mean)?

   37.5% or 0.375 = $P(? < Z < 0) = P(0 < Z < ?)$ – both will be needed this time.

   Looking at Table 12.3:

   $P(0 < Z < 1.15) = 0.3749$

   $P(0 < Z < 1.16) = 0.3770$

   Linear interpolation gives $P(-1.1505 < Z < 0) = P(0 < Z < 1.1505) = 0.375$

   or $Z_{\text{Lower}} = -1.1505$ & $Z_{\text{Upper}} = 1.1505$

   Using Equation 2.2:

   $$-1.1505 = \frac{X_L - 575\text{ mm}}{5\text{ mm}} X_L = 569.2\text{ mm}$$

   and

   $$1.1505 = \frac{X_U - 575\text{ mm}}{5\text{ mm}} X_U = 580.8\text{ mm}$$

2.2 Central Limit Theorem

Most of the statistical techniques that are discussed in this book have the premise that a normal distribution (bell curve) represents the data that is being studied. Although this is generally a good assumption with most manufacturing processes, including injection molding, and products, distributions can sometimes be different. Other than the normal distribution, the most common one that this author has seen is a distribution that is asymmetrical or **skewed**. Figure 2.10 shows an example of a right-skewed distribution, which indicates that there are more values to the right of the distribution's peak than to the left. An example of where this type of distribution might be found is the measurement of warpage, when the parts only warp in one direction. The values can never go below zero so, in effect, the left tail of the distribution is cut short.

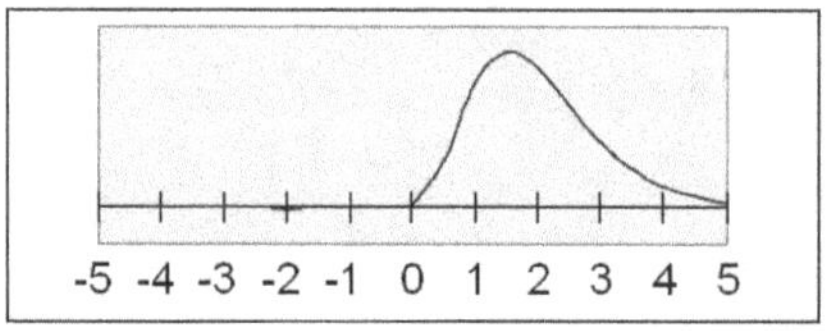

Figure 2.10
Positive or right-skewed distribution

However, regardless of the shape of the distribution of the individual data values, the **central limit theorem** has the effect of changing a non-normal distribution into an approximate normal distribution. It does this by looking at the average of samples of data points. Consider for example, data points taken from a uniform distribution. A uniform distribution is one in which there is an equal chance of obtaining a value within the distribution. For example, if a six-sided die is rolled, there is an equal chance of obtaining a one, two, three, four, five, or six. Figure 2.11 is a histogram of six hundred rolls of a six-sided die.

As can be seen in Figure 2.11, each possible outcome occurred about 100 times. The expected frequency of each outcome can be obtained by multiplying the probability by the number of rolls: $(1/6) \times 600 = 100$ for each possible outcome. The solid line indicates the bell curve that would result by taking the average and sample standard deviation of the 600 rolls and plugging in the values for the mean and standard deviation in Equation 2.1. It should be noted here that the normal distribution is technically intended for continuous variables and not for use with discrete variables (there are 6 discrete values that are possible every time a die is rolled), but this example will demonstrate how discrete distributions can sometimes be approximated by the normal distribution. As can be seen in Figure 2.11, the curve does not fit the histogram well at all. To see the central limit theorem effect, the average of two consecutive rolls or of rolling two dice is calculated and the resulting distribution can be seen in Figure 2.12. Again, the expected outcome can be calculated. For example, for an average of 5.5 there are two possible combinations (5/6 and 6/5) of rolls. The chance of rolling a 5 and then a 6 is $1/6 \times 1/6 = 1/36$.

The chance of rolling a 6 and then a 5 is also 1/36. The expected frequency of an average of 5.5 can then be calculated as (2/36 × 600) = 33.3. An average of 3.5 has the most possible roll combinations and, therefore, has the highest frequency of occurrence. As can be seen in Figure 2.12, the outline of the histogram is already approximated by a bell curve with a sample size of only two from a uniform distribution.

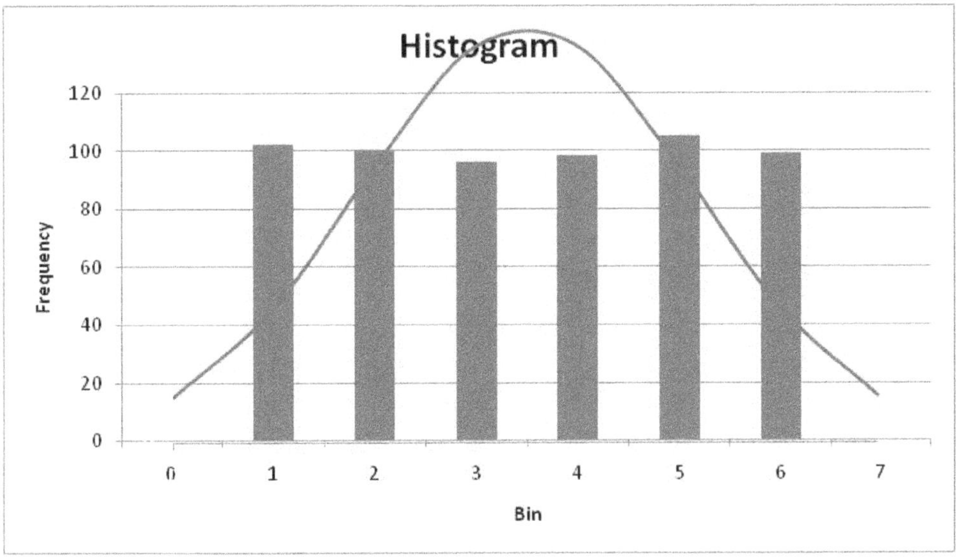

Figure 2.11 Histogram of 600 rolls of a die

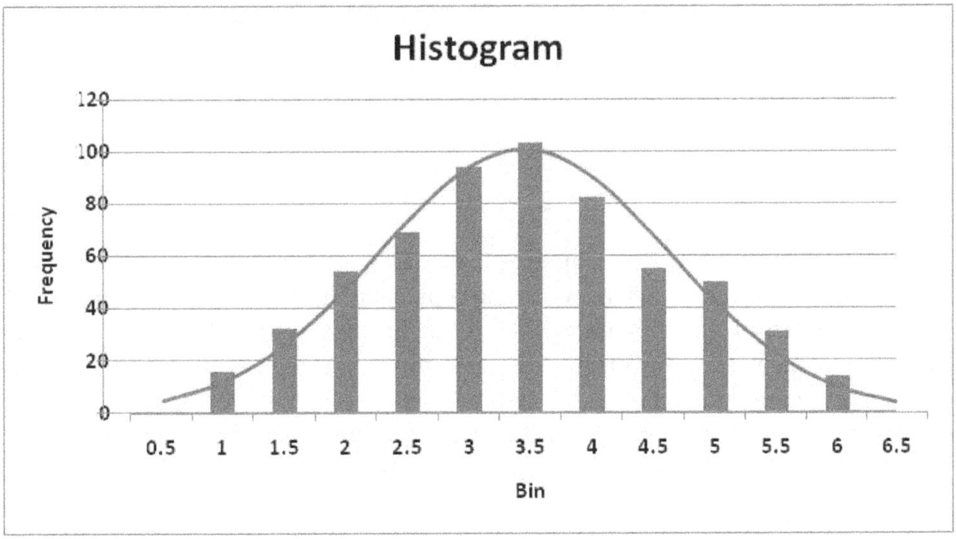

Figure 2.12 Histogram of the average of 600 rolls of two dice

Figure 2.13 shows what the histogram and bell curve look like for a sample size of five. The average of five consecutive rolls or of rolling five dice is what is shown here. Note how the distribution gets higher in the middle and has a narrower spread as the sample size increases. The mean continues to be approximately 3.5 in this example, but the standard deviation decreases as the sample size increases. The relationship between the standard deviation of the individual values, σ_x, and the standard deviation of the average values, $\sigma_{\overline{X}}$, is

$$\sigma_{\overline{X}} = \frac{\sigma_X}{\sqrt{n}} \tag{2.3}$$

Where:

n = sample size

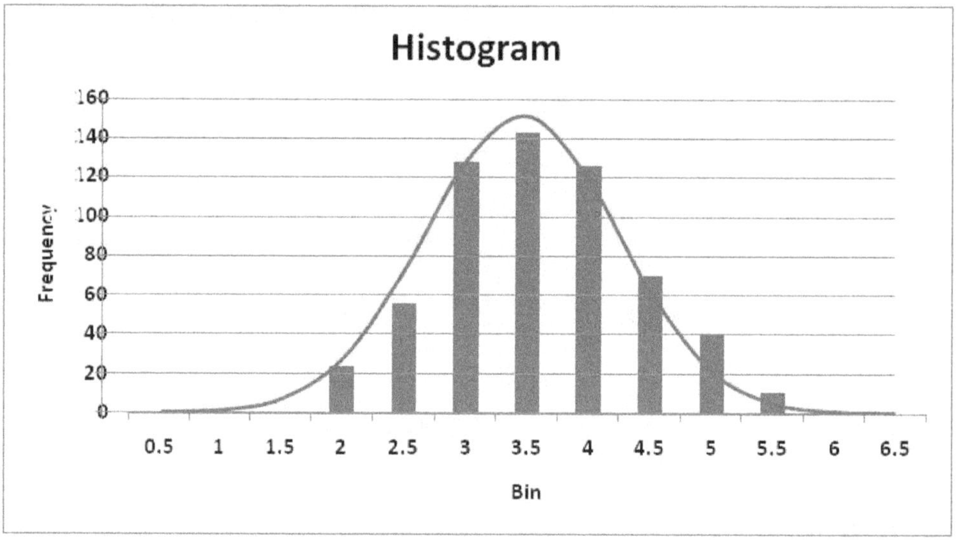

Figure 2.13 Histogram of the average of 600 rolls of five dice

It should also be noted, when comparing Figure 2.12 and Figure 2.13, that the shape of the curve becomes more bell-shaped or more "normal" as the sample size is increased. This should help in understanding why most things are normally distributed. The variation in one variable or product is due to the variation in everything that affects it and, thus, there is an averaging or centralizing effect. For example, the variation in the size of an injection molded part is due to the variation in the molding material, the equipment, the techniques, the ambient conditions, the operators, etc.

2.3 Point Estimates

It has already been pointed out that, to characterize a normal distribution, two parameters are needed; the location parameter or mean, μ_x, and the dispersion parameter or standard deviation, σ_x. These parameters can never be exactly known with a real process, but only estimated using statistics. This book will not review all the possible estimators, but only the two most commonly used for each.

2.3.1 Statistics Used to Estimate the Mean, μ_x

Average (or sample mean):

$$\overline{x} = \frac{1}{n}\sum_{i=1}^{n} x_i \qquad (2.4)$$

Median:

$$\widetilde{x} = 50^{th}\text{percentile of data} \qquad (2.5)$$

The pth percentile is a value such that roughly p% of the data is smaller and $(100 - p)$% of the data is larger. There are three steps for computing a percentile:

1. Sort the data from low to high
2. Count the number of values (n)
3. Select the $p \times (n + 1)$ observation

If $p \times (n + 1)$ is not a whole number, then go half-way between the two adjacent numbers. To find the 50th percentile, $p = 0.5$ or 50%.

2.3.2 Statistics Used to Estimate the Standard Deviation, σ_x

Sample Standard Deviation:

$$s = \sigma_{n-1} = \sqrt{\frac{1}{n-1}\sum_{i=1}^{n}(x_i - \overline{x})^2} \qquad (2.6)$$

Range:

$$R = x_{max} - x_{min} \qquad (2.7)$$

It is important to point out that both sample standard deviation and range are **biased** estimates of standard deviation. An **unbiased** estimator is simply one that doesn't give values that are consistently too large or too small for the parameter being estimated when dealing with data from a normal distribution. Table 2.1 (also Table 12.1 in the Appendix) shows bias correction factors that are needed to convert the biased estimate to an unbiased estimate. These factors are also shown in Table 12.1 for more sample sizes. The biased estimate is divided by the correction factor to give the unbiased estimate.

Table 2.1 Bias Correction Factors for Estimating σ_x

Sample Size	R-Factor d_2	S-Factor c_4
2	1.128	0.7979
3	1.693	0.8862
4	2.059	0.9213
5	2.326	0.9400
6	2.534	0.9515
7	2.704	0.9594
8	2.847	0.9650

For small samples ($n < 8$), it is common practice to use range, R, or average range, $\overline{R}$, to estimate standard deviation:

$$\hat{\sigma}_X = R/d_2 \text{ or } \overline{R}/d_2 \tag{2.8}$$

For large samples, sample standard deviation, s, is generally used and, since c_4 approaches 1.0 as the sample size increases, the biased estimate is sometimes used for simplicity:

$$\hat{\sigma}_X = s/c_4 \text{ or } \overline{s}/c_4 \tag{2.9}$$

2.3.3 Example 2.2: Calculating Point Estimates

1. Calculate the average, median, sample standard deviation, and range for the following data set:

   {12, 5, 13, 9}

Using Equation 2.4 for the average:

$$\bar{x} = \frac{1}{n}\sum_{i=1}^{n} x_i = \frac{1}{4}\sum_{i=1}^{4}\{12, 5, 13, 9\} = \frac{12+5+13+9}{4} = 9.75$$

Using Equation 2.5 for the median:

$\tilde{x}$ = 50th percentile of data

The data sorted in order are:

a. 5

b. 9

c. 12

d. 13

0.5 × (4 + 1) = 2.5, so the median is half-way between 9 and 12 = 10.5

Using Equation 2.6 for the sample standard deviation:

$$s = \sigma_{n-1} = \sqrt{\frac{1}{n-1}\sum_{i=1}^{n}(x_i - \bar{x})^2}$$

$$= \sqrt{\frac{1}{4-1}[(12-9.75)^2 + (5-9.75)^2 + (13-9.75)^2 + (9-9.75)^2]} = 3.594$$

Using Equation 2.7 for the range:

$R = x_{max} - x_{min} = 13 - 5 = 8$

2. Calculate the unbiased estimates of standard deviation using the sample standard deviation and the range calculated in part A.

   Using Equation 2.8, $\hat{\sigma}_X = {}^{R}/_{d_2} = \frac{8}{2.059} = 3.885$

   Using Equation 2.9, $\hat{\sigma}_X = {}^{s}/_{c_4} = \frac{3.594}{0.9213} = 3.901$

In Example 2.2, we found estimates of the mean to be 9.75 and 10.5 using two different statistics. We also found estimates of the standard deviation to be 3.885 and 3.901. The reader may have some questions at this point. Are these good estimates? Are some better than others? How much confidence can be put in these estimates? These questions will be discussed more in the section on interval estimates.

In addition to **unbiasedness**, there are a couple of other important properties to consider when dealing with point estimators. One is **efficiency**. Efficiency refers to the precision of a sample statistic as an estimator of the population parameter. The average is the most efficient estimator of mean, and the sample standard deviation is the most efficient estimator of standard deviation when dealing with normal distributions. This is because each value of the sample contributes to the calculation of the statistic. The relative efficiency of the range compared to the sample standard deviation is shown in Table 2.2.

Table 2.2 Relative Efficiency of R Versus s for Estimating σ_x

Sample Size, n	Relative Efficiency
2	1.000
3	0.992
4	0.975
5	0.955
6	0.930
10	0.850

As can be seen in Table 2.2, range works almost as well as sample standard deviation for small sample sizes and the range method is still commonly used for some of the methods that will be covered later in this book.

Another important property of point estimators is **consistency**. For a statistic to be considered consistent, it should consistently be a closer estimate of the population parameter when the sample size is increased. All four of the statistics (average, median, sample standard deviation, and range) have this property and, thus, would be considered to have good consistency. For each of these statistics, as the sample size increases, confidence in the estimate made increases. Quantifying this confidence level will be discussed in the next section, interval estimates.

2.4 Interval Estimates

Example 2.2 showed that using different statistics as estimators will lead to different estimates of the mean and standard deviation. Even when the same estimator is used, different estimates will be obtained for different samples. Suppose that 10,000 parts were molded in a single cavity mold during the last production run and the customer wants to know the actual size of a certain part characteristic that is critical to quality (CTQ) before the parts are shipped. There are instances when each of those 10,000 parts will need to be measured (at considerable cost) but, in most cases, a random sample will be measured. Two questions that now arise are:

1. How big a sample should be measured?
2. How should the results be reported to the customer?

The case of the 10,000 parts will be considered before these questions are answered. If all 10,000 parts are measured, it is found that the mean is 10.50000 inches, and the standard deviation is 0.00100 inches. This is shown in the graph at the top of Figure 2.14. All the bell curves in Figure 2.14 are drawn to be exactly six estimated standard deviations wide and were constructed by substituting the average and sample standard deviation for the mean and the standard deviation in Equation 2.1. The peak of the bell curve when all 10,000 parts are measured is at 10.50000 inches, the lower "tail" ends at 10.49700 inches, and the upper "tail" ends at 10.50300 inches (6σ or 0.00600 inches wide). The other six graphs in Figure 2.8 are representative of different sample sizes (2, 3, 5, 10, 25, and 100) taken randomly from the 10,000 parts. The two bell curves on each graph represent two independent samples that were taken at different times. A few generalities can be stated about the other six graphs:

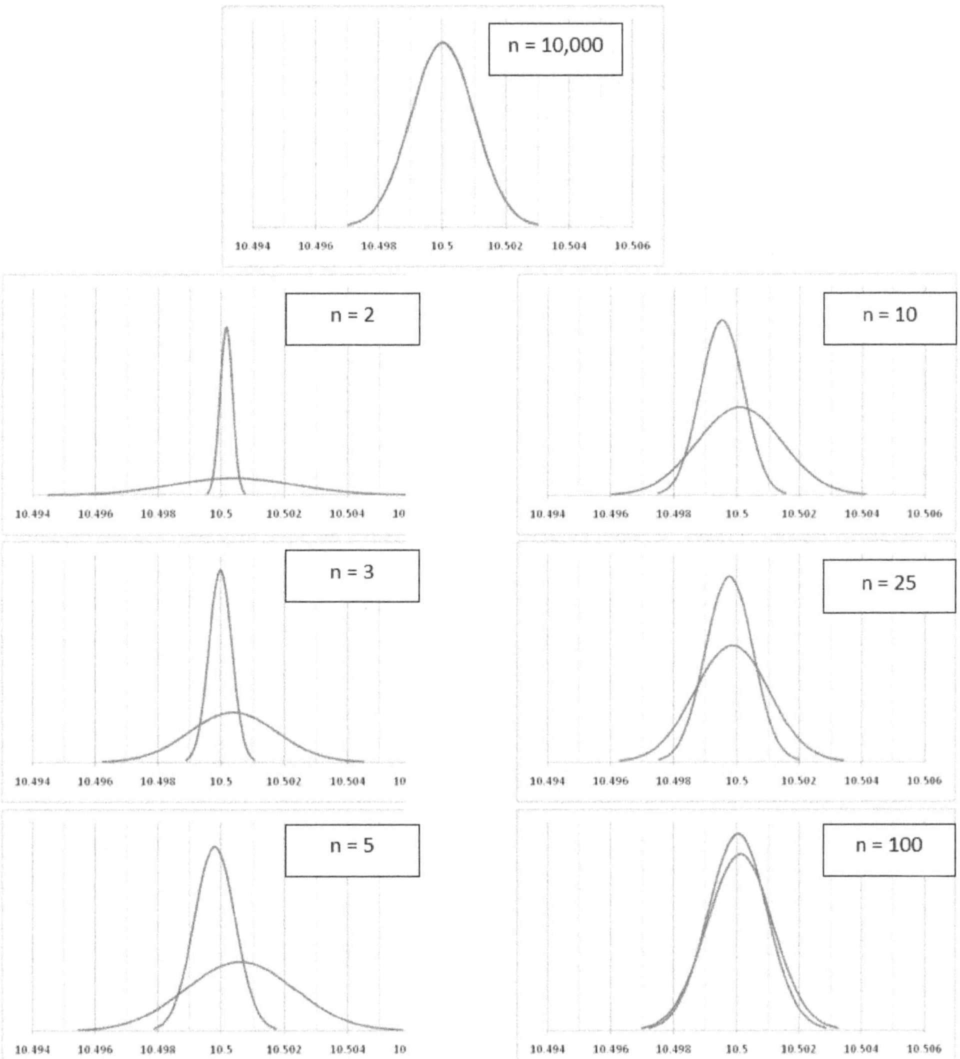

Figure 2.14 Sample size effect on sampling from 10,000 production parts. The x-axis represents a dimension in units of inches

1. None of the peaks are at exactly 10.50000 inches
2. None of the curves are exactly 0.00600 inches wide (six standard deviations)
3. As sample size increases, the two curves are more closely matched
4. As sample size increases, the curves better estimate the entire 10,000 pieces

2.4.1 Confidence Interval Estimate for the Mean, μ_x

An interval estimate or confidence interval (CI) estimate for the mean, μ_x, is simply an interval, for a given sample size, in which the true mean will be found a certain percentage of the time. By far, the most common interval used in industry is the 95% confidence interval. In general, the level of confidence is $(1 - \alpha) \times 100\%$, where α is the proportion in the tails ($\alpha/2$ in the upper tail and $\alpha/2$ in the lower tail) of the distribution that is outside the confidence interval. It would be expected that the estimated confidence interval of the mean does not contain the actual mean $\alpha \times 100\%$ of the time (5% of the time for a 95% CI).

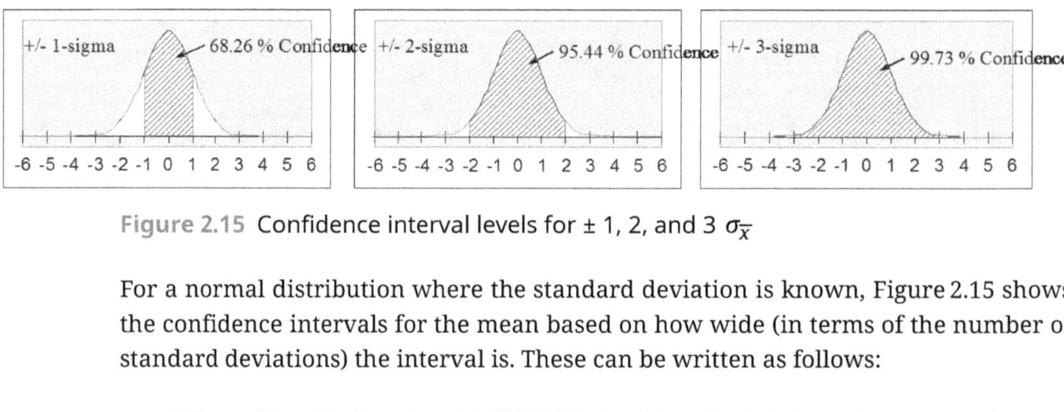

Figure 2.15 Confidence interval levels for ± 1, 2, and 3 $\sigma_{\overline{X}}$

For a normal distribution where the standard deviation is known, Figure 2.15 shows the confidence intervals for the mean based on how wide (in terms of the number of standard deviations) the interval is. These can be written as follows:

For a 68.26% CI: $\overline{X} - \sigma_{\overline{X}} < \mu_X < \overline{X} + \sigma_{\overline{X}}$ or $CI \text{ for } \mu_X = \overline{X} \pm \sigma_{\overline{X}} = \overline{X} \pm \frac{\sigma_X}{\sqrt{n}}$

For a 95.44% CI: $\overline{X} - 2\sigma_{\overline{X}} < \mu_X < \overline{X} + 2\sigma_{\overline{X}}$ or $CI \text{ for } \mu_X = \overline{X} \pm 2\sigma_{\overline{X}} = \overline{X} \pm \frac{2\sigma_X}{\sqrt{n}}$

For a 99.73% CI: $\overline{X} - 3\sigma_{\overline{X}} < \mu_X < \overline{X} + 3\sigma_{\overline{X}}$ or $CI \text{ for } \mu_X = \overline{X} \pm 3\sigma_{\overline{X}} = \overline{X} \pm \frac{3\sigma_X}{\sqrt{n}}$

As you can see, the confidence interval is composed of a band around the sample mean or average that is Z standard deviations of the average wide. The general formula is:

Confidence Interval for a Mean (σ known):

$$CI \text{ for } \mu_X = \overline{X} \pm \frac{Z\sigma_X}{\sqrt{n}} \tag{2.10}$$

Where:

Z = the value corresponding to an area of $(1 - \alpha)/2$ from the center of the standard normal distribution

In practice, the standard deviation isn't always known when doing studies on a process or product. In the early 1900s, a statistician at Guinness Breweries, William S. Gosset, came up with an alternative to the standard normal distribution for cases when the standard deviation is not known. Since Guinness employees weren't permitted to publish under their own names, the distribution that he developed has come to be known as **Student's *t* distribution**. The equation for the confidence interval estimate of the mean using *t*-values is:

Confidence Interval for a Mean (σ unknown):

$$CI \text{ for } \mu_X = \overline{X} \pm t_{n-1}\frac{s}{\sqrt{n}} \tag{2.11}$$

Where:

t_{n-1} = the critical value of the t distribution (Table 12.4 in the Appendix) with $n - 1$ degrees of freedom for an area of $\alpha/2$ in the upper tail.

The concept of degrees of freedom will be needed for several sections in this book. A common way to think about degrees of freedom is the number of independent pieces of information available to estimate or calculate another piece of information. For example, if five numbers have to add up to 100, four of those numbers can be anything; the fifth number is dependent on the sum of the first four. In this case there are $n - 1 = 5 - 1 = 4$ degrees of freedom. With only a few exceptions, the number of degrees of freedom is $n - 1$.

The case of the 10,000 parts from Figure 2.14 will again be used to show the effect of sample size on confidence intervals for a mean.

2.4.1.1 Example 2.3: Calculating Interval Estimates of the Mean

1. Calculate 95% confidence intervals for the mean of the dimension measured from the sample of 10,000 parts when the sample size is two ($n = 2$).
   a. $\overline{X} = 10.50036$ in, $s = 0.001952$ in

      For a 95% CI, $\alpha = 0.05$, so $\alpha/2 = 0.025$

      Degrees of freedom = $n - 1 = 2 - 1 = 1$

      Using Table 12.4, $t_{0.025,1} = 12.7062$, and Equation 2.11,

CI for $\mu_X = \overline{X} \pm t_{n-1}\frac{s}{\sqrt{n}} = 10.50036 \text{ in} \pm \frac{12.7062(0.001953 \text{ in})}{\sqrt{2}}$
$= 10.50036 \text{ in} \pm 0.01755 \text{ in} (X_L = 10.48281 \text{ in}, X_U = 10.51791 \text{ in})$

Note: X_L is the lower confidence limit and X_U is the upper confidence limit.

b. $\overline{X} = 10.50015$ in, $s = 0.000197$ in
CI for $\mu_X = \overline{X} \pm t_{n-1}\frac{s}{\sqrt{n}} = 10.50015 \text{ in} \pm \frac{12.7062(0.000197 \text{ in})}{\sqrt{2}}$
$= 10.50015 \text{ in} \pm 0.00177 \text{ in} (X_L = 10.49838 \text{ in}, X_U = 10.50192 \text{ in})$

2. Make a table (Table 2.3) showing $t_{critical}$, one half the confidence interval width, the lower confidence limit, and the upper confidence limit for both samples in Figure 2.14 when n = 2, 3, 5, 10, 25, and 100. Make a chart (Figure 2.16) to show the different widths of the CIs.

Table 2.3 95% Confidence Intervals of the Means for Figure 2.14 Data

n	$\overline{X}$, in	S, in	$t_{critical}$	δ = ½ CI, in	X_L, in	X_U, in
2	10.50036	0.001952	12.7062	0.01754	10.48282	10.51790
2	10.50015	0.000197	12.7062	0.00177	10.49838	10.50192
3	10.50036	0.00138	4.3027	0.00343	10.49693	10.50379
3	10.49996	0.000359	4.3027	0.00089	10.49907	10.50085
5	10.50059	0.001712	2.7765	0.00213	10.49846	10.50271
5	10.49978	0.000643	2.7765	0.00080	10.49899	10.50058
10	10.50007	0.001348	2.2622	0.00096	10.49911	10.50104
10	10.49951	0.000679	2.2622	0.00049	10.49902	10.49999
25	10.49984	0.001187	2.0639	0.00049	10.49935	10.50033
25	10.49975	0.000746	2.0639	0.00031	10.49944	10.50005
100	10.5001	0.001044	1.9842	0.00021	10.49989	10.50031
100	10.50002	0.000937	1.9842	0.00019	10.49984	10.50021

3. If it was assumed, based on historical Quality department data, that the standard deviation was 0.001 inches, what would the 95% confidence interval be for each sample (make another table and chart; Table 2.4 and Figure 2.17)?

Equation 2.10 will be used in this case, CI for $\mu_X = \overline{X} \pm Z\frac{\sigma_X}{\sqrt{n}} = \overline{X} \pm 1.96\frac{0.001 \, in}{\sqrt{n}}$

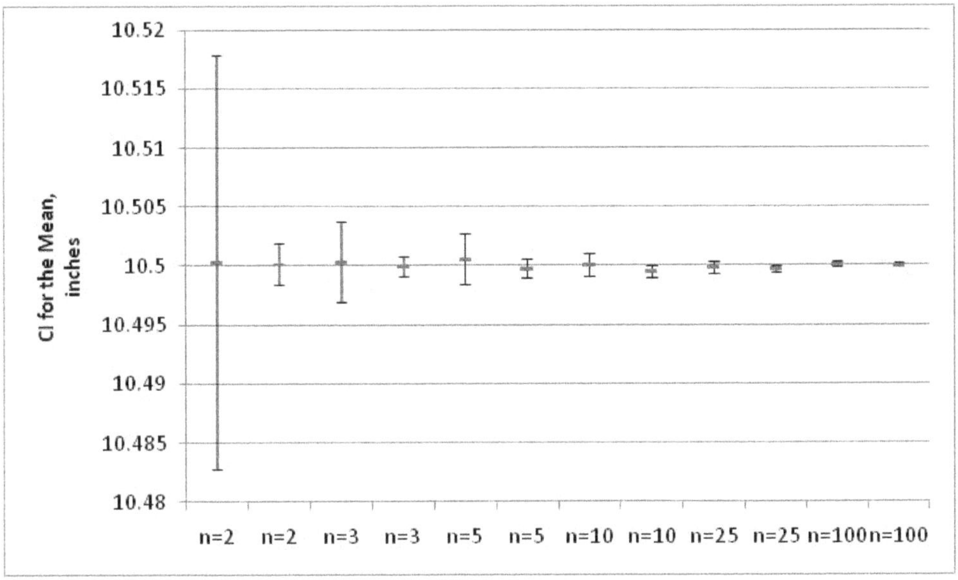

Figure 2.16 95% confidence interval of the means for Table 2.3. The last section in this chapter shows how to make this graph in Excel

Table 2.4 95% Confidence Intervals of the Means for Figure 2.14 Data, Assuming a Standard Deviation of 0.001 inches

n	$\overline{X}$, in	σ_x, in	Z	δ = ½ CI, in	X_L, in	X_U, in
2	10.50036	0.001	1.96	0.00196	10.49897	10.50175
2	10.50015	0.001	1.96	0.00196	10.49876	10.50153
3	10.50036	0.001	1.96	0.00153	10.49923	10.50149
3	10.49996	0.001	1.96	0.00153	10.49883	10.50109
5	10.50059	0.001	1.96	0.00117	10.49971	10.50146
5	10.49978	0.001	1.96	0.00117	10.49891	10.50066
10	10.50007	0.001	1.96	0.00072	10.49945	10.50069
10	10.49951	0.001	1.96	0.00072	10.49889	10.50013
25	10.49984	0.001	1.96	0.00045	10.49945	10.50023
25	10.49975	0.001	1.96	0.00045	10.49935	10.50014
100	10.5001	0.001	1.96	0.00021	10.49990	10.50030
100	10.50002	0.001	1.96	0.00021	10.49983	10.50022

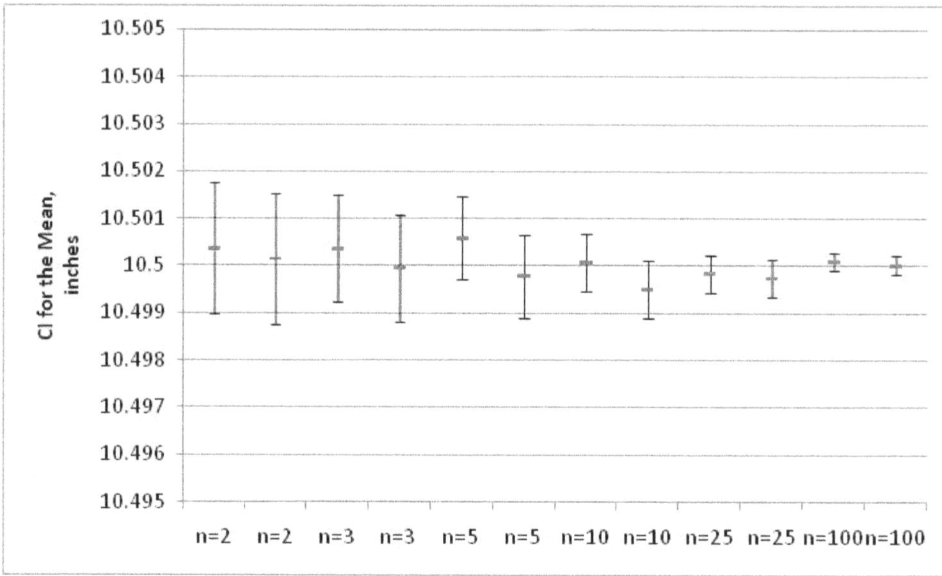

Figure 2.17 95% Confidence interval of the means for Table 2.4

Some general observations of the data from Example 2.3 are:

1. When the sample size increases, the CI narrows
   a. Always when the standard deviation is known
   b. Usually when the standard deviation is not known
2. All but one of the intervals (23/24 or 96%) contains the true mean (10.5000 inches)... it should be expected that 95% would.

So how big should the sample size be? That will depend on how close an estimate the customer is looking for. If the need is to know how big the parts are to the nearest 0.1 inch, then a sample size of two ($n = 2$) would be plenty. If the need is to know to the nearest 0.0001 inch, then a sample size of one hundred ($n = 100$) may not be sufficient. Note the δ column in Table 2.3 and Table 2.4. In that example, δ is how close you can be assured, with a 95% confidence level, that the actual mean is from the average. If you have past data that gives a good idea of the standard deviation, then $\delta = Z\frac{\sigma_x}{\sqrt{n}}$ and Equation 2.12 can be used to determine how big your sample should be:

If $\overline{X}$ is used as the estimate for μ_x, then the sample size n calculated below will ensure a $(1 - \alpha) \times 100\%$ level of confidence that the mean will not vary from $\overline{X}$ by more than δ.

$$n = \left(\frac{Z\sigma_X}{\delta}\right)^2 \quad (2.12)$$

Where:

Z = the value corresponding to an area of $(1 - \alpha)/2$ from the center of the standard normal distribution. The answer should be rounded up if not an integer and a sample size less than 2 is not usually recommended.

2.4.1.2 Example 2.4: Calculating Sample Size for Estimates of the Mean

1. How big of a sample should be measured if the customer wants to know the size of the parts from Example 2.3 to the nearest 0.1 inches, with a 95% CI?

   Using Equation 2.12, $n = \left(\frac{z_{\frac{\alpha}{2}}\sigma}{\delta}\right)^2 = \left(\frac{1.96 \times 0.001\text{ in}}{0.1\text{ in}}\right)^2 = 0.00038$, rounds up to 1, but use a sample size of 2.

2. Make a table (Table 2.5) showing the sample size needed for δ = 0.1, 0.01, 0.001, 0.0005, and 0.0001 inches?

Table 2.5 Data and Results for Example 2.4

δ, in	σ/δ	n (calculated)	n (rounded up)
0.1	0.01	0.00038	1 (2)
0.01	0.1	0.038	1 (2)
0.001	1	3.8	4
0.0005	2	15.4	16
0.0001	10	384.2	385

Note that the sample size requirement increases quite rapidly when the precision of the estimate needs to be less than one standard deviation. Inspection costs would increase rapidly due to the time to inspect so many parts and, in this case, the equipment needed to obtain such precise measurements.

Example 2.4 shows how to calculate what sample size is needed depending on the precision of the estimate that is required, assuming that standard deviation is known from historical data. If standard deviation is not known, then an educated guess of sample size is made and, in most cases, the sample size can be increased if the precision needed is not obtained. The δ column in Table 2.4 shows the precision that is estimated with each sample for Example 2.3. A sample size of twenty-five (n = 25) is usually a good starting point if you have absolutely no idea of what the standard deviation is.

2.4.2 Confidence Interval Estimate for the Standard Deviation, σ_x

When determining a CI for the mean, it was assumed that either the standard deviation was known or not known. It is also possible to calculate a CI for the variability in the data. This is normally done for the variance, σ_X^2, and then may be transformed to the standard deviation, σ_x, by taking the square root of the confidence limits of the variance. Table 12.5 will be used for the chi-square, X^2, distribution which is used to estimate a variance confidence interval. The chi-square distribution is a skewed distribution which becomes more symmetrical as the degrees of freedom increase. The $(1 - \alpha) \times 100\%$ confidence interval is expressed as follows:

Confidence Interval Estimate for the Variance

$$\frac{(n-1)s^2}{X_U^2} \leq \sigma_X^2 \leq \frac{(n-1)s^2}{X_L^2} \tag{2.13}$$

Where:

n = sample size

s^2 = sample variance

X_U^2 = upper critical value of X^2

X_L^2 = lower critical value of X^2

The case of the 10,000 parts from Figure 2.8 will again be used to show the effect of sample size on confidence intervals for a variance.

2.4.2.1 Example 2.5: Calculating Interval Estimates of the Variance

1. Calculate 95% confidence intervals for the variance of the dimension measured from the sample of 10,000 parts when the sample size is twenty-five (n = 25).

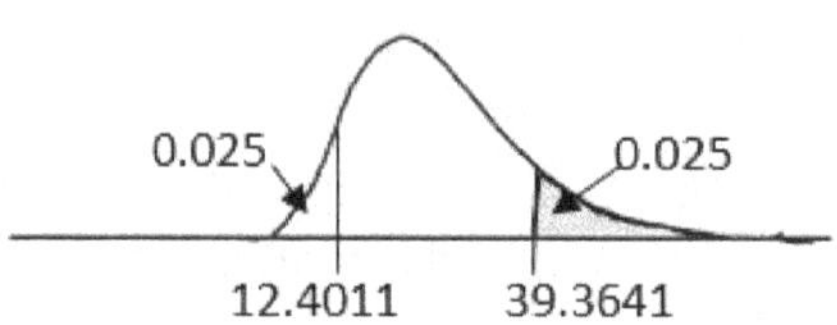

Figure 2.18
Chi-square distribution for Example 2.5

a. $\overline{X}$ = 10.49984 in, s = 0.001187 in

For a 95% CI, α = 0.05, so $\alpha/2$ = 0.025 for the upper (right) critical value of X^2 and $(1 - 0.025) = 0.975$ for the lower (left) critical value of X^2 (Figure 2.18).

Degrees of freedom = $n - 1 = 25 - 1 = 24$

From Table 12.5, $X_U^2 = X_{0.025,24}^2 = 39.3641$

$X_L^2 = X_{0.975,24}^2 = 12.4011$

Using Equation 2.13, $\frac{(n-1)s^2}{X_U^2} \leq \sigma_X^2 \leq \frac{(n-1)s^2}{X_L^2}$

$$\frac{(25-1)(0.0011187 \text{ in})^2}{39.36411} \leq \sigma_X^2 \leq \frac{(25-1)(0.001187 \text{ in})^2}{12.4011}$$

$$8.59038 \times 10^{-7} \leq \sigma_X^2 \leq 2.72679 \times 10^{-6}$$

$$0.000927 \leq \sigma_X \leq 0.001651$$

2. Make a table (Table 2.6) showing X_L^2, X_U^2, the lower confidence limit for standard deviation, and the upper confidence limit for standard deviation for both samples in Figure 2.14 when n = 2, 3, 5, 10, and 25. Make a chart (Figure 2.19) to show the different widths of the CIs.

Table 2.6 95% Confidence Intervals of the Standard Deviation for Figure 2.14 Data

n	$\overline{X}$, in	S, in	X_U^2	X_L^2	Lower Limit σ_X, in	Upper Limit σ_X, in
2	10.50036	0.001952	5.0239	0.0010	0.000871	0.062289
2	10.50015	0.000197	5.0239	0.0010	0.000088	0.006286
3	10.50036	0.00138	7.3778	0.0506	0.000719	0.008673
3	10.49996	0.000359	7.3778	0.0506	0.000187	0.002256
5	10.50059	0.001712	11.1433	0.4844	0.001026	0.004920
5	10.49978	0.000643	11.1433	0.4844	0.000385	0.001848
10	10.50007	0.001348	19.0228	2.7004	0.000927	0.002461
10	10.49951	0.000679	19.0228	2.7004	0.000467	0.001240
25	10.49984	0.001187	39.3641	12.4011	0.000927	0.001651
25	10.49975	0.000746	39.3641	12.4011	0.000582	0.001038

Some general observations of the data from Example 2.5 are:

1. When the sample size is increased, the CI usually narrows.
2. All but one of the intervals (9/10 or 90%) contains the true standard deviation (0.001 inches)... it was expected that 95% would.

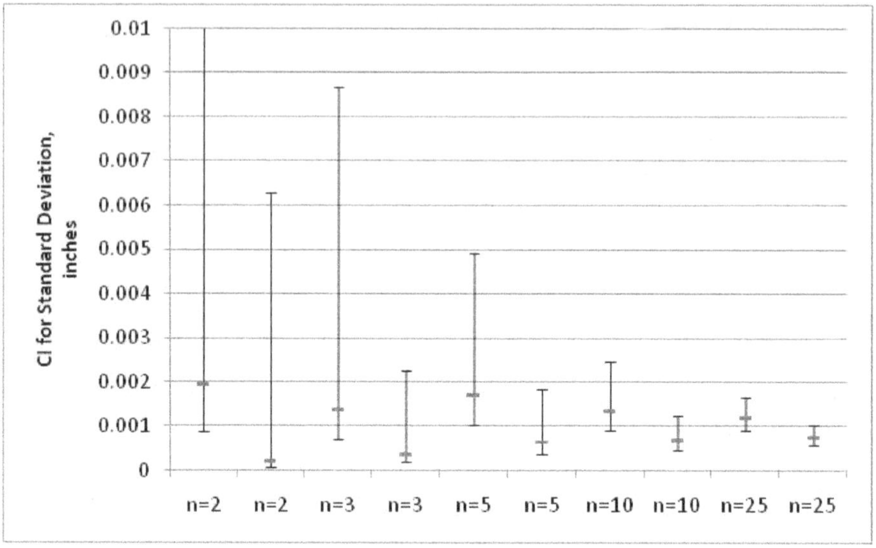

Figure 2.19 95% confidence interval of the standard deviations for Table 2.6. The last section in this chapter shows how to make this graph in Excel

2.4.3 Making Confidence Interval Graphs

Suppose that our 10,000 injection molded parts were made in a four cavity mold, instead of a one cavity mold. For reasons that will be discussed in the next chapter, it would be a good idea to look at each cavity separately. This section will show how to make a graph showing the mean and CIs for each cavity.

2.4.3.1 Example 2.6: Making a Confidence Interval Graph for the Means in Excel & Minitab

After collecting 25 shots (1 part from each cavity), the data was tabulated in Table 2.7.

Table 2.7 Data for Example 2.6. To Calculate the 95% Confidence Interval of the Means Graph in Excel, the Cavity, $\overline{x}$, and δ Columns Are Used

	n	$\overline{x}$, in	S, in	$t_{critical}$	$\delta = 1/2$ CI, in	X_L, in	X_U, in
Cavity 1	25	10.49984	0.001143	2.0639	0.00047183	10.49936	10.50031
Cavity 2	25	10.49959	0.001854	2.0639	0.00076534	10.49882	10.50035
Cavity 3	25	10.50112	0.000878	2.0639	0.00036235	10.50075	10.50148
Cavity 4	25	10.49958	0.001011	2.0639	0.00041745	10.49916	10.49999

Select the cavity column and $\bar{x}$ column, then insert a line chart to get the following:

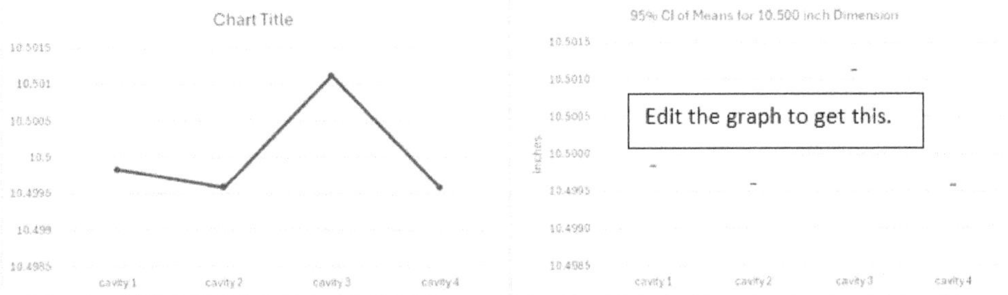

Add the 95% CI limits by going to Chart Elements (comes up when chart is selected) and selecting More Options for the Error Bars, then Custom in the lower right corner of Format Error Bars, and use the δ column for both the positive and negative error value as shown in the following.

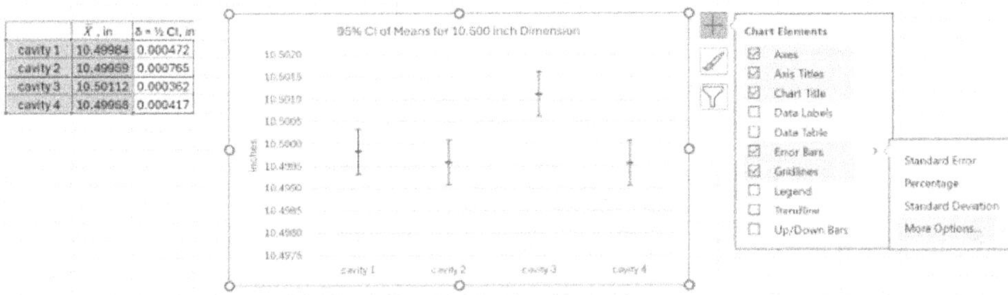

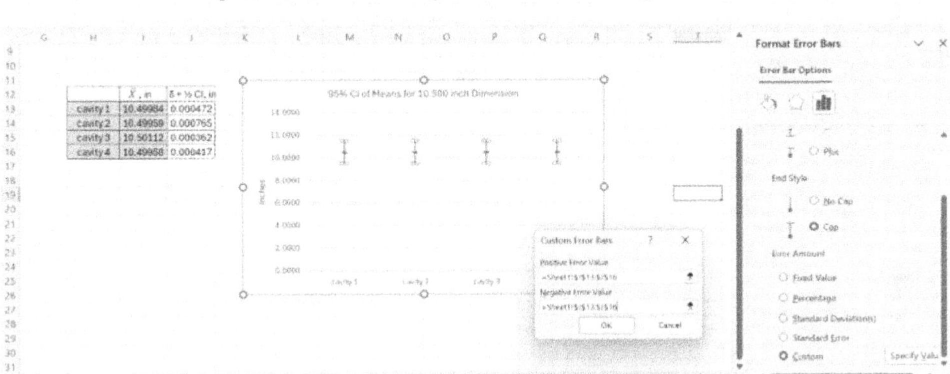

Figure 2.20 Sequence for creating 95% confidence interval for the means in Excel

The results are shown in Figure 2.21.

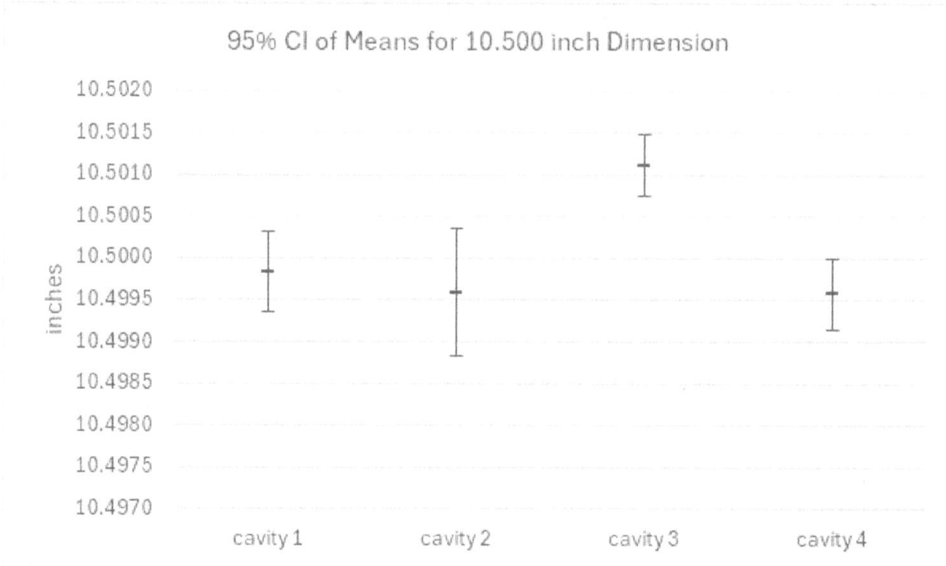

Figure 2.21 95% confidence interval of the means for Example 2.6

Looking at this chart (Figure 2.22), the one thing that can clearly be said, with a 95% confidence level, is that cavity 3's dimension is bigger than the other cavities (not a good thing). This is clear because cavity 3's 95% CI does not overlap any of the others. There is some overlap of the other three cavities. This will be revisited in Chapter 6.

To make the same graph in Minitab, go to Graph, then Interval Plot.

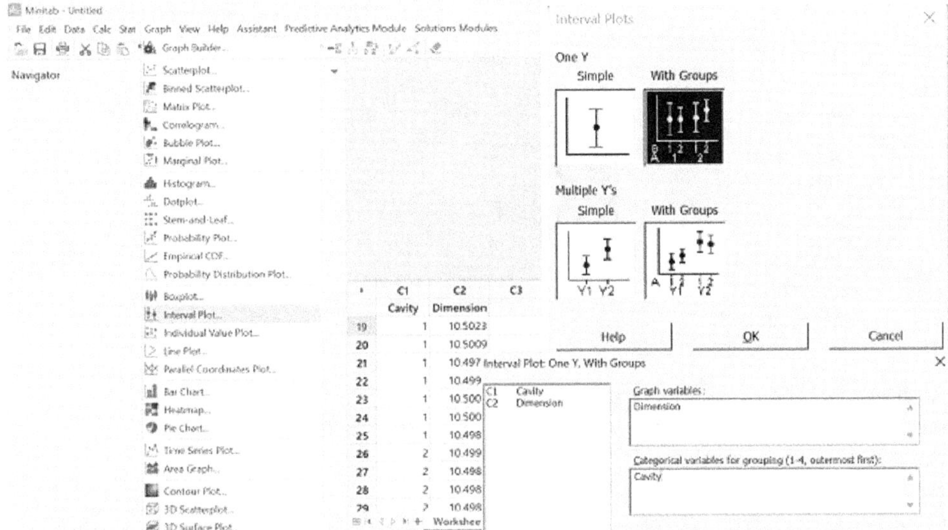

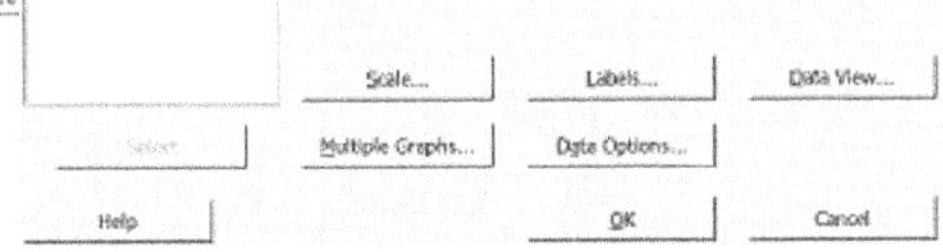

Note that the cavity identifiers are stacked in one column and the twenty-five values for each cavity are stacked in another column. Select the One Y With Groups Option as shown here.

The data column is selected for the graph variables and the Cavity identifiers for the categorical variables for grouping. The graph can be edited by selecting it and using the drop-down menu. The same graph obtained in Excel is the result.

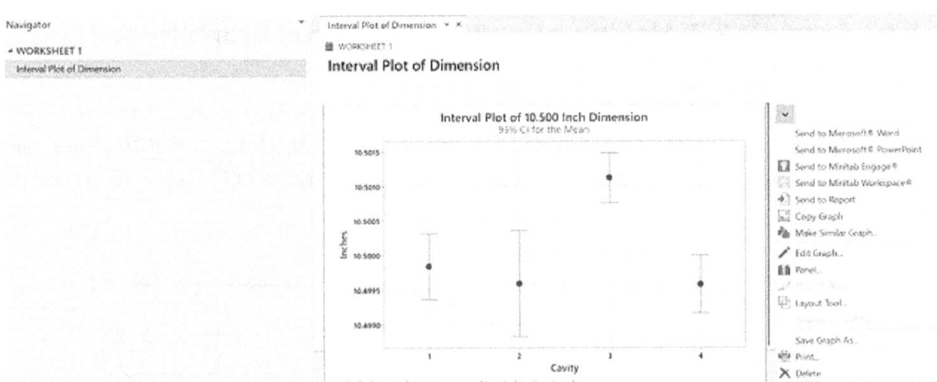

Figure 2.22 Sequence for creating the 95% confidence intervals of the means in Minitab with cavity identifiers

Alternatively, instead of using cavity identifiers in one column, the data for each cavity could be in separate columns and the simple interval plot with multiple Y's would be selected as shown in Figure 2.23.

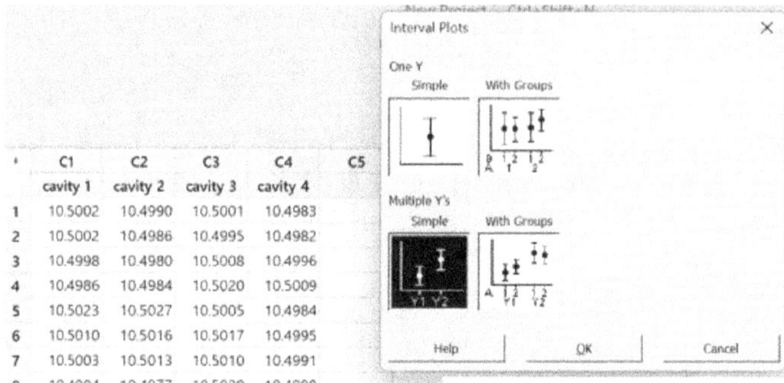

After editing, the resulting graph would be the same except for the x-axis labels being the column headings as seen in the graph.

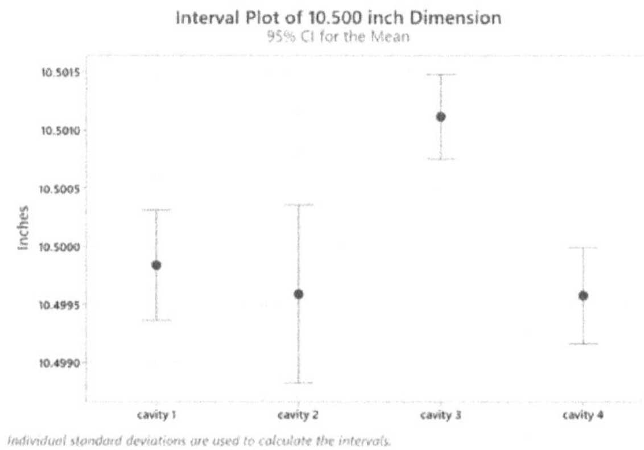

Figure 2.23 Sequence for creating 95% confidence intervals of the means in Minitab with data in separate columns

2.4.3.2 Example 2.7: Making a Confidence Interval Graph for the Standard Deviation in Excel

The following is a continuation of Example 2.6. To create this graph, the lower and upper limits need to first be calculated (Figure 2.24) and, then, the difference that each is from the sample standard deviation (δ) needs to be calculated to make the graph in Excel.

Cavity #	N	S, in	X_U^2	X_L^2	Lower limit σ_x, in	Upper limit σ_x, in	δ_L, in	δ_U, in
1	25	0.001143	39.3641	12.4012	0.00089	0.00159	0.000251	0.000447
2	25	0.001854	39.3641	12.4012	0.00145	0.00258	0.000406	0.000725
3	25	0.000878	39.3641	12.4012	0.00069	0.00122	0.000192	0.000343
4	25	0.001011	39.3641	12.4012	0.00079	0.00141	0.000222	0.000396

Figure 2.24 Data for Example 2.7 to make 95% confidence interval graph of the standard deviations in Excel

Although this book will not go through how to make the standard deviation confidence interval graphs based on the chi-square distribution in Minitab, if Minitab is available, the values in the previous table can be checked as follows. The data will need to be arranged with each cavity in different columns (Figure 2.25).

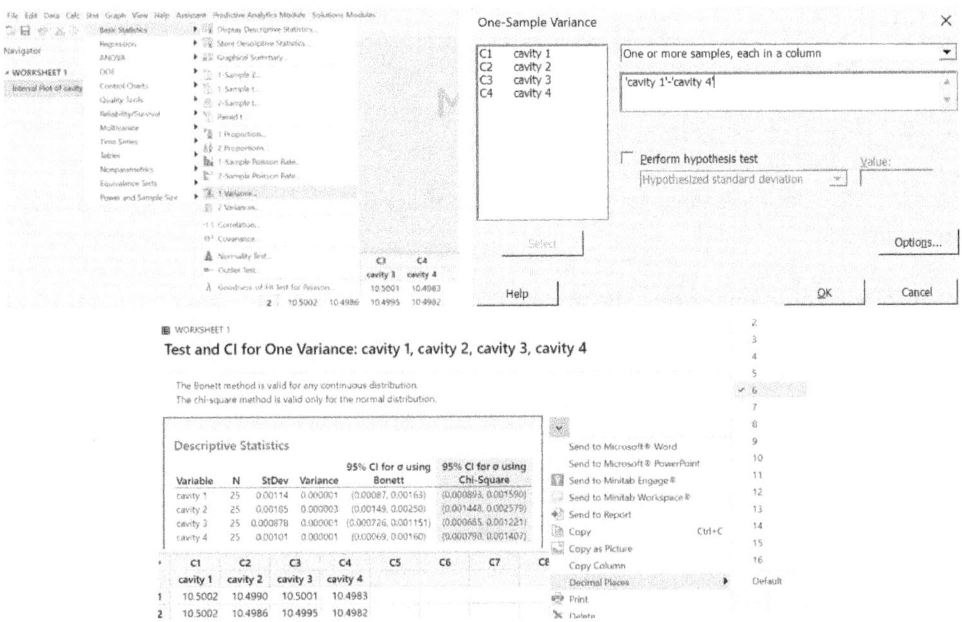

Figure 2.25 Sequence for determining 95% confidence intervals of the standard deviations in Minitab

The chi-square confidence intervals obtained in Minitab are the same as those from Excel. Also shown is how to change the number of decimal points that Minitab displays in the descriptive statistics.

Now that we have confirmed that the numbers are the same in Excel and Minitab, the 95% confidence intervals for the standard deviation can be made in Excel, as shown in Figure 2.26.

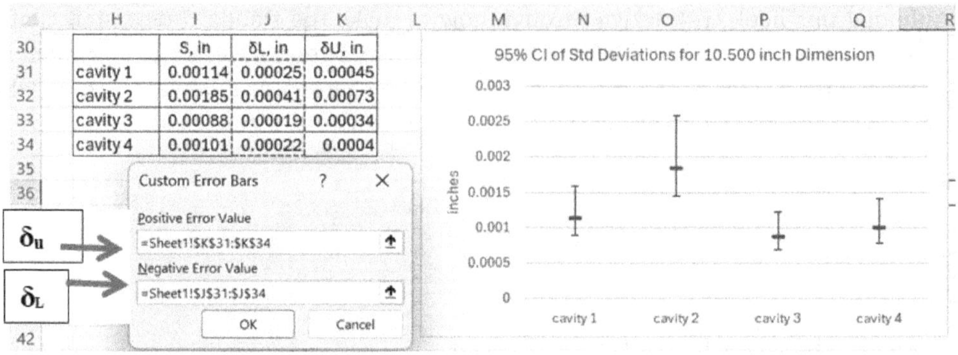

Figure 2.26 Data to select for making 95% confidence intervals of the standard deviations in Excel

It is important to check that the intervals match what was calculated once the graph has been made. Based on the confidence interval for cavity 2 being higher and not overlapping the intervals of 3 and 4, it can be said that cavity 2 has more variation than 3 or 4. We will revisit how to make the determination of whether 2 has more variation than 1 when we get to the Hypothesis Testing chapter (Chapter 6).

3 Statistical Process Control

Most of the examples in Chapter 2 dealt with taking samples of a product after the completion of a production run. It was assumed that the product characteristics being measured displayed a normal distribution; that all parts molded in one run from one cavity would display a nice bell-curve distribution. In practice, some processes do not work that way. Walter Shewart, the father of SPC, discovered that processes displayed two types of variation; controlled variation and uncontrolled variation.

1. Controlled variation – This type of variation is characterized by a stable and consistent pattern of variation over time. Also called random, chance, predictable, natural, or common cause variation. It is also sometimes referred to as the noise in the process.
2. Uncontrolled variation – This type of variation is characterized by a pattern of variation that changes over time. Also called non-random, assignable, unpredictable, unnatural, or special cause variation. It is also sometimes referred to as the signals that are present in the process.

There are many sources of variation in the injection molding process (and in most other processes). Even with a single cavity mold, there will be variation in plastic material properties (viscosity is one that has a large effect on injection molding), screw tip performance, temperatures (barrel, mold, drying, feed throat, oil, ambient, etc.), mold venting, clamping, etc. If an operator is controlling the mold open time because of insert placement, part removal, or some other process step, they become a significant source of variation. To compound all these sources of variation is the issue of measurement variation which will be discussed in a later chapter. With multi-cavity molds, more sources of variation are present (see Figure 3.1).

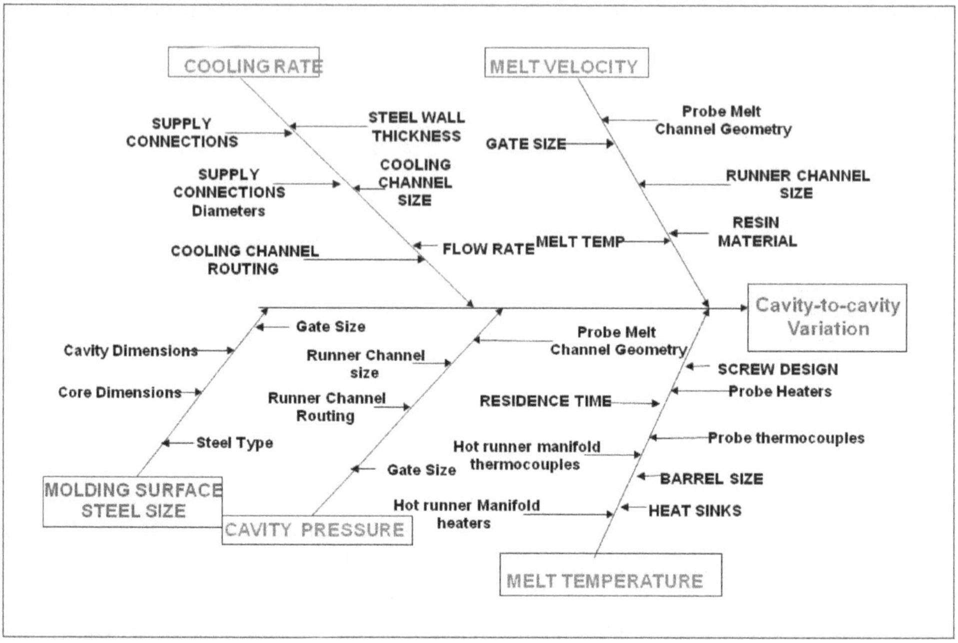

Figure 3.1 Possible sources of variation in multi-cavity molds (courtesy of Beckton Dickinson)

All these sources of variation (and many more which aren't listed here) contribute to the common cause variation and can be the source of a special cause of variation. How some of these sources of variation would show up on a control chart if they were to become special causes of variation is shown in Figure 3.2.

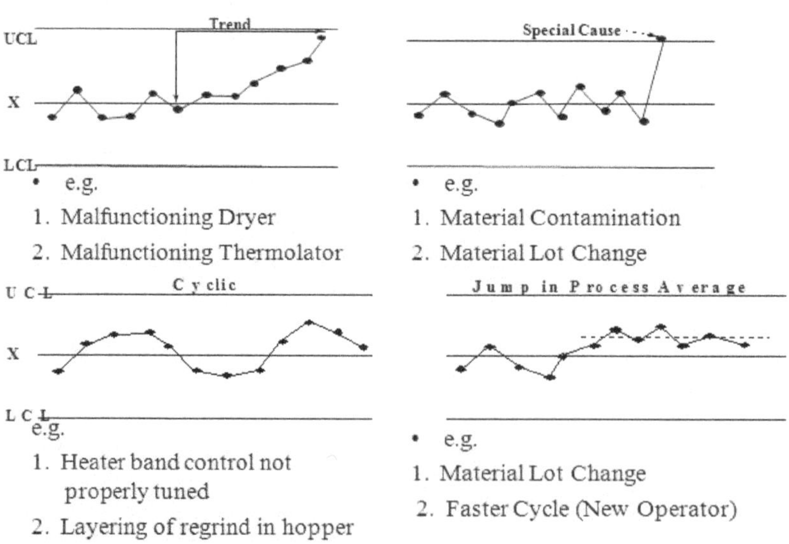

Figure 3.2 Examples of uncontrolled (special cause) variation

The screw tip is a good example of common cause versus special cause variation. On any machine with a screw tip, there will be some, hopefully very small, variation in how much plastic leaks back past the screw tip before it completely shuts off and acts like a plunger when it comes forward. This would be considered common cause or controlled variation if the variation remains relatively constant over time. This controlled variation is the primary reason that the injection velocity is slowed when the part is 95–99% full, instead of 99.999% full (see Chapter 11). There are different screw tip designs which will have more or less variation for a given process. However, if a check ring cracks, becomes excessively worn, or solid pellets or contamination prevent the screw tip from functioning correctly, then the material leakage becomes inconsistent or increases excessively, and the resulting part variation increases. This would be an example of special cause or uncontrolled variation. The main purpose of control charts is to catch the special causes of variation before a lot of inconsistent, and perhaps out-of-specification, product is made. However, control charts can, and this author would say should, also be used to track the common cause variation and be used to continuously improve processes by reducing common cause variation. How to judge which processes have too much common cause variation will be discussed in the next chapter.

SPC involves measuring a characteristic or characteristics of a part as it is being made, calculating statistics, and comparing those statistics to what they have been in the past to determine if the variation present is common cause or special cause. In the case of injection molded parts, there will be somewhat of a time lag, since the parts are usually given time to stabilize before measurements are made. When starting with a new product, there will be no past or historical data with which to compare, so a baseline needs to be established. This is typically done after the process has been set up using recommended procedures (see Chapter 11). Subgroups of consecutive shots (typically 5 shots) are collected for a period of typically four to twenty-four hours, until at least twenty-five subgroups have been collected. The statistics, usually the average and the range, are calculated for each subgroup and then graphed as a time series to see whether the process is stable or in-control (only common cause variation present). If patterns like those seen in Figure 3.1 are found when doing the baseline study, then it is important that the source of the special or uncontrollable variation be found and made controllable. Even if all the parts meet specifications, but there is special cause variation present in the data, there is a high probability that the special cause of variation will cause problems in future production.

Before discussing more of the concepts that apply to SPC, examples of how to produce the following charts will be provided as a frame of reference.

1. Average (X-Bar)/Range charts
2. Individuals/Moving Range charts
3. Three-Way charts
4. Process Aim charts

3.1 Average/Range Charts

The Average and Range charts are by far the most common type of control chart. The average sample standard deviation is sometimes used instead of the average range, but for the typical control chart with a subgroup size of five, the Range chart is basically equivalent to the sample standard deviation and will be used exclusively in this book.

3.1.1 Example 3.1: Average/Range Charts in Excel and Minitab

It is desired produce control charts for the baseline process for the distance between two snap fits on an injection molded part. This characteristic is a critical-to-quality (CTQ) feature, due to its effect on assembly. If the distance between the two is too small, then insertion force tends to be high and the plastic arm will sometimes crack. If the distance is too large, then the part holds the mating piece too loosely, causing vibration problems. For this example, assume that a one-cavity mold is being used and that thirty subgroups of five have been collected for a CTQ dimension. The results are shown in Table 3.1 (all readings are in inches).

Table 3.1 Table of Data for Example 3.1

Subgroup						Subgroup Average	Subgroup Range
1	3.5012	3.4962	3.4976	3.4998	3.4999	3.4989	0.0050
2	3.4971	3.4989	3.4961	3.5049	3.4956	3.4985	0.0094
3	3.5001	3.4988	3.4979	3.4992	3.5010	3.4994	0.0031
4	3.5027	3.4977	3.5006	3.4979	3.4998	3.4997	0.0050
5	3.4973	3.4975	3.5032	3.5037	3.5013	3.5006	0.0064
6	3.5004	3.5021	3.4965	3.4976	3.5044	3.5002	0.0079
7	3.5000	3.4993	3.5018	3.4988	3.5024	3.5005	0.0036
8	3.5011	3.5012	3.5011	3.4983	3.4999	3.5003	0.0030
9	3.5036	3.4997	3.5024	3.5015	3.5016	3.5018	0.0040
10	3.5003	3.4995	3.4936	3.5009	3.5048	3.4998	0.0113
11	3.5007	3.4967	3.4955	3.5021	3.4975	3.4985	0.0066
12	3.4997	3.5041	3.4984	3.5009	3.5020	3.5010	0.0057

Subgroup						Subgroup Average	Subgroup Range
13	3.4989	3.5017	3.5022	3.5014	3.4988	3.5006	0.0033
14	3.5000	3.5000	3.5005	3.5018	3.5012	3.5007	0.0019
15	3.5001	3.4994	3.5008	3.5026	3.4999	3.5006	0.0032
16	3.4971	3.4992	3.4984	3.5033	3.5031	3.5002	0.0062
17	3.4977	3.4977	3.4996	3.5022	3.5018	3.4998	0.0045
18	3.5006	3.5028	3.5024	3.5063	3.4996	3.5023	0.0066
19	3.4999	3.4966	3.5029	3.5035	3.4999	3.5006	0.0069
20	3.4996	3.4967	3.5014	3.4968	3.5003	3.4990	0.0047
21	3.5027	3.5045	3.4965	3.4986	3.5045	3.5014	0.0080
22	3.5019	3.4980	3.5006	3.4969	3.5026	3.5000	0.0057
23	3.4995	3.5024	3.4972	3.4993	3.4994	3.4995	0.0052
24	3.5012	3.4993	3.5009	3.5047	3.4999	3.5012	0.0054
25	3.5029	3.4979	3.4989	3.4985	3.4976	3.4992	0.0053
26	3.5006	3.4991	3.4988	3.4965	3.4979	3.4986	0.0041
27	3.4975	3.4991	3.5034	3.5036	3.5033	3.5014	0.0061
28	3.4998	3.4975	3.5028	3.4981	3.5008	3.4998	0.0053
29	3.5012	3.4984	3.4988	3.4949	3.4974	3.4981	0.0063
30	3.5039	3.5017	3.4977	3.5017	3.5014	3.5013	0.0062

Note that the average and the range for the five parts in each subgroup have also been calculated in Table 3.1. Next, the average range and the control limits for the Range chart need to be calculated using Equation 3.1 to Equation 3.3.

Equations for the limits and centerline for the Range control chart

Centerline of the Range chart = Average Range:

$$\overline{R} = \sum_{i=1}^{k} {}^{R_i}/_{k} \tag{3.1}$$

Upper control limit of the Range chart:

$$UCL_R = D_4\overline{R} \tag{3.2}$$

Lower control limit of the Range chart:

$$LCL_R = D_3\overline{R} \tag{3.3}$$

Where:

D_4 and D_3 = control chart constants found in Table 12.2 of the Appendix

k = number of subgroups

With a subgroup size of five (n = 5), Table 12.2 shows that D_4 = 2.114 and D_3 = 0. Taking the average of the thirty ranges (Equation 3.1), it is found that $\overline{R}$ = **0.0055 inches.** Using Equation 3.2 and Equation 3.3, the control limits are:

- $UCL_R = D_4R$ = 2.114 × 0.0055 inches = 0.0117 inches
- $LCL_R = D_3\overline{R}$ = 0 × 0.0055 inches = 0 inches

The Range chart can then be produced (Figure 3.3).

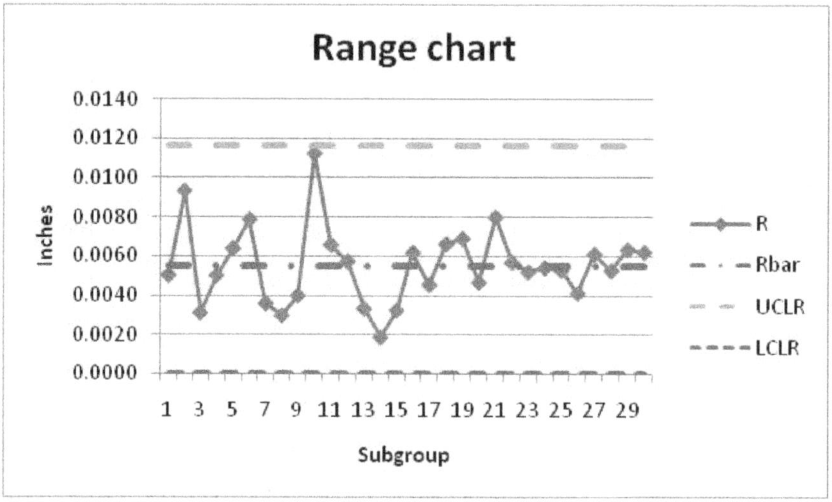

Figure 3.3 Example of Range chart for subgrouped data

This chart gives a look at the **within subgroup** variation. Since none of the points are outside the control limits, it can be said that the within-subgroup variation or the shot-to-shot variation is consistent.

Next the Grand Average and the control limits for the Average chart need to be calculated using Equation 3.4 to Equation 3.6.

Formulas for the limits and centerline of the Average control chart

Centerline of Average chart = Grand Average:

$$\overline{\overline{X}} = \sum_{i=1}^{k} \overline{x}_i / k \tag{3.4}$$

Upper control limit of Average chart:

$$UCL_{\overline{X}} = \overline{\overline{X}} + A_2\overline{R} \tag{3.5}$$

Lower control limit of Average chart:

$$LCL_{\overline{X}} = \overline{\overline{X}} - A_2\overline{R} \tag{3.6}$$

Where:

A_2 = control chart constant found in Table 12.2 of the Appendix

k = number of subgroups

With a subgroup size of five (n = 5), Table 12.2 shows that A_2 = 0.577. Taking the average of the thirty averages (Equation 3.4), it is found that $\overline{\overline{X}}$ = 3.5001 inches. Using Equation 3.5 and Equation 3.6, the control limits are:

- $UCL_{\overline{X}} = \overline{\overline{X}} + A_2\overline{R}$ = 3.5001 inches + 0.577 × 0.0055 inches = 3.5033 inches
- $LCL_{\overline{X}} = \overline{\overline{X}} - A_2\overline{R}$ = 3.5001 inches − 0.577 × 0.0055 inches = 3.4969 inches

The Average chart can then be produced (Figure 3.4).

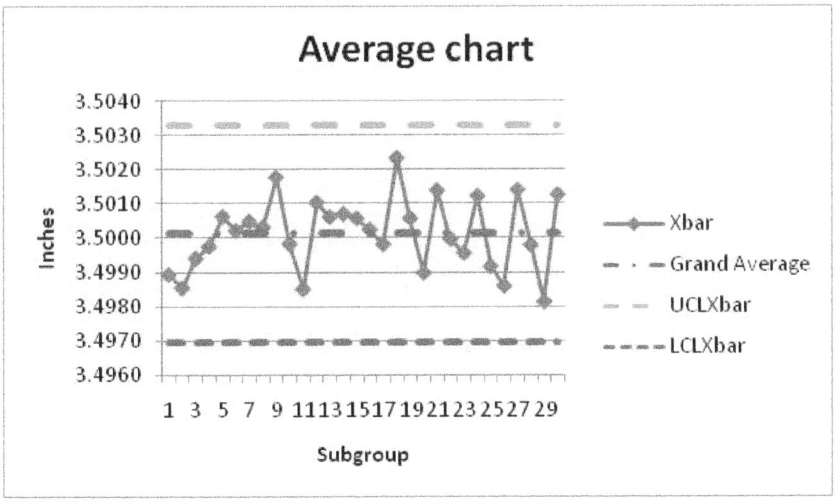

Figure 3.4 Example of an Average chart for subgrouped data

This chart shows the differences **between subgroups**. Since none of the points are outside the control limits, it can be said that there weren't any detectable differences between subgroups or, in this case, over time. It should be noted that the Average chart could be deemed out of control even if no points are outside the control limits (see Figure 3.2) – this will be discussed later in this chapter.

The Average and Range charts can be produced in Minitab, as shown in Figure 3.5.

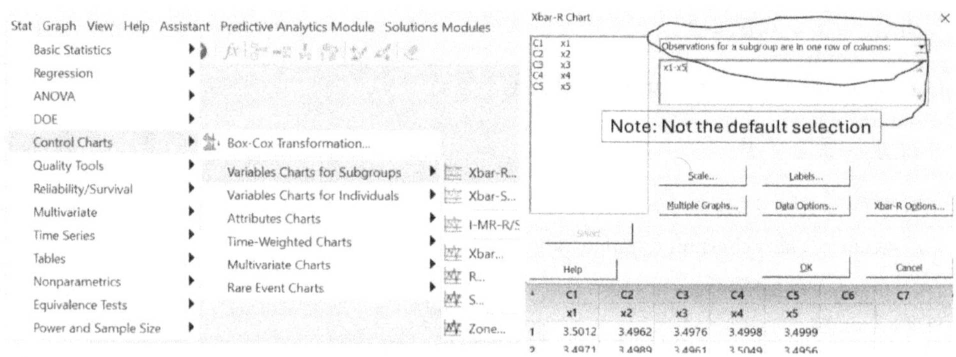

Figure 3.5 Sequence to produce the Average and the Range charts in Minitab

The following graphs match those made in Excel (Figure 3.6).

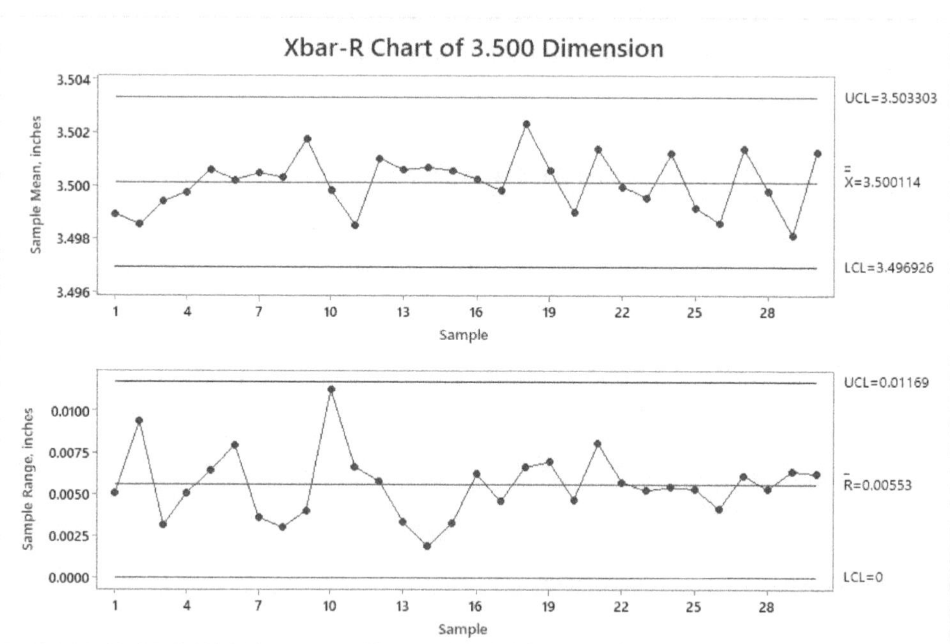

Figure 3.6 Average and Range charts in Minitab for Example 3.1

3.2 Individuals/Moving Range Charts

Although the Average and Range charts from Example 3.1 are generally preferred, there are times when the subgroup size will be one ($n = 1$). Examples of when this is desirable are when a destructive test is done to obtain the data and the parts are expensive (large parts, parts with expensive inserts, etc.), when batches of product (such as a batch of compounded plastic) are being made and the main concern is batch-to-batch variation, when inspection costs need to be minimized, or when every part is being inspected by means of automation. In some cases, it is desirable to monitor continuously (every shot from a molding process) to pick up patterns that may not be seen in intermittently taken data.

3.2.1 Example 3.2: Individuals/Moving Range Charts in Excel and Minitab

This example will look at making control charts to investigate the amount of mismatch of an injection molded briefcase that is assembled in an automation cell right next to the molding machine. The mismatch is checked automatically before the briefcases are packed. Minimizing the amount of mismatch is a key to the perceived quality of the briefcases and thus the mismatch is a CTQ dimension. The data shown in Table 3.2 is for thirty consecutive briefcases (all dimensions are in millimeters).

Table 3.2 Data for Example 3.2

Briefcase	Mismatch	Moving Range
1	0.54	
2	–0.08	0.62
3	0.47	0.55
4	0.24	0.23
5	0.60	0.36
6	0.36	0.24
7	1.16	0.80
8	0.39	0.77
9	1.08	0.69
10	0.00	1.08
11	0.26	0.26

Table 3.2 Data for Example 3.2 (*continued*)

Briefcase	Mismatch	Moving Range
12	0.32	0.06
13	0.57	0.25
14	0.36	0.21
15	0.32	0.04
16	0.63	0.31
17	0.31	0.33
18	0.91	0.60
19	0.77	0.14
20	0.79	0.01
21	0.73	0.06
22	0.17	0.56
23	0.04	0.13
24	0.22	0.18
25	0.51	0.29
26	1.14	0.63
27	-0.11	1.25
28	0.56	0.67
29	0.32	0.24
30	0.43	0.11

Note that the Moving Range has also been calculated using Equation 3.7.

Equation for Moving Range:

$$mR = |X_i - X_{i+1}| \tag{3.7}$$

Next, the Average Moving Range and the control limits for the Moving Range chart can be calculated using Equation 3.8 toEquation 3.10.

Equations for the limits and centerline for the Moving Range control chart

Centerline of the Moving Range chart = Average Moving Range:

$$m\overline{R} = \sum_{i=1}^{k} {}^{mR_i}/_{k} \tag{3.8}$$

Upper control limit of the Moving Range chart:

$$UCL_{mR} = D_4 m\overline{R} \tag{3.9}$$

Lower control limit of the Moving Range chart:

$$LCL_{mR} = D_3 m\overline{R} \tag{3.10}$$

Where:

D_4 and D_3 = control chart constants found in Table 12.2 of the Appendix

k = number of moving ranges

For Individuals/Moving Range charts, a subgroup size of two (n = 2) is typically used. Therefore, D_4 = 3.268 and D_3 = 0. Taking the average of the twenty-nine moving ranges (Equation 3.8), it is found that $m\overline{R}$ = 0.40 mm. Using Equation 3.9 and Equation 3.10, the control limits are:

- $UCL_{mR} = D_4 m\overline{R}$ = 3.268 × 0.40 mm = 1.31 mm
- $LCL_R = D_3\overline{R}$ = 0 × 0.40 mm = 0 mm

The Moving Range chart can then be produced (Figure 3.7).

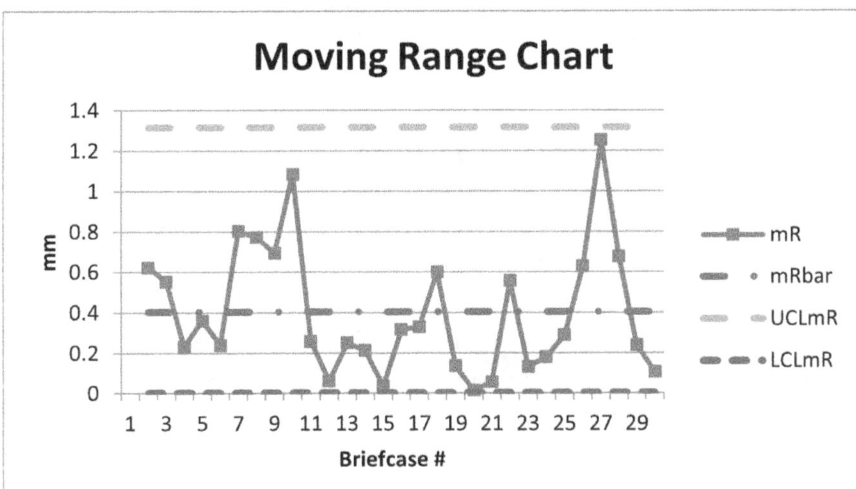

Figure 3.7 Example of a Moving Range chart

Since none of the points are outside the control limits, it can be said that the short-term variation or the shot-to-shot variation is consistent.

Next, the average and the control limits for the Individuals chart need to be calculated using Equation 3.11 to Equation 3.13.

Equations for the limits and centerline of the Individuals control chart

Centerline of Individuals chart = Average:

$$\overline{X} = \sum_{i=1}^{k} x_i/k \tag{3.11}$$

Upper control limit of Individuals chart:

$$UCL_X = \overline{X} + 2.66m\overline{R} \tag{3.12}$$

Lower control limit of Individuals chart:

$$LCL_X = \overline{X} - 2.66m\overline{R} \tag{3.13}$$

Where:

k = number of individual values

Taking the average of the thirty individual values (Equation 3.11), it is found that $\overline{X}$ = 0.47 mm. Using Equation 3.12 and Equation 3.13, the control limits are:

- $UCL_X = \overline{\overline{X}} + 2.66\ m\overline{R}$ = 0.47 mm + 2.66 × 0.40 mm = 1.54 mm
- $LCL_X = \overline{\overline{X}} - 2.66\ m\overline{R}$ = 0.47 mm − 2.66 × 0.40 mm = −0.60 mm

The Individuals chart can then be produced (Figure 3.8).

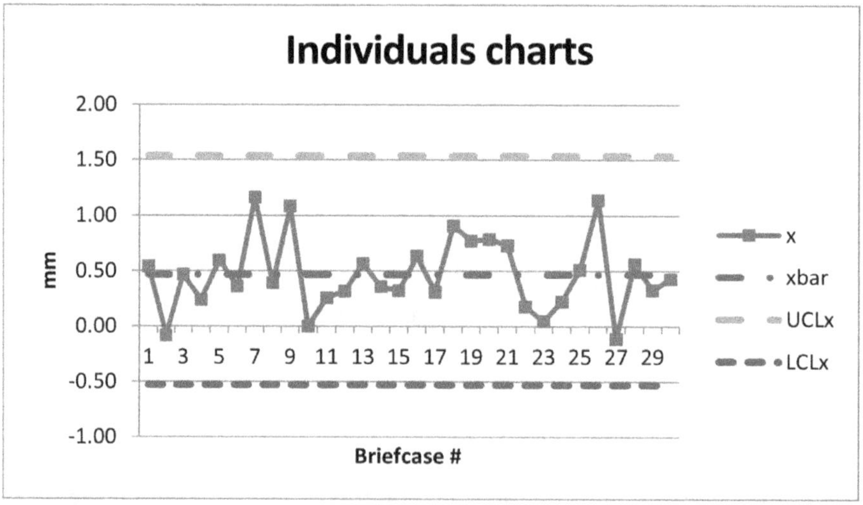

Figure 3.8 Example of an Individuals chart

The same charts can be produced in Minitab as shown in Figure 3.9 and Figure 3.10.

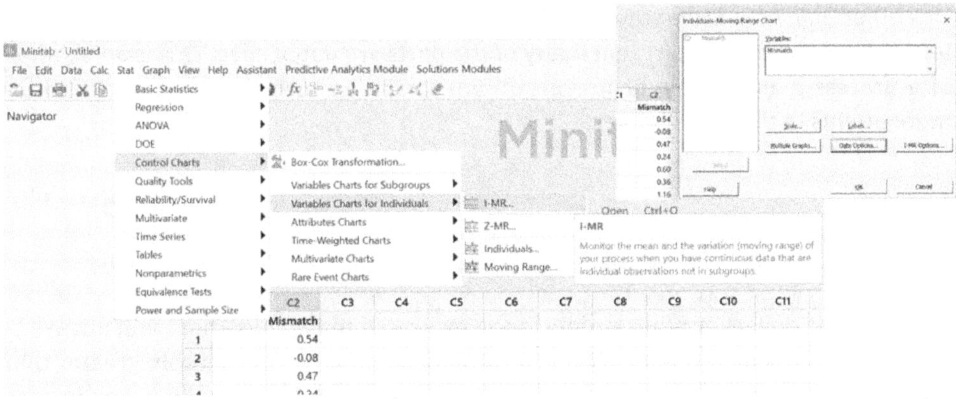

Figure 3.9 Sequence for producing Individuals/Moving Range charts in Minitab

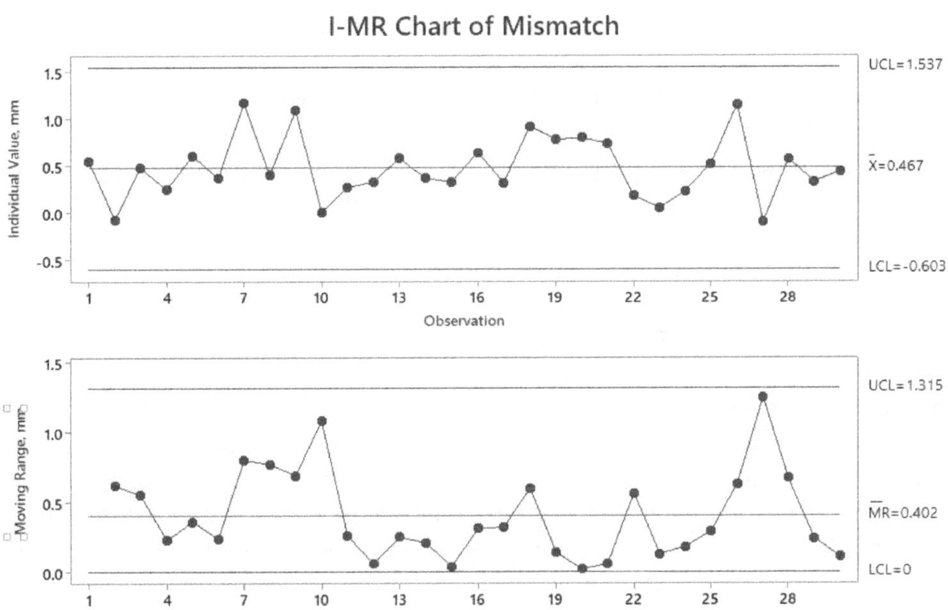

Figure 3.10 Individuals/Moving Range charts for Example 3.2 produced in Minitab

Since none of the points are outside the control limits, it can be said that there weren't any detectable differences between subgroups or, in this case, over the time of the thirty consecutive shots. It should be noted that the Individuals chart could be deemed out of control even if no points are outside the control limits (see Figure 3.2) – this will be discussed later in this chapter.

One other note here is that the specification and tolerance for mismatch is 0 ± 0.5 mm. It is easy to see on the chart that many of the parts are out of spec. This is an example of a process that is stable or in-control, but unable to meet design specifications – more on this in the next chapter.

3.3 Three-Way Charts

When plastic pellets or other products are produced in batches, there will always be variation from batch to batch. This variation is generally considerably greater than the amount of variation within a batch. Three-Way control charts allow the monitoring of both sources of variation without unrealistically expecting the Average chart to be in-control based on control limits calculated using the within-batch variation ($\overline{R}$ from the Range chart). A subgroup is taken from each batch and the average for each is calculated. Each average is then treated as an individual value to produce a Moving Range chart which checks whether the between-batch variation remains consistent. The Average Moving Range is then used to determine the control limits for the Average chart. The Average chart checks to see if there are detectable differences over time based on the average variation between batches. The standard Range chart is then used to check whether the within-batch variation remains consistent.

3.3.1 Example 3.3: Three-Way Charts

A company is compounding virgin polyethylene with scraps from several diaper manufacturers, shredded tires, and other types of recycled materials, in addition to UV inhibitors, colorants and several other additives. The blend is then injection molded into its proprietary product and sold at a premium price. To try to understand all the sources of variation that it sees, the company has decided to use SPC. It generally takes about two hours to extrude and pelletize one batch and the company will take a sample at the beginning and then every half hour in order to get a subgroup size of five (n = 5). A tensile bar is molded from each sample and the tensile strength (along with several other properties) is measured. The data from twenty-five consecutive batches is presented in Table 3.3 (all measurements are in psi).

Table 3.3 Data for Example 3.3

Batch						Range	Average	*mR*
1	4004	4006	3975	4019	4011	44	4003	
2	4143	4138	4101	4131	4131	42	4129	126
3	4105	4104	4121	4118	4139	35	4117	11
4	4407	4394	4398	4359	4372	47	4386	269
5	4419	4381	4409	4410	4398	38	4403	17
6	4258	4282	4247	4258	4261	36	4261	142
7	4318	4309	4291	4291	4288	29	4300	38
8	4074	4086	4060	4048	4099	50	4073	226
9	4045	4021	4041	4019	4036	27	4032	41
10	4043	4026	4044	4057	4033	30	4041	8
11	4024	4026	4022	4028	4045	24	4029	12
12	4153	4181	4176	4147	4167	34	4165	136
13	5067	5066	5063	5033	5042	34	5054	890
14	4250	4238	4260	4257	4269	31	4255	800
15	4269	4271	4256	4251	4273	21	4264	9
16	4125	4155	4151	4142	4138	29	4142	122
17	4081	4104	4085	4088	4086	23	4089	53
18	4011	4037	4013	4034	4008	29	4021	68
19	4310	4295	4324	4305	4320	28	4311	290
20	4389	4382	4383	4359	4361	30	4375	64
21	3731	3710	3732	3754	3741	44	3734	641
22	3334	3336	3327	3292	3301	44	3318	416
23	4513	4509	4531	4512	4529	22	4518	1201
24	4047	4036	4015	4025	4024	32	4029	489
25	3726	3742	3738	3738	3768	42	3742	287

Note that the average and the range have been calculated for each subgroup, along with the moving ranges of the averages.

Taking the average of the twenty-four moving ranges (Equation 3.8), it is found that $m\overline{R}$ = 265 psi. Using Equation 3.9 and Equation 3.10, the control limits are:

- $UCL_{mR} = D_4 m\overline{R}$ = 3.268 × 265 mm = 866 psi
- $LCL_R = D_3\overline{R}$ = 0 × 0.40 mm = 0 mm

The Moving Range chart can then be produced (Figure 3.11).

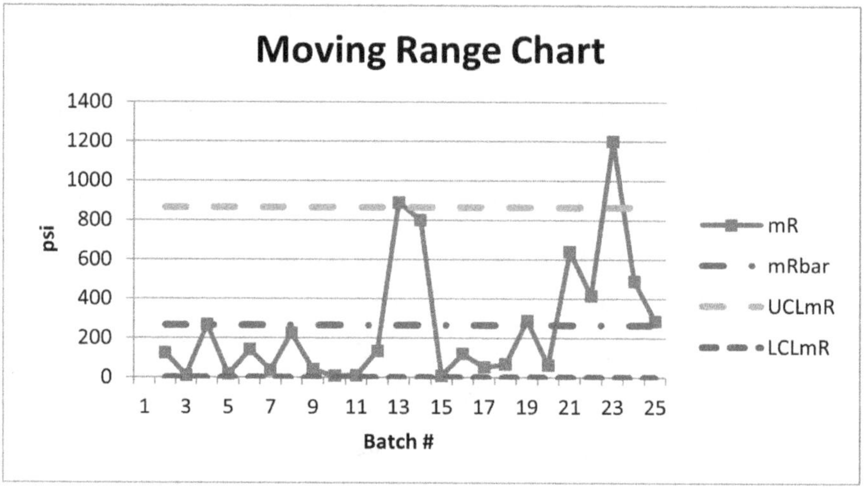

Figure 3.11 Moving Range chart for Example 3.3

There are a couple of points (13 and 23) that are out of control and thus the between-batch variation is not consistent.

Next, the Individuals chart using the averages as individual values should be produced. Taking the average of the twenty-five readings (Equation 3.11), it is found that $\overline{X}$ = 4152 psi. Using Equation 3.12 and Equation 3.13, the control limits are:

- $UCL_X = \overline{X} + 2.66 m\overline{R}$ = 4152 psi + 2.66 × 265 psi = 4856 psi
- $LCL_X = \overline{X} - 2.66 m\overline{R}$ = 4152 psi − 2.66 × 265 psi = 3447 psi

The Average chart can then be produced (Figure 3.12).

As can be seen, batches 13 and 22 are outside the control limits and so there are detectable differences in tensile strength between batches. If the goal is to maximize tensile strength, then batch 13 should be investigated more closely to duplicate what was done differently. On the other hand, whatever was done differently for batch 22 should be avoided in the future.

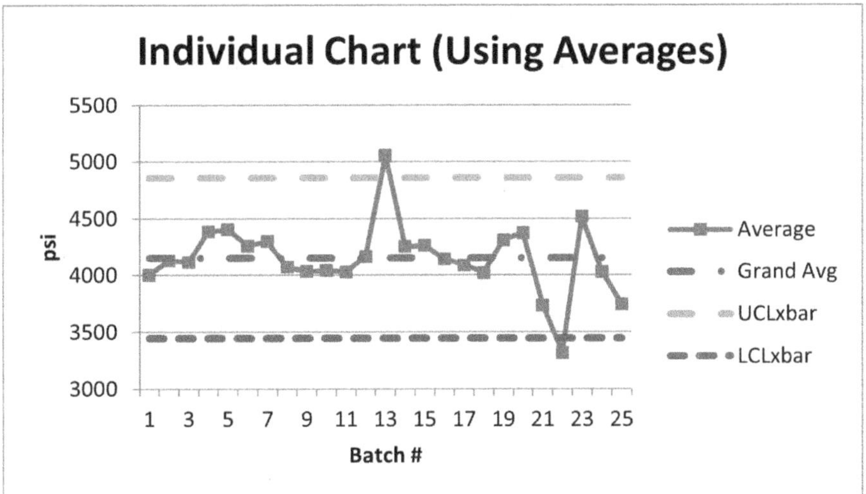

Figure 3.12 Individuals chart using batch averages for Example 3.3

Next, the standard Range chart should be produced. With a subgroup size of five (n = 5), Table 12.2 shows that D_4 = 2.114 and D_3 = 0. Taking the average of the twenty-five ranges (Equation 3.1), it is found that $\overline{R}$ = 34 psi. Using Equation 3.2 and Equation 3.3, the control limits are:

- $UCL_R = D_4\overline{R}$ = 2.114 × 34 psi = 72 psi
- $LCL_R = D_3\overline{R}$ = 0 × 34 psi = 0 psi

The Range chart can then be produced (Figure 3.13).

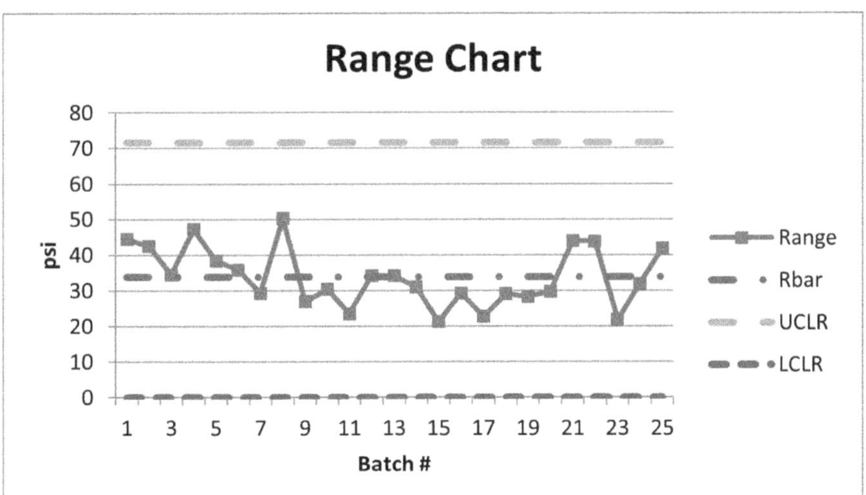

Figure 3.13 Range chart for Example 3.3

This chart shows that the within-batch variation is consistent.

The Three-Way charts can be produced in Minitab, as shown in Figure 3.14 and Figure 3.15.

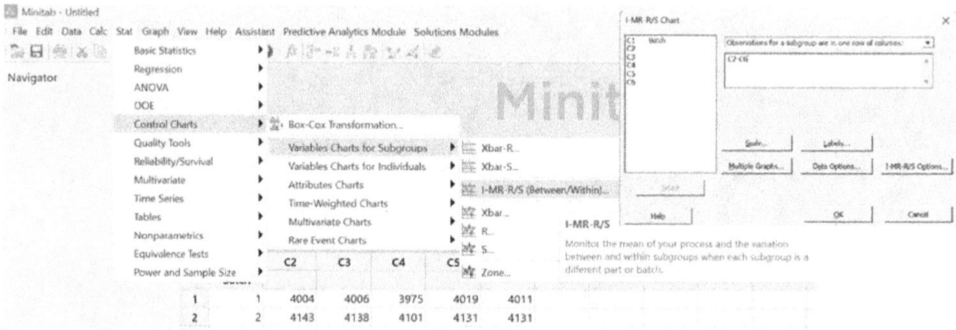

Figure 3.14 Sequence for producing Three-Way control charts in Minitab

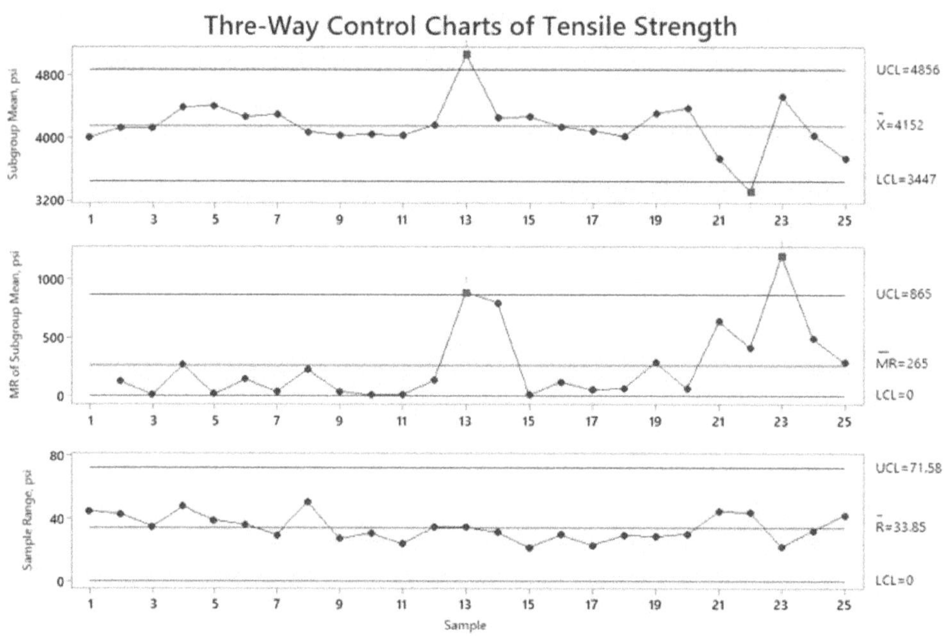

Figure 3.15 Three-Way control charts for Example 3.3 produced in Minitab

3.4 Process Aim Charts

It is common in the injection molding industry for a mold to sit on a shelf for several weeks in between production runs. When the mold is returned to production, it may or may not be run in the same molding machine with the same auxiliary equipment (dryers, temperature control unit, etc.). It will probably start up using a different batch of plastic than it was run with the last time. Even using good procedures to ensure that the machine settings are such that should give rise to constant plastic conditions, it is not uncommon that the parts will be different. The purpose of the Process Aim chart is to get the CTQ characteristics close to the target at the beginning of a production run.

Regardless of whether or not the Process Aim chart is used at the beginning of a production run, it can be a useful tool when an uncontrollable factor (generally a new batch of material) causes the process to go out of control and need adjustment. It will get the process back on course quicker than simply making an adjustment and waiting to see if the process is in-control after that. The steps of this procedure are as follows:

1. Using historical data, calculate $\hat{\sigma}_X$. This will generally be $\bar{R}/d_2$ if using control charts for subgrouped data
2. Make a Process Aim chart with the following decision lines:
   a. Centerline at target value – usually either the Grand Average from the last production run or the nominal value (center of design tolerance)
   b. Two lines at a distance of one sigma ($\hat{\sigma}_X$) from the centerline, on each side
   c. Two lines at a distance of two sigma ($\hat{\sigma}_X$) from the centerline, on each side
   d. Two lines at a distance of three sigma ($\hat{\sigma}_X$) from the centerline, on each side
3. Take ten consecutive shots (numbered sequentially) from the process using settings that will most likely give the same parts as the last production run
4. Plot the ten data points on the Process Aim chart in order
5. Check if any points show a lack of control, using the following four detection rules:
   a. Any point outside the three-sigma limits
   b. Two out of three points outside either two-sigma limit, on the same side
   c. Four out of five points outside either one-sigma limit, on the same side
   d. Eight consecutive points on the same side of the centerline
6. If a lack of control is found, make a process adjustment based on past experience or DOE results and go back to step 3
7. Once a lack of control is not found, switch from process aim mode back to SPC mode and start using the regular control charts

3.4.1 Example 3.4: Process Aim Charts

Assume the mold from Example 3.1 has been out of production for a few weeks and the nominal print dimension is 3.5000 inches. The average range for example 3.1 was 0.0055 inches. Using Equation 2.8, the standard deviation of the individual values, $\hat{\sigma}_X$, is:

$$\hat{\sigma}_X = \bar{R}/d_2 = 0.0055\frac{\text{inches}}{2.326} = 0.0024\,\text{inches}$$

The Process Aim chart with a nominal line, ± 1 sigma, ± 2 sigma, and ± 3 sigma can now be produced (Figure 3.16).

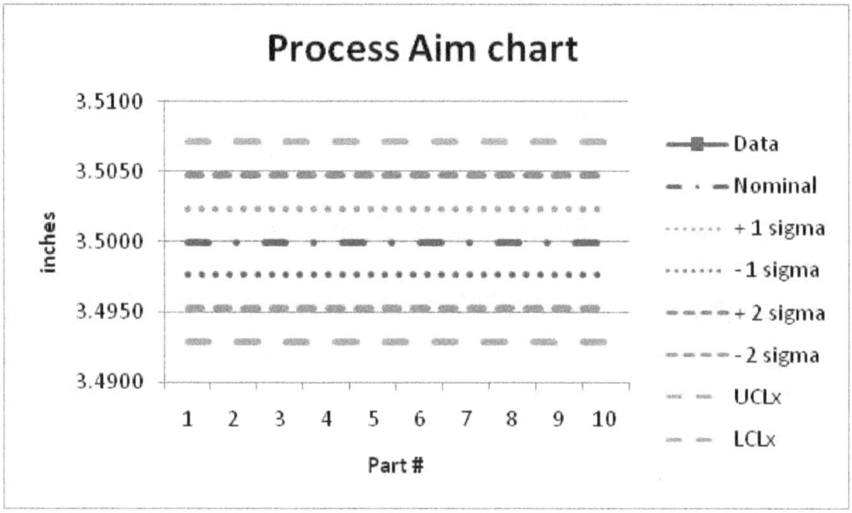

Figure 3.16 Process Aim chart for Example 3.4

When the process is started back up, a sample of ten parts is taken which measure and chart as shown in Figure 3.17.

As can be seen, the first four out of five parts are below the −1 sigma line and, therefore, the target has not been hit. If adding the data to the chart as the parts are measured, the measurements could stop at that point and the process could be adjusted in a way that would increase the size of the dimension. In many cases, there will be more than one CTQ characteristic, so the process change may depend on other measurements. Let's assume that this is the only CTQ characteristic, the process is adjusted, and a new sample of ten parts are taken which measure and chart as shown in Figure 3.18.

Part	
1	3.4968
2	3.4973
3	3.4968
4	3.4990
5	3.4950
6	3.4978
7	3.4932
8	3.4987
9	3.4962
10	3.4967

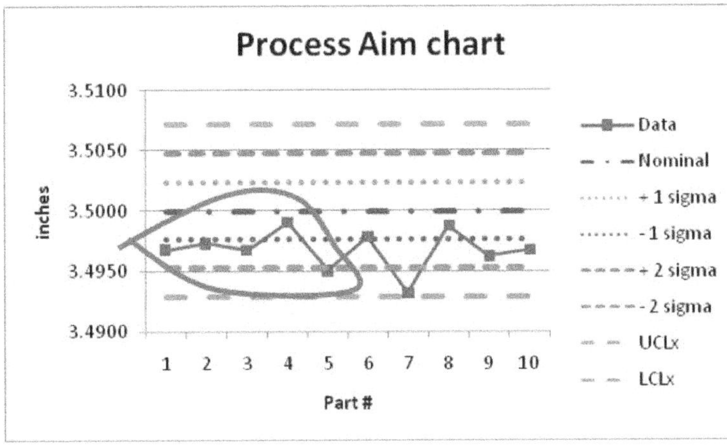

Figure 3.17 Process Aim chart for Example 3.3 that is out of control

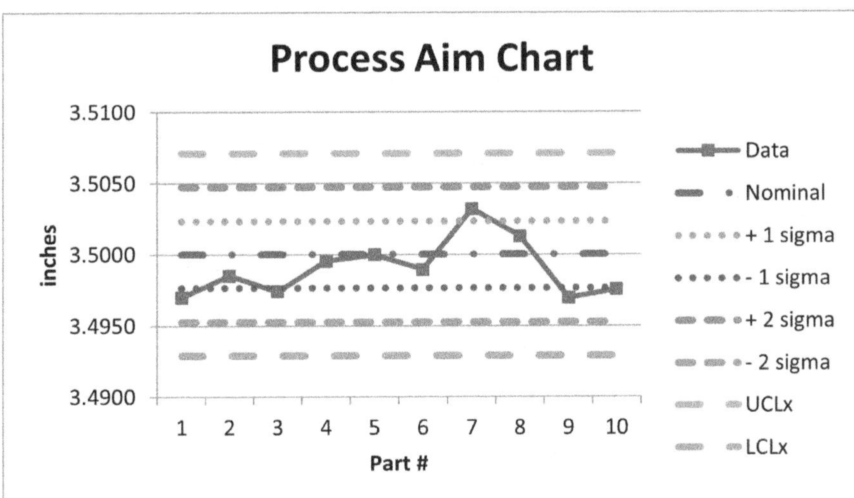

Figure 3.18 Process Aim chart for Example 3.3 that is in-control

As this chart does not show any signs of lack of control, the process settings will not be adjusted further, and the switch will be made back to the standard control chart mode.

The Process Aim chart can also be produced in Minitab, as shown in Figure 3.19 and Figure 3.20.

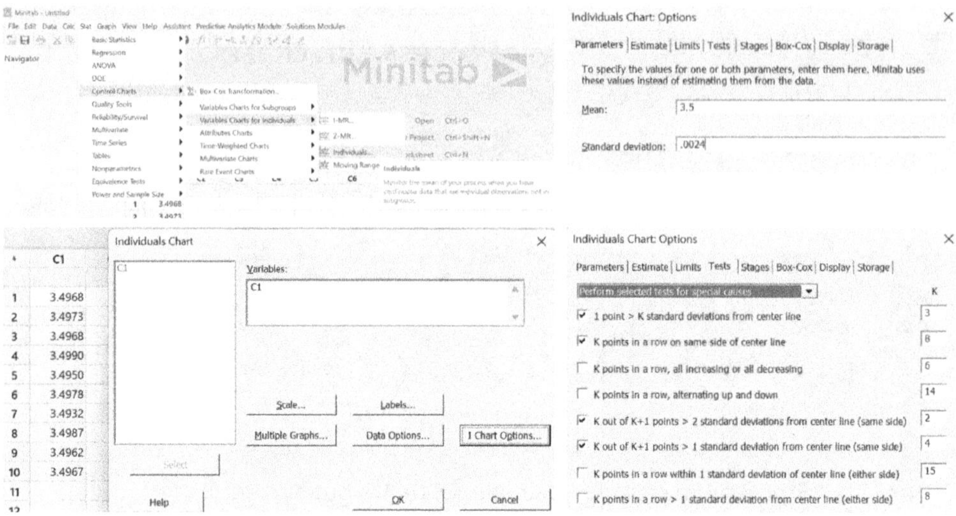

Figure 3.19 Sequence for producing a Process Aim chart in Minitab

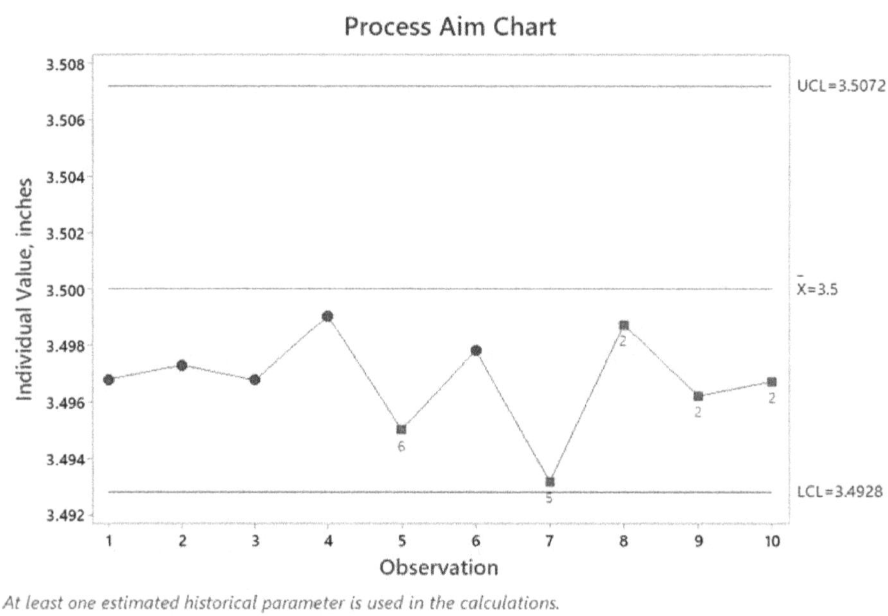

Figure 3.20 Process Aim chart made in Minitab that is out of control

Producing the Process Aim Chart in Minitab does not show the one-sigma and two-sigma lines but, choosing the Western Electric Zone Tests (discussed later in this chapter) in the Options does show that a rule was violated at value 5, and the chart is out of control.

The data can then be deleted and the chart updated when more parts are run, as shown in Figure 3.21. The chart could be updated after each part was measured instead of waiting for all ten.

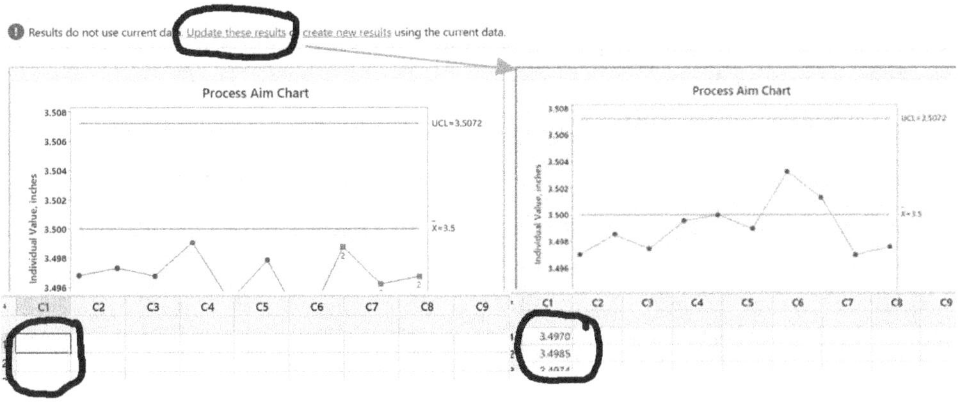

Figure 3.21 Process Aim chart made in Minitab that is in-control

It should be noted here that an alternate way to determine whether the sample of ten was close enough to the target would be to use Equation 2.11 to find the 95% CI for the mean. If this was done for this example, the results shown in Table 3.4 would be obtained.

Table 3.4 95% Confidence Interval Data for Out-of-Control Process Aim Data

n	$\overline{X}$, in	S, in	$t_{critical}$	δ = ½ CI, in	X_L, in	X_U, in
10	3.4968	0.00173	2.2622	0.00123	3.4955	3.4980

The nominal 3.5000 is not within the CI, so a process adjustment would be made (Table 3.5).

Table 3.5 95% Confidence Interval for In-Control Process Aim Data

n	$\overline{X}$, in	S, in	$t_{critical}$	δ = ½ CI, in	X_L, in	X_U, in
10	3.4990	0.00204	2.2622	0.00146	3.4976	3.5005

This time, the nominal 3.5000 is within the CI, so the process would not need to be adjusted further.

The results for the Process Aim chart will generally give the same results as the CI method and has the advantage of being a more visual method.

3.5 Important Considerations when Using Shewart's Control Charts

The remainder of this chapter will discuss these three important considerations that should always be followed when using control charts for variables data from CTQ product characteristics.

1. Use control limits set at three-sigma units on each side of the centerline
2. Use an average dispersion statistic when computing three-sigma control limits
3. Use rational sampling and subgrouping when collecting data for control charts

3.5.1 Three-Sigma Limits

All of the four examples earlier in this chapter have had control limits based on a centerline plus or minus three estimated standard deviations. The following summarizes how the limits were derived for each type of chart.

1. Average $\left(\overline{X}\right)$ chart

   Using Equations 2.3 and 2.8:

   $$\overline{\overline{X}} \pm 3\sigma_{\overline{X}} = \overline{\overline{X}} \pm 3\frac{\sigma_X}{\sqrt{n}} = \overline{\overline{X}} \pm 3\frac{\overline{R}/d_2}{\sqrt{n}} = \overline{\overline{X}} \pm A_2\overline{R}$$

   where

   $$A_2 = \frac{3}{A_2\sqrt{n}}$$

2. Individuals (X) chart

   Using Equation 2.8:

   $$\overline{X} \pm 3\sigma_X = \overline{X} \pm 3\frac{m\overline{R}}{d_2} = \overline{X} \pm 3\frac{m\overline{R}}{1.128} = \overline{X} \pm 2.66m\overline{R}$$

   when $n = 2$ for mR chart

3. Range (R) chart

   Using the relationship $\sigma_R = {d_3\overline{R}}/{d_2}$:

   $$\overline{R} \pm 3\sigma_R = \overline{R} \pm {d_3\overline{R}}/{d_2} = \overline{R}\left(1 \pm {d_3}/{d_2}\right)$$

so

$$UCL_R = D_4\overline{R}$$

and

$$LCL_R = D_3\overline{R}$$

Where:

$D_4 = (1 + d_3/d_2)$

$D_3 = (1 - d_3/d_2)$

The reason that plus or minus three estimated standard deviations are used is because they work. Shewart chose these limits because, in practice, they provided reasonable action limits – not because with a perfectly normal data distribution 99.73% of the data would be within the limits. Regardless of what limits are chosen, there are two possible types of error when interpreting the charts (see Figure 3.22). A Type I or alpha error occurs when a data point is outside the limits and the process does not have any special causes or uncontrollable sources of variation present. This is also sometimes called the producer's risk because the producer will waste resources trying to figure out what changed in the process. A Type II or beta error occurs when a special cause of variation is present, but the control chart does not catch it. This is sometimes called the consumer's risk because, if there is a special cause present, there is a higher risk that the product that the consumer gets will be different.

		Decision Based on Control Charts	
		Process In-Control (Only "chance" causes of variation present)	Process Out-of-Control ("Assignable" causes of variation present)
Actual State of Process	In-Control	Good Decision	Type I error α error Producer's Risk
	Out-of-Control	Type II error β error Consumer's Risk	Good Decision

Figure 3.22 Types of errors that occur with control charts

A hundred years of practice have proven three-sigma limits to be good action limits for production processes by giving a good balance of the alpha and beta errors. To pick up certain special causes earlier, additional rules can be applied to control charts, as was done in the case of the Process Aim chart. These rules are sometimes called the

Western Electric Zone Tests and they state that the process is deemed to be out of control if:

- Rule 1: Any point is outside the three-sigma limits
- Rule 2: Two out of three points are outside either two-sigma limit, on the same side
- Rule 3: Four out of five points are outside either one-sigma limit, on the same side, and
- Rule 4: Eight consecutive points are on the same side of the centerline

Another common rule is seven or eight points in a row either increasing or decreasing. Using additional rules increases the alpha error (and decreases the beta error) while also reducing what is called the Average Run Length (ARL). The ARL is the theoretical number of data points that are collected between false alarms or alpha errors. If Rule 1 is the only detection rule used and the distribution is perfectly normal, it would be expected that 0.27% (100% – 99.73%) of the time a point would be outside the control limits without a special cause of variation being present. This is the same as one point for every 370 (1/0.0027) points, so the ARL is 370. If all four of the Western Electric Zone Tests are used, the ARL drops to 92.

The Western Electric Zone Tests can be set up in Minitab, as shown in Figure 3.23.

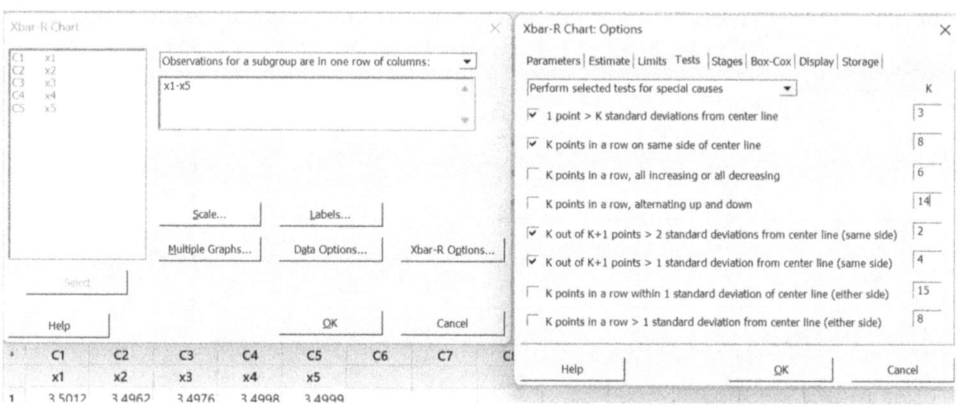

Figure 3.23 Setting up Western Electric control chart rules in Minitab

When monitoring the CTQ characteristic directly, it is recommended that three-sigma limits always be used. When monitoring a process variable or other part characteristic that doesn't perfectly correlate with a CTQ, it is sometimes necessary to use wider limits to keep from having too many false alarms – this will be discussed further in the Process Monitoring chapter (Chapter 10).

3.5.2 Average Dispersion Statistic for Estimating Standard Deviation

The average range ($\overline{R}$ or $m\overline{R}$) is an example of an average dispersion statistic. It is the statistic that was used to calculate the width of all the control limits in Examples 3.1–3.4. To show why these limits are applied, an example of data with a special cause of variation present will be used.

3.5.3 Example 3.5: Methods of Estimating Sigma for Control Limits

A process that, without special causes of variation present, normally has a characteristic with a mean of 10 and a standard deviation (σ_X) of 1 is interrupted with a special cause of variation that causes the characteristic to increase by 1.5 every hour. Estimate the standard deviation using the data given and the following five methods:

1. Average Range of each subgroup
2. Average Sample Standard Deviation of each subgroup
3. Average Moving Range of each subgroup Average
4. Overall Standard Deviation of all data points
5. Standard Deviation of the Averages

Ten subgroups of five are collected every hour and the measurements shown in Table 3.6 are obtained.

Table 3.6 Data for Example 3.5

Hour	X1	X2	X3	X4	X5	X-bar	*R*	*s*	*mR*
0	8.567	10.167	12.122	10.269	7.646	9.754	4.476	1.724	
1	12.387	11.180	10.796	10.143	11.263	11.154	2.245	0.819	1.400
2	13.779	12.818	12.642	14.249	15.069	13.711	2.427	1.010	2.558
3	13.253	15.091	16.147	14.161	14.042	14.539	2.895	1.111	0.827
4	16.904	16.834	15.547	15.385	16.132	16.161	1.519	0.704	1.622
5	18.246	16.506	17.740	18.260	17.330	17.616	1.754	0.732	1.456
6	18.717	19.434	17.863	17.866	19.418	18.659	1.572	0.782	1.043
7	21.968	20.758	20.322	19.260	23.011	21.064	3.751	1.459	2.404
8	23.081	21.532	21.653	22.437	22.899	22.320	1.549	0.706	1.256
9	23.674	22.212	23.567	23.562	22.588	23.121	1.461	0.672	0.801
Average						**16.810**	**2.365**	**0.972**	**1.485**

Note that the average, range, and sample standard deviation for each subgroup have been calculated. The moving ranges of the subgroup averages have also been calculated. The Grand Average, Average Range, Average Sample Deviation, and Average Moving Range have been calculated and are shown under the column that they have averaged.

1. Using the Average Range of each subgroup

   For a subgroup size of five ($n = 5$), $d_2 = 2.326$. Using Equation 2.8:

   $$\hat{\sigma}_X = \overline{R}/d_2 = 2.365/2.326$$

2. Using Average Sample Standard Deviation of each subgroup

   For a subgroup size of five ($n = 5$), $c_4 = 0.94$. Using Equation 2.9:

   $$\hat{\sigma}_X = \overline{s}/c_4 = 0.972/0.94$$

3. Using the Average Moving Range of each subgroup Average

   Using a subgroup size of two ($n = 2$), $d_2 = 1.128$. Using Equation 2.8 (substituting $m\overline{R}$ for $\overline{R}$):

   $$\hat{\sigma}_X = m\overline{R}/d_2 = 1.485/1.128 = 1.317$$

4. Using the Overall Standard Deviation of all data points

   Using all fifty individual values, $c_4 = 0.9949$ with a sample size of fifty ($n = 50$), and Equations 2.6 and 2.9:

   $$\hat{\sigma}_X = s/c_4 = \sqrt{\frac{1}{n-1}\sum_{i=1}^{n}(x_i - \overline{x})^2}/c_4 = \frac{4.493}{0.9949} = 4.516$$

5. Using the Standard Deviation of the Averages

   Using the ten sample standard deviations, $c_4 = 0.9727$ for a sample size of ten ($n = 10$), and Equations 2.6 and 2.9 (for averages):

   $$\hat{\sigma}_{\overline{X}} = s_{\overline{X}}/c_4 = \sqrt{\frac{1}{n-1}\sum_{i=1}^{n}\left(\overline{x}_i - \overline{\overline{X}}\right)^2}/c_4 = \frac{4.586}{0.9727} = 4.715$$

   Using Equation 2.3 with a subgroup size of five ($n = 5$) gives:

   $$\hat{\sigma}_X = \hat{\sigma}_{\overline{X}}\sqrt{n} = 4.715\sqrt{5} = 10.543$$

In the preceding example, using either the average range or the average sample standard deviation of the subgroups is the best way to estimate standard deviation with subgrouped data. If the data is perfectly normal, then any method would be equivalent to any of the other methods shown in this example. However, if a special cause of variation is present between the subgroups, then the estimate of standard deviation will be inflated when using the other three methods and the control charts will rarely show an out-of-control condition when a special cause is present. It is not advisable to use these methods unless the data is normally distributed, with no special causes of variation present when the control limits are set up.

3.5.4 Rational Sampling and Subgrouping

A control chart is only as good as the data that is plotted on it. It is important that some thought go into how the data is collected. When doing this, the sources of variation should always be kept in mind. The following six principles are a good guideline (Wheeler, 1992):

1. Never knowingly subgroup unlike things together
2. Minimize the variation within each subgroup
3. Maximize the opportunity for variation between the subgroups
4. Average across noise, not across signals
5. Treat the chart in accordance with the use of the data
6. Establish operational definitions for the sampling procedure

3.5.5 Never Knowingly Subgroup Unlike Things Together

This principle basically means that the parts to be subgrouped together need to come from the same bell curve (same process). The most obvious way of doing this incorrectly with injection molding is when parts from different cavities are put in the same subgroup. This is debatable when the cavities are not statistically different, but this author rarely recommends this practice. The following example illustrates why subgrouping parts from different cavities together can be a problem.

3.5.5.1 Example 3.6: Incorrect Four-Cavity Subgrouping Method

A four-cavity insert-molded part with a snap-fit latch on one side of the part is being made. The warpage of the side latch is a CTQ characteristic. The mean warpage and standard deviation of each of the four cavities are shown in Figure 3.24.

Cavity #	Mean, in.	Std. Dev., in.
1	-0.010	0.00167
2	-0.005	0.00167
3	0.005	0.00167
4	0.010	0.00167

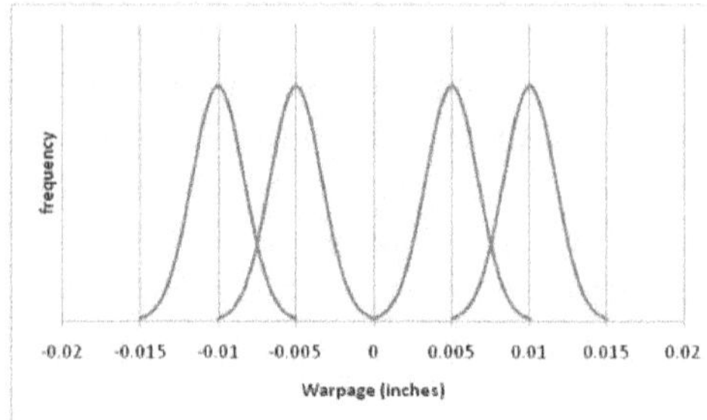

Figure 3.24 Bell curves for Example 3.6

As can be seen, each cavity has the same amount of variation, but the means are all different. First, we will look at an incorrect way of subgrouping to see what problem it causes. The data in Table 3.7 is composed of measurements from one shot per hour for twenty-five hours (all dimensions are in inches).

Table 3.7 Data for Example 3.6

Hour	Cavity 1	Cavity 2	Cavity 3	Cavity 4	X-Bar	*R*
1	−0.0106	−0.0019	0.0029	0.0104	0.0002	0.0210
2	−0.0101	−0.0031	0.0055	0.0121	0.0011	0.0222
3	−0.0114	−0.0049	0.0084	0.0083	0.0001	0.0198
4	−0.0104	−0.0058	0.0024	0.0077	−0.0015	0.0181
5	−0.0098	−0.0034	0.0041	0.0097	0.0002	0.0195
6	−0.0094	−0.0030	0.0049	0.0104	0.0007	0.0198
7	−0.0082	−0.0055	0.0063	0.0069	−0.0001	0.0150
8	−0.0117	−0.0036	0.0047	0.0110	0.0001	0.0227
9	−0.0102	−0.0051	0.0082	0.0108	0.0009	0.0211
10	−0.0112	−0.0072	0.0061	0.0126	0.0001	0.0238
11	−0.0084	−0.0042	0.0040	0.0095	0.0002	0.0178
12	−0.0078	−0.0080	0.0044	0.0099	−0.0004	0.0179
13	−0.0109	−0.0033	0.0067	0.0109	0.0008	0.0219
14	−0.0109	−0.0043	0.0041	0.0114	0.0001	0.0223
15	−0.0102	−0.0050	0.0044	0.0096	−0.0003	0.0198

Hour	Cavity 1	Cavity 2	Cavity 3	Cavity 4	X-Bar	*R*
16	-0.0078	-0.0054	0.0034	0.0095	-0.0001	0.0173
17	-0.0069	-0.0067	0.0049	0.0092	0.0001	0.0162
18	-0.0100	-0.0070	0.0055	0.0106	-0.0002	0.0206
19	-0.0121	-0.0061	0.0048	0.0096	-0.0010	0.0217
20	-0.0114	-0.0056	0.0042	0.0092	-0.0009	0.0206
21	-0.0114	-0.0055	0.0055	0.0100	-0.0003	0.0214
22	-0.0084	-0.0080	0.0051	0.0112	0.0000	0.0196
23	-0.0079	-0.0021	0.0054	0.0091	0.0011	0.0170
24	-0.0084	-0.0040	0.0082	0.0087	0.0011	0.0171
25	-0.0108	-0.0033	0.0063	0.0110	0.0008	0.0218

The average and range of each subgroup have also been calculated in the preceding data. The Grand Average is 0.0001 inches, and the Average Range is 0.0198 inches.

Note: It is obvious, from looking at the bell curves, that the Grand Average would be about 0 inches and the Average Range about 0.020 inches.

For a subgroup size of four ($n = 4$), $D_4 = 2.282$, $D_3 = 0$, and $A_2 = 0.729$. Using Equation 3.1 – Equation 3.6, the charts in Figure 3.25 can be produced.

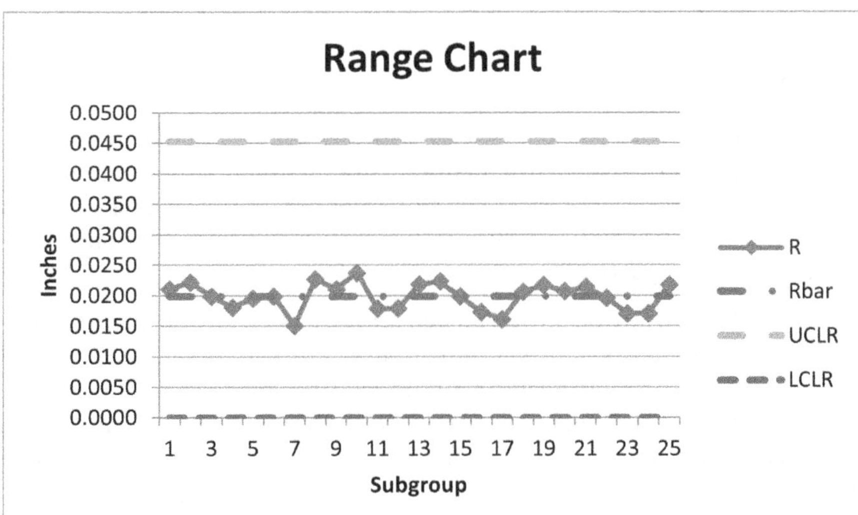

Figure 3.25 Example of stratification caused by subgrouping cavities together

It is obvious, from looking at the charts from Figure 3.25, that there is something wrong. None of the points are even more than one standard deviation from the centerline. This is called stratification and it occurs whenever multiple means and or variation occur in the subgroup samples. In addition to putting multiple cavities in a subgroup, another common mistake is not waiting a sufficient length of time after molding before measuring parts and being inconsistent in that amount of time. Depending on the thickness and material, it is usually recommended that a cooling time of twenty minutes to two hours be allowed before measuring the parts. If the cooling time is not long enough, then the first part measured will almost always be the biggest and the last part measured, the smallest.

Admittedly, Example 3.6 is an extreme case of stratification. However, the cases that aren't as extreme would be more difficult to detect by looking at the charts; the control limits would be inflated, but the points wouldn't hug the centerline as tightly (be wary when someone says that the cavities have been "statistically proven" to be the same – more on this in Chapter 6).

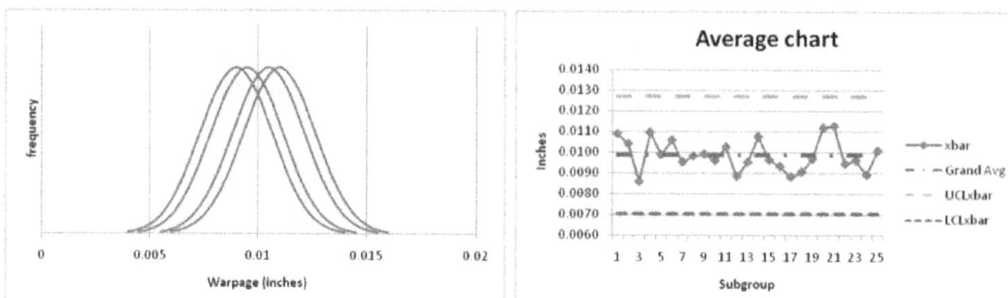

Figure 3.26 Example of a small amount of stratification

The bell curves and Average chart shown in Figure 3.26 are an example of this. The chart isn't that much different from one done correctly, but some of the ability to recognize special causes of variation has been lost by putting all four cavities in one subgroup and on one chart (the total width of the control limits is about 0.001 inches too wide on this Average chart).

So, how then should multiple cavity molds be handled? Ideally, from the ability-to-learn-about-your-process point of view, Average and Range charts would be produced for each cavity, typically with a subgroup size of five. The problem with this is the amount of measurement time that is involved. For Example 3.6, if five parts were taken for each cavity, twenty parts would need to be measured every time subgroups were collected. There are times when this would be necessary, but normally that amount of inspection cost is not tolerated for very long. For Example 3.6, one option would be to do Individuals and Moving Range charts. This would yield better information from the same amount of data. However, for higher-cavitation-level molds (8,

16, 32, etc.), inspection costs can become a major concern. One common method is to look for indicator cavities. Once it is known which process factors (temperatures, pressures, times, etc.) affect the CTQ characteristic, the cavities that are most likely to see variation in that factor would be monitored. More on how to determine which process factors are important is presented in Chapters 7 to 9. For example, if a DOE shows that mold temperature was the major contributing factor to variation in a CTQ characteristic, the first and last cavities in the cooling circuit (if a series circuit) could be monitored. Other things such as runner layout and shear history of the plastic going to each cavity could be considerations when determining indicator cavities.

3.5.6 Minimize the Variation within each Subgroup

As has already been discussed and shown in the previous examples, it is the within-subgroup variation that defines the common cause variation and determines how wide the bell curve and control limits are. By minimizing the variation within the subgroups, control charts will be created which are sensitive to the presence of special causes of variation. In injection molding, taking five consecutive shots usually will give a minimum amount of variation. In rare cases, the subgroups are too homogeneous. Example 3.3, where the within-batch variation was not a reasonable estimator of between-batch variation, was such a case.

3.5.7 Maximize the Opportunity for Variation between the Subgroups

Anything that is known to be a possible source of variation should not be purposely included in the same subgroup. A major way of accomplishing this is by not subgrouping unlike things, such as different cavities, together (Subgrouping Principle 1). There are other things that are known to vary during production that will always be on the same charts. Some examples are batches of material, product from different operators, and product from different times of the day. Parts from different batches of material, different operators, and different times of the day should not be placed in the same subgroup in order that the opportunity for variation between subgroups may be maximized and the most useful charts generated.

3.5.8 Average across Noise, Not across Signals

Again, noise is another term for common cause variation and signal is another term for special cause of variation. If the first three subgrouping principles are followed, then this principle should be the natural result. The purpose of the subgroup average is to average out the common cause variation and give an accurate estimate of the mean at that point. This would indicate that the larger the subgroup, the better, since, as was shown in Chapter 2, a larger sample size will give an estimated mean with the smallest CI. However, if the sample size is large, it increases the chance that a special cause will creep into the subgroup. Again, a subgroup size of five has proven to provide a nice balance.

3.5.9 Treat the Chart in Accordance with the Use of the Data

Some logic needs to be applied when coming up with a strategy for control charts. For example, how often subgroups are collected should depend on how often they need to be collected. When first starting up a new process, the subgroups may be collected more often than after the process has exhibited long periods of control and vice versa if special causes of variation are frequently disturbing a process. Another common practice, which is a good one, is to collect one hundred consecutive shots during the baseline study of a new mold. A running record of individual values can reveal things that periodic subgroups do not. The cyclic pattern shown in Figure 3.27 is a common problem that probably wouldn't be picked up with periodic subgroups.

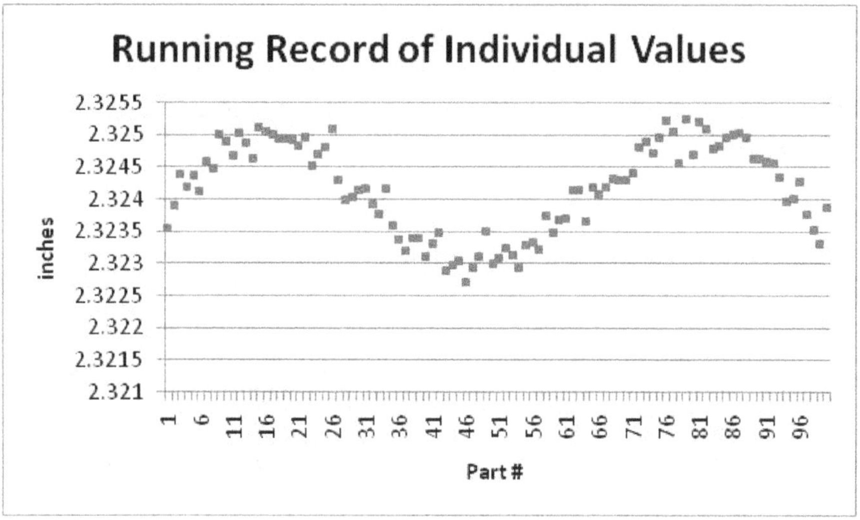

Figure 3.27 Example of cyclic data

This type of pattern is common if temperature controls are not tuned correctly or material that needs to be mixed is not being mixed well. From Figure 3.27, it appears that the cycling is causing about 0.002 inches of range in the part dimension and that, if the cycling were eliminated, the range would be reduced to about 0.0005 inches. This example would most likely be detected in standard control charts since the variation of five consecutive parts (n = 5) is so much less than the variation over time. The only way it wouldn't be detected is if the frequency of the sampling happened to coincide with that of the sinusoidal pattern, i.e. if subgroups were taken about every 64 (or some multiplier of 64) parts. But even if detected, the pattern would most likely not be as easy to see in charts made with periodically collected subgroups.

Another consideration that should be kept in mind with injection molded parts is that a mold will typically be run in several different molding machines in production. It is not uncommon for different control limits to be needed for different molding machines, since the common cause variation associated with each machine will be different.

Updating of control limits will also vary with the application. For example, a characteristic will sometimes shift up or down when a new batch of material (special cause of variation) is introduced. If the shift takes the new average too far from nominal or target for that characteristic, then the process should be adjusted or the material rejected. A Process Aim chart can be useful when this occurs. However, if the shift is very small in comparison to the spec tolerance width, then it may be acceptable to recalculate new control limits for the new batch of material.

In this book, several ways of using Average and Range charts will be reviewed other than those for monitoring a production process. Among them will be measurement system analysis and DOE analysis. When using these charts, it is always important to keep in mind the key questions that each chart asks.

- Range chart – Is the within-subgroup variation consistent?
- Average chart – Given the average variation within subgroups, are there detectable differences between subgroups?

What is within and between subgroups will vary with the application.

3.5.10 Establish Operational Definitions for the Sampling Procedure

In addition to how often the parts are collected, a control plan should reference systematic procedures for how soon after molding, and exactly how, the parts are to be measured. This will be discussed further in the Measurement chapter (Chapter 5). How to handle out-of-control conditions should also be defined.

4 Process Capability

Up to this point, the design specifications of the parts have not been mentioned very often. It is important to understand that the control limits discussed in Chapter 3 are not the design specification (spec) limits or tolerance limits. Hopefully, the control limits are much narrower than the spec limit width or tolerance width. Control limits are intended to help you learn about your process and make it as consistent as possible. Spec limits are needed to ensure that parts fit correctly with other parts and function correctly. Four states of every production process have been described (Wheeler, 1992):

1. **The Ideal State**

   This state is the goal of every process. Every CTQ characteristic is in control, at the nominal or center of the spec limits, and has common cause variation that is small compared to tolerance width.

2. **The Threshold State**

   In this state, the process is in control, but is making some product that is out of spec. The molded briefcases in Example 3.2 were in this state. Figure 3.7 shows that a spec of 0 mm (± 0.5 mm) is not met, as many parts are over 0.5 mm. This state can sometimes be met by centering the process. In the case of the briefcases, an adjustment to the mold steel or the assembly process may be able to shift the average mismatch closer to 0 mm from 0.42 mm. However, a closer inspection of Figure 3.7 shows that the total spread in data points is more than 1 mm (± 0.5 mm). Therefore, in order to meet the spec with all parts, it will also be necessary to find a method of reducing the common cause variation. This may involve changing the assembly tooling or procedure.

3. **The Brink of Chaos**

   This state is the reverse of the Threshold State. All the parts being made are in spec, but the control charts show that special causes of variation are present. This is a disaster waiting to happen. If whatever is causing the special cause of variation (cracked check ring, sticking hydraulic valve, fouled water lines, etc.) gets worse, then a lot of out-of-spec product will probably be produced. This is where control charts could be utilized to identify the problem and convert this process to the Ideal State.

4. **The State of Chaos**

   In this state, the process is out of control and the parts are out of spec. This author has been to several molding shops where processing was treated as "art" and every process was an original. They had been operating on the Brink of Chaos with many of those processes but, as spec limits got tighter on new products, they entered the State of Chaos. This is because of the way they operated their machinery and the number of special causes of variation present, many of which were self-imposed because of the way the process was set up in the first place. The same fix for a problem never seemed to work twice, but they all had their own special fixes that they didn't like to share (mostly because they had no idea why they sometimes worked). Changing the shot size, changing the injection velocity, or changing hot runner tip temperatures, etc., was always the answer (pick one) for every problem they had. Maintenance was never considered until parts couldn't be made at all. In cases such as these, training in the correct methods and procedures is needed for the company to survive.

   It should also be pointed out that chaos is the state that nature prefers. A universal force called entropy is always driving processes toward wear and breakdown. It takes knowledge of your processes to keep the forces of entropy under control.

The question then becomes: how is the state of the process quantified? As discussed in the last chapter, a baseline study is normally performed before a new mold is sent to production. The more common name for this baseline study is a process capability study. Different indices are used to quantify how the actual process compares to the spec. It is important to point out that these indices have very little meaning unless the process is in control, so if the control charts made for the baseline study are not in control (Brink of Chaos or State of Chaos), the indices should not be calculated and the special causes of variation should be remedied first. The three indices that are most used and the equations used to calculate them are shown in Table 4.1.

Table 4.1 Process Capability Equations

Index Name	Equation
Process Potential Index	$$C_p = \frac{USL - LSL}{6\hat{\sigma}_X} \quad (4.1)$$
Process Capability Index	$$C_{pk} = \frac{\min(Z_l \& Z_u)}{3} \quad (4.2)$$ where $Z_l = \frac{\hat{\mu}_X - LSL}{\hat{\sigma}_X} \& Z_u = \frac{USL - \hat{\mu}_X}{\hat{\sigma}_X}$
Taguchi Capability Index	$$C_{pm} = \frac{USL - LSL}{6\sqrt{MSD_\tau}} \quad (4.3)$$ where $MSD_\tau = \hat{\sigma}_X^2 + (\hat{\mu}_X - \tau)^2$
USL = Upper Spec Limit **LSL** = Lower Spec Limit τ = Target or (USL + LSL)/2	$\hat{\sigma}_X = \overline{R}/d_2$ $\hat{\mu}_X = \overline{\overline{X}}$ or Grand Average ***MSD_τ*** = *Mean Square Deviation about the Target*

One important note is that some industries use P_p, P_{pk}, and P_{pm} instead of, or in addition to, C_p, C_{pk}, and C_{pm}. The *P* indices are usually called Process Performance indices. The only difference in the calculation is in the way that the estimate of standard deviation, $\hat{\sigma}_X$, is calculated. Instead of using an average dispersion statistic, the sample standard deviation of all the samples is used (see Method 4 in Example 3.5). In most cases, sample size is large enough and so the bias correction factor is not included; consequently, the equation used is $\hat{\sigma}_X = s$. If the data that is used for the capability studies is normally distributed, the *P* indices will be about the same as the *C* indices and, if not, the *P* indices will be smaller because the estimate of standard deviation will be inflated. Sometimes the *P* indices will be calculated for a more extended capability run that will purposely include causes of special variation that may be present during production. This would include things like different batches of material, different molding machines, slight changes in the process settings, limits of the process window determined with a DOE and so on (more on this in the DOE chapters). Although the pure statistician typically does not approve of this practice (all the data is not from one bell curve), the Process Performance indices are a more severe test of process capability. If this type of modified capability study is done to produce both *C* and *P* indices, the following two rules must be followed:

1. An equal number of subgroups should be run for each process variation (material batch, machine, settings, etc.) combination to be run.
2. No subgroup should include data from more than one process variation (follow subgrouping rules).

4.1 Process Potential Index

The C_p is called the Process Potential Index because, for a given process, it is the highest that either C_{pk} or C_{pm} can be if the Grand Average is centered exactly in the spec limits or is equal to the target. The higher the number that is calculated for any of the three statistical indices, the better. The numerator in Equation 4.1 represents the design specifications and the total width of the spec tolerance band. The denominator represents the actual process variation and is 99.73% of the width of the theoretical bell curve. The smaller the actual process variation is compared to the tolerance band, the less likely the process is to make parts that are out of spec. However, the equation for the index formula does not contain the spec limits or the target, so it doesn't guarantee that any of the parts are made to print, just that it has the potential to. A Process Potential Index of one (C_p = 1.0) would indicate that, at best (if the process is exactly centered and perfectly normal), 99.73% of the product would be within the spec limits – this is not typically acceptable.

4.2 Process Capability Index

Like C_p, the Process Capability Index, C_{pk}, also compares the spec tolerance width to the process variation. However, C_{pk} also has a measure of whether parts are being made to spec, since the spec limits are included in the equations. It also allows an estimation of how many parts will be produced which are out of spec, since Z-value calculations are part of the C_{pk} calculation. C_{pk} will be less than C_p if the Grand Average is not centered exactly between the spec limits. A negative value for C_{pk} indicates that the Grand Average is outside of the spec limits. It should be noted that, in Equation 4.2, the Z-value for the distance to the LSL is calculated as $\hat{\mu}_X - LSL$ instead of $LSL - \hat{\mu}_X$. This reversal keeps the Z-value positive whenever the Grand Average is within the spec limits. If customers require that their suppliers have certain characteristics to meet a minimum process capability index, C_{pk} is what is usually specified. Motorola started asking their suppliers to meet a minimum Process Capability Index of two (C_{pk} > 2.0) in the mid 1980s and coined the terms "six sigma process" and "six sigma supplier". With a six sigma process, the closest the Grand Average is to any spec limit is six sigmas (standard deviations). Depending on how important the characteristic involved is, minimum C_{pk} values of 2.0, 1.67, 1.5, and 1.33 are commonly specified requirements.

4.3 Taguchi Capability Index

The Taguchi Capability Index, C_{pm}, has an added feature compared to C_{pk}. For companies that have truly adopted Taguchi's "On target with minimum variance" definition of World Class Quality discussed in Chapter 1, C_{pm} will drive them toward that goal. The premise behind Taguchi's definition of World Class Quality is that any deviation from the target, τ, will increase costs. Taguchi's Loss Function, $\mathcal{L}(x)$, shows one way to quantify that cost (Equation 4.4):

Taguchi's Loss Function

$$\mathcal{L}(x) = K(x - \tau)^2 \quad (4.4)$$

Where:

K = constant

τ = target value or (USL+LSL)/2

Figure 4.1 shows an example of the Taguchi Loss Function, Equation 4.4, and the Normal Curve Distribution, Equation 2.1, for Example 3.1 on the same graph. Here, it is assumed that the spec is 3.505 inches ± 0.015 inches. To calculate the loss ($) associated with not being exactly on target with zero variance, the loss ($/unit) at each value of x (distance between snap fits) is multiplied by the frequency (number of units) of that value of x occurring. This can be done by integrating Equation 4.5.

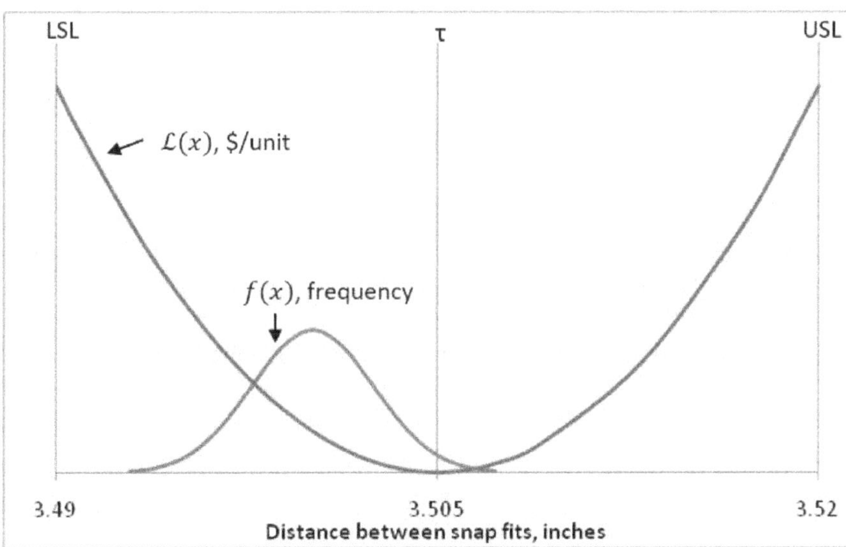

Figure 4.1 Example of a Taguchi Loss Function superimposed on a bell curve

$$\text{Loss} = \int_{\text{allx}} \mathcal{L}(x)f(x)dx = K[\sigma_x{}^2 + (\mu_x - \tau)^2] = K(\text{MSD}_\tau) \tag{4.5}$$

It can be seen, both in Equation 4.5 and by looking at Figure 4.1, that the loss will be minimized by centering the bell curve on the target ($\mu x = \tau$) and minimizing the standard deviation, σ_x. The only values for x that have no cost associated are those that are exactly on target. The C_{pk} index works under the old school of thought where any value between the spec limits is deemed to have no associated cost. The C_{pm} index gives the added incentive of centering the process by putting MSD_τ in the denominator of the equation.

Figure 4.2 shows why the C_{pm} index is best at encouraging companies to hit the target. All three lines on each graph would be on top of each other if the nominal spec were equal to the estimated mean. The graph in which the nominal spec (3.5025-inch nominal) is closest to the estimated mean (3.5001 inches) is the graph in which the C_{pm} and C_{pk} lines most closely match. On this graph, the nominal spec is one estimated standard deviation (0.0024 inches) away from the estimated mean. On the other hand, the graph in which the nominal spec (3.5100-inch nominal) is farthest from the estimated mean (3.5001 inches) has a C_{pm} index that doesn't improve much even if the tolerance width is increased considerably. The nominal spec is about four estimated standard deviations from the estimated mean on this graph. Note that C_p and C_{pk} increase at the same rate (have the same slope) on all three graphs.

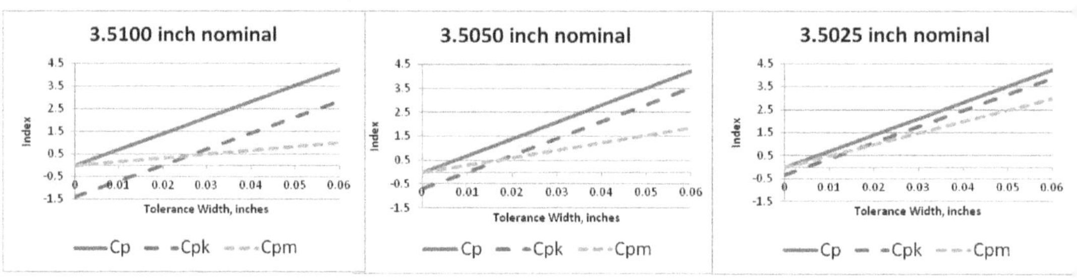

Figure 4.2 Process capability indices for Example 3.1 (where $\hat{\mu}_X$ =3.5001 inches and $\hat{\sigma}_X$= 0.0024 inches) and with different specs (nominal and tolerance)

Example 3.1 will now be revisited to show how each index is calculated.

4.4 Example 4.1: Calculating Process Capability Indices

To calculate the indices for Example 3.1 which was a baseline study of the distance between snap fits using Average and Range charts, the summary statistics of Average Range and Grand Average are needed. The Average Range was 0.055 inches, and the

Grand Average was 3.5001 inches. In addition, for this example it is assumed that the spec is 3.510 inches ± 0.015 inches (total tolerance width of 0.030 inches). The Grand Average is the estimated mean, $\hat{\mu}_X$, and the Average Range is converted into the estimated standard deviation, $\hat{\sigma}_X$, using Equation 2.8, with d_2 = 2.326 for a subgroup size of five (n = 5).

$$\hat{\sigma}_X = \frac{\overline{R}}{d_2} = \frac{0.0055\text{ inches}}{2.326} = 0.0024\text{ inches}$$

The Process Potential index is calculated using Equation 4.1.

$$C_p = \frac{USL - LSL}{6\hat{\sigma}_X} = \frac{3.525\text{ inches} - 3.495\text{ inches}}{6 \times 0.0024\text{ inches}} = \frac{0.030\text{ inches}}{0.0144\text{ inches}} = 2.10$$

This shows that if the process was exactly centered C_{pk} and C_{pm} would be 2.10.

For C_{pk}, the Z-values are calculated first.

$$Z_l = \frac{\hat{\mu}_X - LSL}{\hat{\sigma}_X} = \frac{3.5001\text{ inches} - 3.495\text{ inches}}{0.0024\text{ inches}} = 2.15$$

$$Z_u = \frac{USL - \hat{\mu}_X}{\hat{\sigma}_X} = \frac{3.525\text{ inches} - 3.5001\text{ inches}}{0.0024\text{ inches}} = 10.47$$

Note that Table 12.3 in the Appendix can be used to estimate how many parts would be out of spec. For a Z_l-value of 2.15, the value from Table 12.3 is 0.4842. Therefore, the proportion of parts that would be expected to be out of spec is 0.5 – 0.4842 = 0.0158 or 1.58% or 15,800 ppm. The Z_u-value of 10.47 is way off of Table 12.3 and it can be assumed that the ppms are close to zero.

Equation 4.2 can then be used to calculate the C_{pk} index.

$$C_{pk} = \frac{\min(Z_l \& Z_u)}{3} = \frac{2.15}{3} = 0.72$$

For C_{pm}, the MSD_τ is calculated first.

$$MSD_\tau = \hat{\sigma}_X^2 + (\hat{\mu}_X - \tau)^2 = (0.0024\text{ inches})^2 + (3.5001\text{ inches} - 3.510\text{ inches})^2$$
$$= 0.000104\text{ inches}^2$$

Equation 4.3 can then be used to calculate the C_{pm} index.

$$C_{pm} = \frac{USL - LSL}{6\sqrt{MSD_\tau}} = \frac{3.525\text{ inches} - 3.495\text{ inches}}{6\sqrt{0.000104\text{ inches}^2}} = 0.49$$

The preceding shows how all the values in Figure 4.2 were calculated. Figure 4.2 can be also used to look at scenarios of different spec tolerance widths and different distances between the estimated mean and the target.

Just as the confidence intervals were determined for estimated means and estimated standard deviations, they can also be calculated for the statistical indices discussed in this chapter. The lower 95% CI limit is what would be of interest when trying to meet a minimum index number. This book will not go through the equations to calculate the CI limits for these indices, but they can be found in the help menus of Minitab. Confidence intervals can be shown when doing the analysis in Minitab or other statistical packages.

4.5 Example 4.2: Using Minitab to Do Capability Analysis

The numbers calculated in Example 4.1 can be obtained in Minitab. In addition, the confidence intervals and a graphical representation of the analysis is given there. The lower confidence interval limit of C_{pk}, C_{pm}, or P_{pk}, is what is commonly required to meet a minimum number (2.0 for a six sigma process, for example). It should also be noted that what Minitab reports as C_{pm} is really P_{pm} since it uses s instead of $\overline{R}$ to calculate estimated standard deviation.

First, the raw data from Example 3.1 is copied from Excel to Minitab. Each row represents one subgroup of five and the thirty rows of data are copied and pasted into Minitab. The Capability Analysis is done as shown in Figure 4.3:

When using confidence limits, it is generally the lower 95% confidence limits that are reported. The lower 95% confidence limits (Lower CL) for this example are: $C_p = 1.82$, $C_{pk} = 0.61$, and $C_{pm} = 0.47$ (which is really P_{pm}).

What action should be taken when the calculated indices do not meet the required minimum number? If the Process Potential Index, C_p, is below the required minimum, then the following two options are available to improve it.

1. Reduce the variation, σ_x. By making the denominator of Equation 4.1 smaller, the index will get better. If the capability study was done using normal, best operating procedures, then this option will prove to be very difficult. Running the parts on a more consistent molding machine, running a more consistent material mix, changing the gate location, and automating the cycle if an operator is currently running semi-automatic are all examples of steps that might be needed. Most of these steps would also have an associated price tag. If, however, the capability study was done under subpar conditions, then this option may be possible without adding cost to the final production operation. Redoing a capability study does have a significant cost, so this mistake is not repeated very often. One possible scenario for subpar conditions is that the new mold is run on the machine that needs maintenance for the capability study because all the "good" machines currently have "hot" production jobs in them. However, most customers require capability studies to be run on the machine(s) that will be used in production.

Figure 4.3 Sequence for capability analysis in Minitab

2. Increase the spec tolerance width. By making the numerator of Equation 4.1 bigger, the index will get better. This is usually a touchy proposition, due to the customer/supplier relationship. The best time to iron out spec limits that are unrealistic is before accepting a job from a customer and before building a mold. There will be times, however, where the design spec is not based on sound engineering analysis but, instead, made artificially tight to avoid the time needed to do the analysis. This option will have the possible cost of damaging the customer/supplier relationship and, possibly, the cost of any required analysis to prove that the spec limits were too tight in the first place.

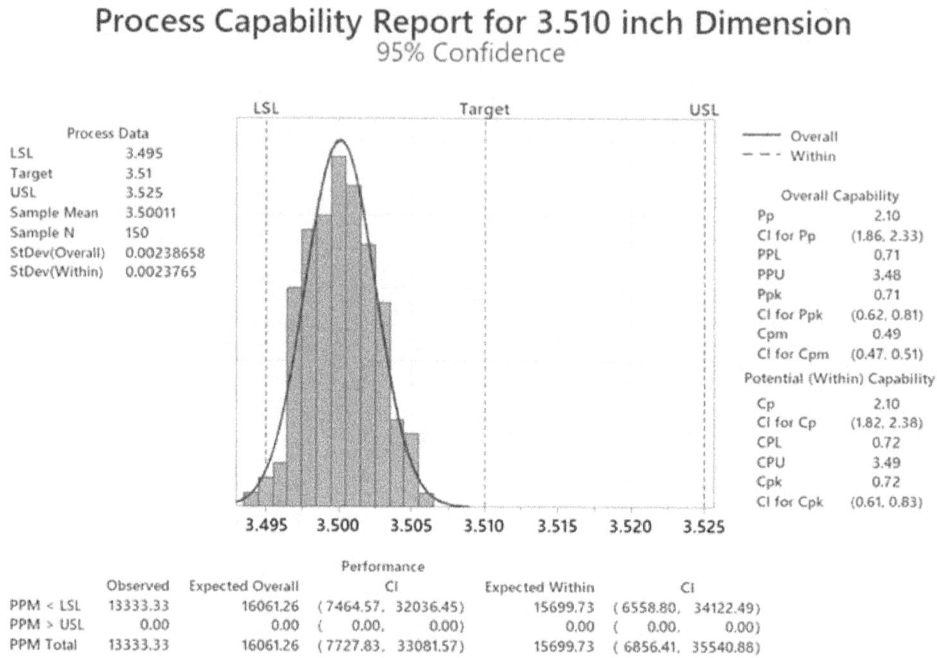

The actual process spread is represented by 6 sigma.

Figure 4.4 Results for Example 4.1 in Minitab

If the Process Potential Index, C_p, is acceptable, but C_{pk} and/or C_{pm} are below the required minimum, then the following step is the preferred option (steps 1 and 2 would also help).

3. Center the process; move μ_x closer to τ. It has already been pointed out that with the ideal, $\mu_x = \tau$, all three indices are equal. With injection molding, it is important not to use up the process window when doing this. For example, if a dimension is made smaller by reducing the hold pressure, this may substantially increase the possibility of getting short shots during production when the viscosity of a new batch of material is slightly higher. When working with dimensions, it is usually desirable to center the dimensions by modifying the mold steel so that the process will have as wide a process window as possible. There may be other times when the gate location or the thickness of the part needs modified. Of course, there will always be exceptions to adjusting the steel instead of the process settings, based on the economics of the situation. If the cost to modify the mold is higher than the cost of poor quality that results from the smaller process window, then it may be better to modify the process settings and live with the smaller process window.

5 The Measurement Process

One of the sources of variation, when dealing with quantifiable results, is measurement variation. The quality characteristics that can be studied using the statistical methods in this book all have to be measured. Measurement processes are used often when a new plastic part is being developed. The mold maker uses measurements to ensure that the mold is built to the design. The size of the components and the properties of the steel or other material are examples of things that can be measured. If the measurement equipment is not correct, the mold may not even be able to be assembled and, even if it can be assembled, the parts that come out of it won't be right. The equipment used to make the mold and to mold the plastic parts employs measurement systems to control the processes. After the plastic parts are molded, measurement ensures that the parts are the right size and meet all the other requirements (strength, color, surface finish, etc.).

Like any process, the measurement process can be characterized by a location parameter and a dispersion parameter. An error in the estimated location or mean of the measurements can be very costly. If, when making a mold, all the steel dimensions are off by 0.10 inches, or some other amount, the mold may go together fine, but all the dimensions will be about 0.10 inches off. Problems such as these are avoided by keeping all measurement equipment calibrated to a known standard. It is also important that the error or variation (dispersion) in measurements is small compared to the actual variation in the parts when tracking the variation in the parts and to get a good estimate of the mean with a small number of measurements. If one part inspector or appraiser has more variation in measurements than another, control charts are more likely to go out of control when the operator with the high variation is checking the parts. Unfortunately, one of the main special causes of variation that explains why a control chart goes out of control is a measurement problem. Method of measurement, fixture design, and operator training are keys to minimizing measurement variation.

Another way to think about location and dispersion is what happens when a rifle is fired at a target. Figure 5.1 shows four targets that had 5 bullets fired at it. Accuracy is the ability to hit the target. If measuring a dimension, the target can be thought of as the actual dimension. Precision is the ability to have little variation in the shots (a tight grouping). Sighting in or calibration of the rifle is necessary for good accuracy. A fixture to hold the rifle still or increased training and practice should improve precision. If the precision difference shown in Figure 5.1 is just due to two different shooters, it should be obvious to the reader that the shooter with good precision would not have to take as many shots before making a calibration adjustment. Due to the shooter's good precision or low amount of variation, a high confidence level can be obtained with fewer shots.

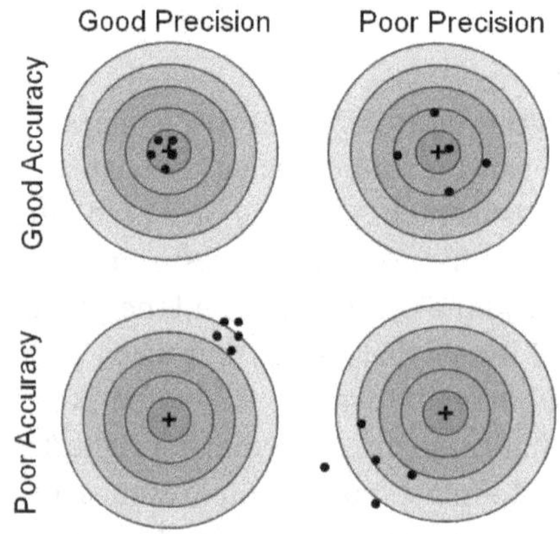

Figure 5.1
Precision and accuracy

When checking the accuracy of a piece of measuring equipment, the following are three of the issues that generally need to be addressed:

1. Measurement bias
2. Measurement stability
3. Measurement linearity

5.1 Measurement Bias

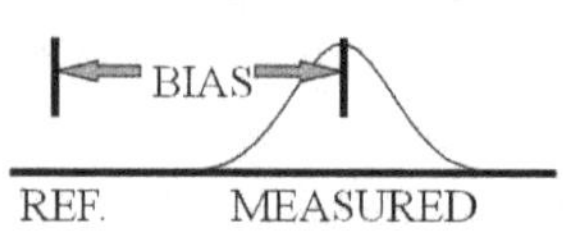

Figure 5.2
Measurement bias

Bias is merely a shift from the correct value. It is important that a good measuring procedure has been developed and a value has been certified for a reference sample before checking for bias. Bias is usually checked by one appraiser checking a reference sample ten times. If the bias is unusually large, it is wise to check for the following before recalibrating:

1. Instrument used improperly by the appraiser
2. Instrument damaged or worn
3. Error in the reference sample
4. Instrument made to the wrong dimensions
5. Instrument measuring the wrong characteristic

5.2 Measurement Stability

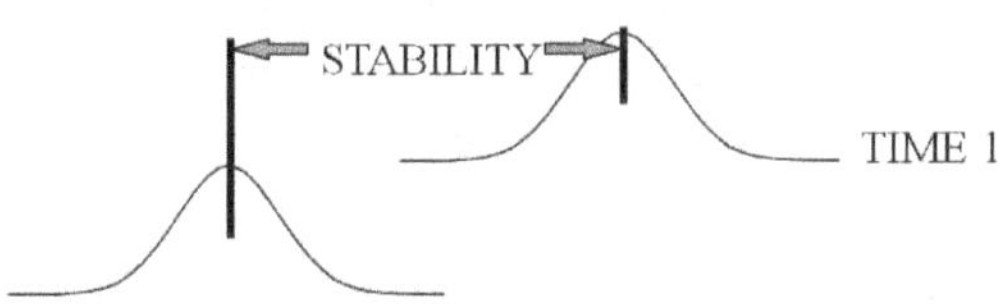

Figure 5.3
Measurement stability

A lack of stability is indicated if measurements of the reference sample show special causes of variation over time. A good way to check for stability is by measuring the reference sample three to five times on a periodic basis (daily, weekly). Control limits can be set up on Average and Range charts to look for special causes of variation. An example of a special cause of variation would be large temperature shifts in the room where the parts are being measured. This is why the rooms where the parts get measured are almost always air conditioned and the temperature is controlled tightly.

5.3 Measurement Linearity

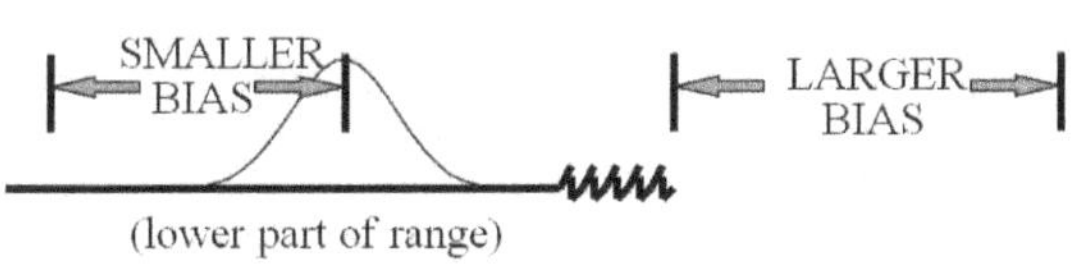

Figure 5.4
Measurement linearity

A lack of linearity is indicated when the bias is different through the expected operating range of the measuring equipment. This is usually checked by measuring two or three reference samples. If, for example, a reference 1.00 gram weight gave a reading of 0.98 grams and a 10.0 gram reference gave a reading of 10.1 grams, there would be nonlinearity present. With newer equipment, the fix for this is generally an adjustment in the software.

5.4 Gage Repeatability & Reproducibility

When checking the precision or variation of a measurement system, the following two components are examined.

1. **Gage Repeatability or Equipment Variation (EV)**

   This component is the amount of variation that would be obtained when one person measures the same characteristic on one part several times. Most people who are trying to measure something will automatically do an informal check of equipment variation by measuring the same thing two or three times to see if they can repeat the same number or get close to the same number.

2. **Gage Reproducibility or Appraiser Variation (AV)**

   This is the variation that occurs between inspectors or appraisers. Different people may put parts in fixtures differently, visually line things up differently, follow procedures differently, etc. Training the appraisers is a key to reducing this variation. More automation in the inspection procedure can also reduce this component of variation.

These two components of measurement variation are then combined into a gage R & R (Repeatability & Reproducibility). This measurement variation or R & R is normally then compared to the total variation in the parts and/or the spec tolerance width. The total variation in the parts is the sum of the measurement variation and the actual part-to-part variation. An important rule to understand before working through this example is that standard deviations cannot be added together, but variances can.

Totaling Variation Components

$$\begin{aligned} \sigma_T &\neq \sum_{all\,i} \sigma_i \\ \sigma_T^2 &= \sum_{all\,i} \sigma_i^2 \\ \sigma_T &= \sqrt{\sum_{all\,i} \sigma_i^2} \end{aligned} \tag{5.1}$$

Where:

σ_T = total variation

σ_i = a component of the total variation

When doing a gage R & R study, several parts (typically five to ten) are measured by several appraisers (typically two to four) several times (typically three to five). The parts should be measured randomly and, ideally, should be numbered in such a way that the appraiser does not know which part they are measuring until after the reading is taken. The three sources of variation in such a study and the equations used to quantify the variation are presented in Table 5.1.

Table 5.1 Gage R & R Equations

Source of Variation	Standard Deviation Calculation
Repeatability (EV)	$$\hat{\sigma}_{EV} = \sum_{i=1}^{a}\sum_{j=1}^{b}\frac{R_{ij}}{ab}/d_2^* \quad (5.2)$$ Where: R_{ij} = range of measurements by appraiser j for part i; $n = c$; $k = ab$ for d_2^* (Table 12.8)
Reproducibility (AV)	$$\hat{\sigma}_{AV} = \overline{X}_{DIFF}/d_2^* \quad (5.3)$$ Where: $\overline{X}_{DIFF} = \max\left(\overline{X}_1,\ldots,\overline{X}_b\right) - min\left(\overline{X}_1,\ldots,\overline{X}_b\right)$, $\overline{X}_i$ = average of all readings for appraiser i; $n = b$; $k = 1$ for d_2^* (Table 12.8)
Part-to-Part (P)	$$\hat{\sigma}_P = R_P/d_2^* \quad (5.4)$$ Where: $R_P = \max\left(\overline{X}_1,\ldots,\overline{X}_a\right) - \min\left(\overline{X}_1,\ldots,\overline{X}_a\right)$; $\overline{X}_i$ = average of all readings for part i; $n = a$; $k = 1$ for d_2^* (Table 12.8)

a = Number of parts, b = number of appraisers, c = number of measurements per part per operator

Note that all the estimated standard deviation equations in Table 5.1 are variations of Equation 2.8. The numerator is the range for the source of error in question. The denominator uses d_2^* instead of d_2. d_2^* is sometimes used when the number of subgroups, k, is small ($d_2^* = d_2$ when $k = \infty$).

Using Equation 5.1, the estimated standard deviation of the measurement variation can be calculated (Equation 5.5).

$$\hat{\sigma}_{R\&R} = \sqrt{\hat{\sigma}_{EV}^2 + \hat{\sigma}_{AV}^2} \quad (5.5)$$

The total variation (TV) is a combination of the measurement variation and the actual part-to-part variation. Again, using Equation 5.1, it is calculated as shown in Equation 5.6.

$$\hat{\sigma}_{TV} = \sqrt{\hat{\sigma}^2_{R\&R} + \hat{\sigma}^2_p} \tag{5.6}$$

Ideally, the amount of measurement error is small when compared to both the total variation and to the spec tolerance width. These comparisons are usually reported as % R & R and are calculated as shown in Equation 5.7 and Equation 5.8.

Based on total variation:

$$\% \, R\&R = \frac{\hat{\sigma}_{R\&R}}{\hat{\sigma}_{TV}} \times 100\% \tag{5.7}$$

Based on tolerance:

$$\% \, R\&R = \frac{6\hat{\sigma}_{R\&R}}{\text{spec tolerance width}} \times 100\% \tag{5.8}$$

It should be noted that some industries use 5.15 in the numerator of Equation 5.8 instead of 6. A bell curve that is 5.15 standard deviations wide contains 99% of the data if perfectly normally distributed versus 99.73% for 6 standard deviations. This book will use 6 standard deviations to maintain consistency with other chapters and the default in Minitab is 6 standard deviations.

The smaller the % R & R based on either total variation or tolerance, the better. Assuming the spec tolerance width is based on sound engineering analysis, the course of action will differ if one of these % R & Rs is much bigger than the other. If, based on tolerance, the R & R is acceptable, but it is large based on total variation, then resources should be spent on improving the measurement system if it is important to track the variation of that characteristic. The EV and AV components should be reviewed to determine whether the improvement needs to be in the equipment or appraisers or both. If, based on total variation, the R & R is acceptable, but it is large based on tolerance, then resources should be spent on improving the part-to-part variation first. If the part-to-part variation is reduced to the point where the measurement system can no longer detect it, then resources should be spent on improving the measurement process.

What, then, is an acceptable % R & R? The following are the rules that are most often applied to the results of R & R studies:

1. Less than 10% – Always acceptable
2. 10% to 30% – Sometimes acceptable, depending on the application and cost to improve
3. Greater than 30% – Usually not acceptable

If a capability study, using Average and Range charts, has been carried out and the parts were measured with the same measurement system as used in the R & R study, then $\hat{\sigma}_{TV}$ in Equation 5.7 should be replaced by $\overline{R}/d_2$ for an estimate of total variation based on more data than just the data from the R & R study.

The preceding R & R study that has been described is called the Method One Average and Range method. There are two other commonly employed methods, which generally yield very similar results:

1. Method Two Average and Range method – The only difference is that the following is substituted for Equation 5.3.

   $$\hat{\sigma}_{AV} = \sqrt{\left(\overline{X}_{DIFF}/d_2^*\right)^2 - \hat{\sigma}_{EV}^2/ac}$$

   This equation basically has the premise that appraiser variation is influenced by the equipment variation.

2. Analysis of Variance (ANOVA) method – ANOVA is a technique that will be looked at when analyzing DOEs in later chapters. It has a couple of advantages in that it uses all the data in the calculations (not just ranges) and it can be used to look for interactions between appraisers and parts. These calculations will not be covered here. The equations can be found in the help menus in Minitab.

5.5 Example 5.1: R & R Study

An R & R study was performed on a gauge used to measure the length of an injection molded part. The gauge consisted of a digital indicator that was zeroed when a 17.00 inch reference standard was put in the fixture used to hold the part. Thus, the readings represent how much longer each part is than 17.00 inches. The design spec for the length is 17.100 ± 0.030 inches (0.060 inch wide spec tolerance). Determine the % R & R based on both total variation in the study and tolerance.

Figure 5.5 shows the data collected from the four operators.

Note that the data is rearranged in two different ways to show how some of the intermediate statistics are calculated. Using subgroups consisting of the three measurements of each part by each operator, the first thing to do is to construct the Average and Range charts (Figure 5.6).

operator	part	1	2	3		xbar	R
Mike	part 1	0.0760	0.0740	0.0740		0.0747	0.0020
	part 2	0.0760	0.0775	0.0755		0.0763	0.0020
	part 3	0.0890	0.0890	0.0890		0.0890	0.0000
	part 4	0.0945	0.0950	0.0930		0.0942	0.0020
	part 5	0.0905	0.0900	0.0910		0.0905	0.0010
				Mike's xbar =	0.0849		

operator	part	1	2	3		xbar	R
Keith	part 1	0.0755	0.0750	0.0755		0.0753	0.0005
	part 2	0.0770	0.0775	0.0775		0.0773	0.0005
	part 3	0.0900	0.0900	0.0905		0.0902	0.0005
	part 4	0.0945	0.0945	0.0945		0.0945	0.0000
	part 5	0.0915	0.0915	0.0910		0.0913	0.0005
				Keith's xbar =	0.0857		

operator	part	1	2	3		xbar	R
Tom	part 1	0.0740	0.0735	0.0745		0.0740	0.0010
	part 2	0.0750	0.0765	0.0760		0.0758	0.0015
	part 3	0.0890	0.0880	0.0870		0.0880	0.0020
	part 4	0.0930	0.0925	0.0945		0.0933	0.0020
	part 5	0.0900	0.0900	0.0915		0.0905	0.0015
				Tom's xbar =	0.0843		

operator	part	1	2	3		xbar	R
Gary	part 1	0.0740	0.0740	0.0735		0.0738	0.0005
	part 2	0.0755	0.0765	0.0760		0.0760	0.0010
	part 3	0.0885	0.0880	0.0885		0.0883	0.0005
	part 4	0.0925	0.0930	0.0925		0.0927	0.0005
	part 5	0.0895	0.0895	0.0910		0.0900	0.0015
				Gary's xbar =	0.0842		

Overall Grand Average = 0.0848 Rbar = 0.00105

	Mike			Keith			Tom			Gary			
	1	2	3	1	2	3	1	2	3	1	2	3	Average
part 1	0.0760	0.0740	0.0740	0.0755	0.0750	0.0755	0.0740	0.0735	0.0745	0.0740	0.0740	0.0735	0.0745
part 2	0.0760	0.0775	0.0755	0.0770	0.0775	0.0775	0.0750	0.0765	0.0760	0.0755	0.0765	0.0760	0.0764
part 3	0.0890	0.0890	0.0890	0.0900	0.0900	0.0905	0.0890	0.0880	0.0870	0.0885	0.0880	0.0885	0.0889
part 4	0.0945	0.0950	0.0930	0.0945	0.0945	0.0945	0.0930	0.0925	0.0945	0.0925	0.0930	0.0925	0.0937
part 5	0.0905	0.0900	0.0910	0.0915	0.0915	0.0910	0.0900	0.0900	0.0915	0.0895	0.0895	0.0910	0.0906

Figure 5.5 Data and intermediate statistics for an R & R Study

The groupings of four sets of data points represent the operators on each chart. The five data points within group represent the five parts. As was discussed in Chapter 3, the Range chart answers the question, "Is the within-subgroup variation consistent?" Since the chart is in-control, the answer is yes – the measurement-to-measurement variation is consistent. The Average chart answers the question, "Given the variation within the subgroups, are there detectable differences between subgroups?" Since this chart is out of control, the answer is also yes – there are differences from part-to-part that exceed what can be explained by the common cause measurement-to-measurement variation. When the Average chart is out of control and the pattern of part averages is similar for each operator, as it is in this case, the experienced eye will realize that the measurement system is likely performing well before reviewing the numbers. An out-of-control Average chart in an R & R study is a good thing, provided that the pattern matches across operators, because it indicates that the part-to-part variation (the spread of the data points) is greater than the measurement variation (the width of control limits on the Average chart).

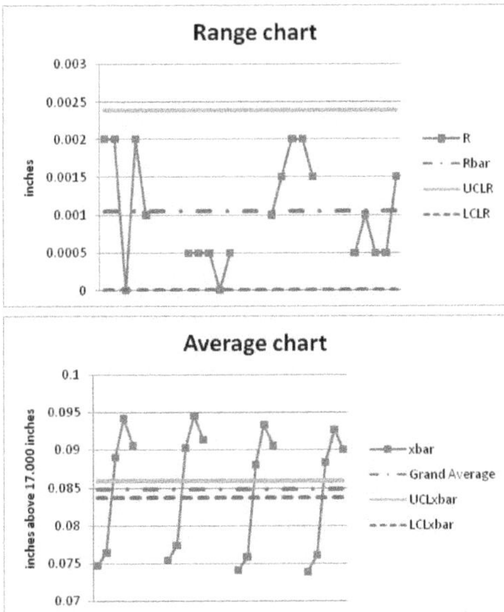

Figure 5.6
Average and Range charts for Example 5.1

The numbers can then be calculated. Using Equation 5.2–Equation 5.4, the estimated standard deviation for each source of variation is obtained.

Source of Variation	Standard Deviation Calculation
Repeatability (EV)	$\hat{\sigma}_{EV} = \sum_{i=1}^{a}\sum_{j=1}^{b}\frac{R_{ij}}{ab}/d_2^* = \frac{0.0210\text{ inches}}{5\times 4}/1.705$ $= \frac{0.00105\text{ inches}}{1.705} = 6.158\times 10^{-4}\text{ inches}$ $n = c = 3$ & $k = ab = 5 \times 4 = 20$ for $d_2^* = 1.705$ (Table 12.8)
Reproducibility (AV)	$\hat{\sigma}_{AV} = \overline{X}_{DIFF}/d_2^* = (0.0857\text{ inches} - 0.0842\text{ inches})/2.237$ $= 6.821\times 10^{-4}\text{ inches}$ $n = b = 4$ & $k = 1$ for $d_2^* = 2.237$ (Table 12.8)
Part-to-Part	$\hat{\sigma}_P = R_P/d_2^* = \frac{(0.0937\text{ inches} - 0.0745\text{ inches})}{2.477}$ $= 7.755\times 10^{-3}\text{ inches}$ $n = a = 5$ & $k = 1$ for $d_2^* = 2.477$ (Table 12.8)

From Equation 5.5, the estimated standard deviation of the measurement variation can be calculated.

$$\hat{\sigma}_{R\&R} = \sqrt{\hat{\sigma}_{EV}^2 + \hat{\sigma}_{AV}^2} = \sqrt{(6.168\times 10^{-4}\text{ inches})^2 + (6.705\times 10^{-4}\text{ inches})^2}$$
$$= 9.189\times 10^{-4}\text{ inches}$$

Total variation (TV) is calculated from Equation 5.6.

$$\hat{\sigma}_{TV} = \sqrt{\hat{\sigma}^2_{R\&R} + \hat{\sigma}^2_p} = \sqrt{(9.189 \times 10^{-4}\ \text{inches})^2 + (7.755 \times 10^{-3}\ \text{inches})^2}$$
$$= 7.809 \times 10^{-3}\ \text{inches}$$

The % R & R based on total variation is calculated with the aid of Equation 5.7.

$$\%\ R\&R\ \text{based on TV} = \frac{\hat{\sigma}_{R\&R}}{\hat{\sigma}_{TV}} \times 100\% = \frac{9.189 \times 10^{-4}\ \text{inches}}{7.809 \times 10^{-3}\ \text{inches}} \times 100\% = 11.8\%$$

The % R & R based on tolerance is calculated from Equation 5.8.

$$\%\ R\&R\ \text{based on tolerance} = \frac{6\hat{\sigma}_{R\&R}}{\text{spec tolerance width}} \times 100\%$$
$$= \frac{6 \times 9.189 \times 10^{-4}\ \text{inches}}{0.060\ \text{inches}} \times 100\% = 9.2\%$$

Assume that a capability study is run and that the Average and Range charts show good process control with an Average Range of 0.0138 inches with a subgroup size of 5. A better number for % R & R can be calculated as follows:

$$\%\ R\&R\ \text{based on TV} = \frac{\hat{\sigma}_{R\&R}}{\bar{R}/d_2} \times 100\% = \frac{9.189 \times 10^{-4}\ \text{inches}}{0.0138\ \text{inches}/2.326} \times 100\%$$
$$= \frac{9.189 \times 10^{-4}\ \text{inches}}{5.933 \times 10^{-3}\ \text{inches}} \times 100\% = 15.5\%$$

Although not less than 10%, there is a good chance that the 15.5% based on total variation would be accepted (10–30% range). The 9.2% based on tolerance is acceptable.

5.6 Example 5.2: R & R Study with Minitab

The following will repeat Example 5.1 in Minitab. Put the numbers in one column, use another column to identify the part number and a third column to identify the appraiser. Next, go to Stat, then Quality Tools, then Gage Study, then Gage R & R Study (Crossed) as shown in Figure 5.7.

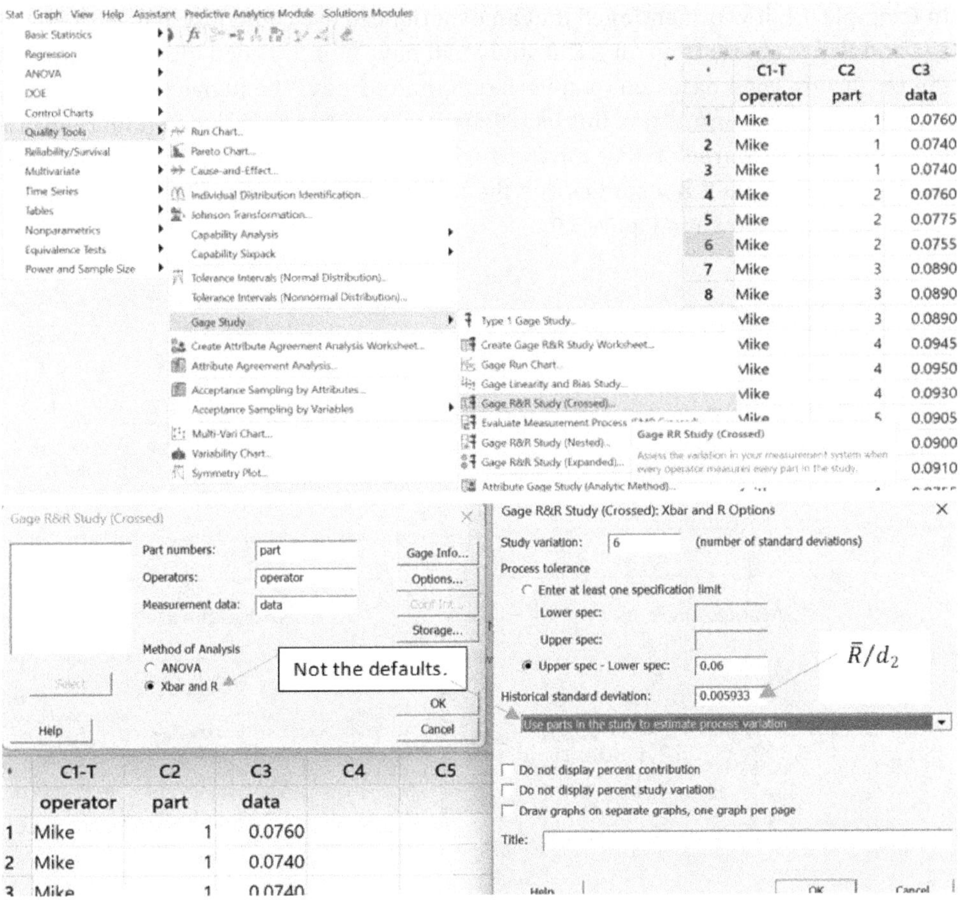

Figure 5.7 Sequence for Gage R & R analysis in Minitab

The results and graphs are the same as calculated in the previous example (Figure 5.8).

Gage Evaluation

Source	StdDev (SD)	Study Var (6 × SD)	%Study Var (%SV)	%Tolerance (SV/Toler)	%Process (SV/Proc)
Total Gage R&R	0.0009187	0.0055121	11.78	9.19	15.48
Repeatability	0.0006161	0.0036968	7.90	6.16	10.38
Reproducibility	0.0006814	0.0040886	8.74	6.81	11.49
Part-To-Part	0.0077414	0.0464485	99.30	77.41	130.48
Total Variation	0.0077957	0.0467745	100.00	77.96	131.40

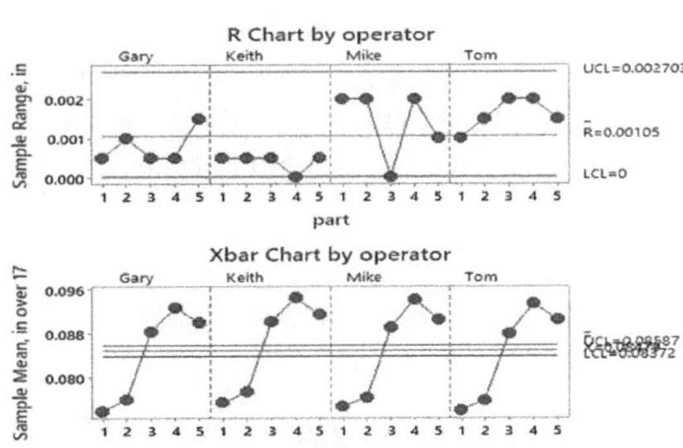

Figure 5.8 Minitab results for Example 5.2

In Example 5.1, it was mentioned that an experienced eye could just look at the Average and Range charts from an R & R study and have a pretty good idea as to whether the results are good based on total variation in the study. The Range chart should always be in-control and, if not, this usually indicates either a typo or that more operator training is required or that a fixture is inconsistently holding the parts. Several Average charts from R & R studies and the resultant % R & R based on total variation in the study are shown in Figure 5.9.

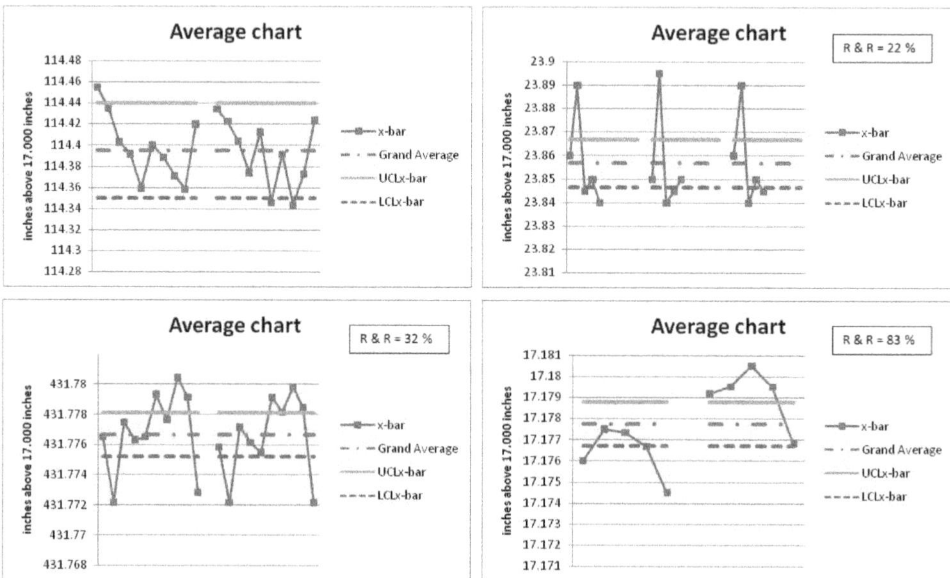

Figure 5.9 Judging % R & R from the Average chart

Three of the preceding graphs showed very little AV variation. This is evident from the pattern of the part groupings. If they are close, then the reproducibility across different appraisers is good. However, poor reproducibility can be seen in the graph with an R & R of 83%, where the second appraiser's readings were much higher than the first. For the three graphs that had good reproducibility, you can see that those with better repeatability (lower EV variation) the ones that have tighter control limits relative to the spread of the data points (part-to-part variation).

5.7 Inadequate Measurement Units

One final measurement consideration will be reviewed in this chapter. The issue of measurement resolution is important when deciding on how to measure parts. The cost of measurement equipment can sometimes increase by nearly one order of magnitude when measurements are taken to one more decimal place, so it is important to have enough resolution – but not more than necessary. An easy rule to follow is to ensure that the Range chart from the process capability study has at least six (6) possible values below the Upper Control Limit (Wheeler, 1992). Example 3.1 will again be used to show what happens when the measurement system does not have high enough resolution.

5.7.1 Example 5.3: Inadequate Measurement Units

The original parts in Example 3.1 were measured to the nearest 0.0001 inches. The Upper Control limit of the Range chart was calculated to be 0.0117 inches. Therefore, for the next subgroup taken from the process, there would be 117 possible values for the range below the Upper Control limit. These values would be 0, 0.0001, 0.0002, ..., or 0.0116 inches. The original Average and Range charts are shown in Figure 5.10.

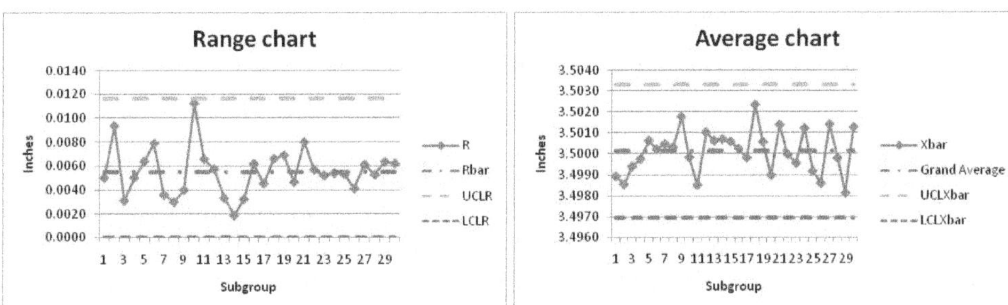

Figure 5.10 Original Average and Range charts for Example 5.3

Now if the original data is rounded to the nearest 0.001 inch, the charts in Figure 5.11 are obtained.

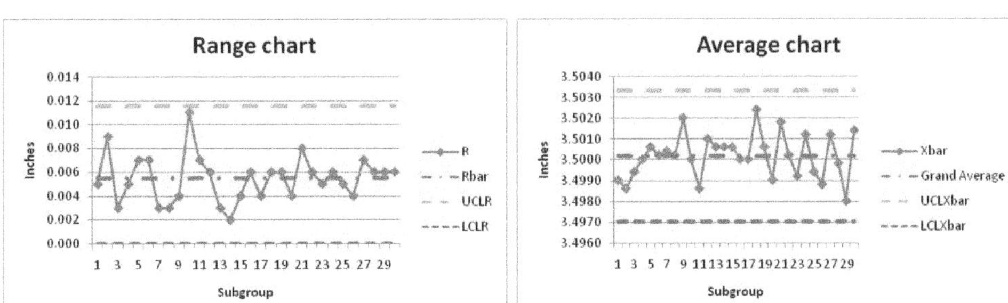

Figure 5.11 Average and Range charts for Example 5.3 with data rounded to 0.001 inches

Comparing these charts to the original, very little has changed. You can see that a few points have shifted very slightly, but the lines and points remain almost identical. There are still adequate measurement units since there are more than six (6) possibilities below the Upper Control limit of the Range chart. With three-decimal-place resolution, the next subgroup has 12 possible values: 0, 0.001, 0.002, ..., up to 0.011 inches. Three decimal places would therefore provide enough resolution for this dimension.

Figure 5.12 shows what would happen if our measurement resolution were to the nearest 0.005 inches.

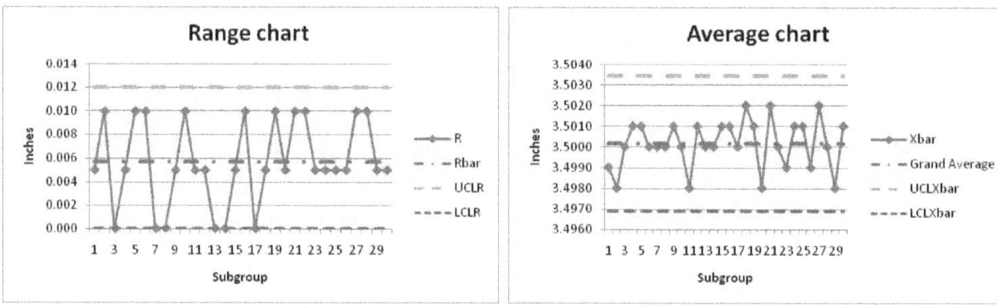

Figure 5.12 Average and Range charts for Example 5.3 with data rounded to 0.005 inches

Looking at these charts, even though the control limits did not change much, the pattern of the points changed significantly. With this level of resolution, the next subgroup would only have 3 possible values under the Upper Control limit: 0, 0.005, or 0.010 inches. Since this is fewer than six (6), this measurement resolution would not be adequate. Even if the rule of six (6) possible range values were not known, a Range chart that contains many zero values and repeated values as the one shown in Figure 5.12, is an indication that the measurement resolution needs to be improved.

6 Hypothesis Testing

The subject of this chapter is how to make statistical comparisons of two parameters. The following are examples of questions that might be answered with the comparisons:

1. Which of two materials gives parts with better physical properties?
2. Which of two materials gives parts with better dimensional stability?
3. Which of two machines molds parts with less variability?
4. Can a new process set-up reduce the variability in the molded parts?
5. Can a new gate location or fewer (or more) gates make better parts?
6. Can a different assembly method make assemblies that last longer?

To make comparisons, samples will have to be taken and parameters (mean and standard deviation) will need to be estimated. As can be imagined, when two samples are taken, it is very rare for the estimated means or standard deviations to be exactly the same. This chapter covers methods for determining whether two estimates are statistically different or not.

Control charts (Chapter 3) are a form of hypothesis testing. The hypothesis is that the characteristic being monitored is in control. Unless there is sufficient evidence that the characteristic is out of control, it is assumed that the process is in control. However, it is important to point out that there is still the possibility that the process is out of control even if the evidence doesn't show that it is. That would be an example of a Type II or β error and was shown in Figure 3.22. That chart can be modified for more general hypothesis testing as shown in Figure 6.1.

		Decision Based on Calculated Statistics	
		Accept Hypothesis (Only "chance" causes of variation present – Parameters the same)	Reject Hypothesis ("Assignable" causes of variation present – Parameters different)
Actual State	Parameters the same	Good Decision	Type I error α Error
	Parameters different	Type II error β Error	Good Decision

Figure 6.1 Types of errors for hypothesis tests

As was discussed in Chapter 3, the amount of α error can be set. An α error of 0.05 or 5% is typically used and is the default in Minitab and most statistical packages. The β error cannot be set as easily. It is a result of the set α error, the type of test or distribution that is being used, the sample size, and how much the actual mean deviates from the assumed mean, δ.

For an Average chart with $\pm 3\hat{\sigma}_{\overline{X}}$ control limits, the α error is 0.27% (100% – 99.73%) which means 0.27% of the subgroup Averages will be outside the control limits when there aren't any special or assignable causes of variation present. Figure 6.2 shows what a distribution of the subgroup Averages would look like.

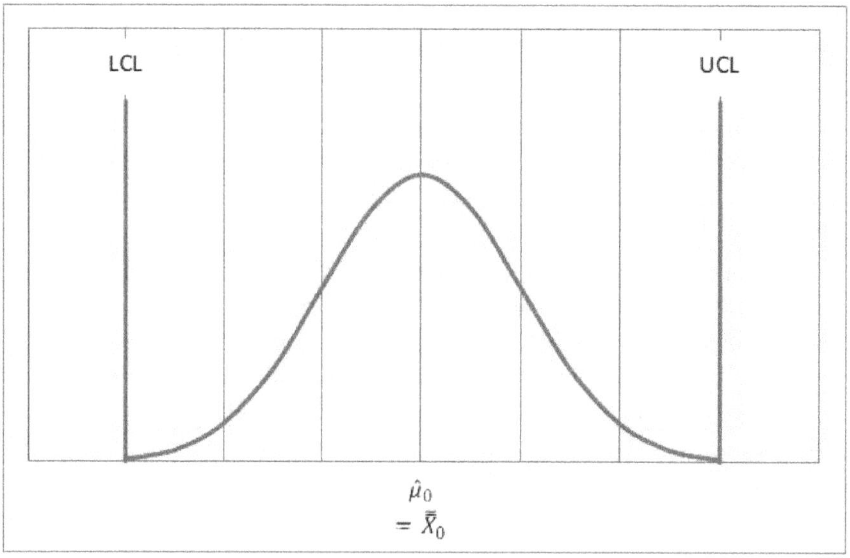

Figure 6.2 Normal distribution of Averages from a control chart

The control chart will be used as an example of how to calculate the β error.

6.1 Example 6.1: Calculating the Beta Error

The β error deals with whether the control chart will catch a special cause of variation. Consider, for example, that the process shifts by an amount equal to exactly three standard deviations of the Averages as shown in Figure 6.3 (only God would truly know that the process shifted by this amount).

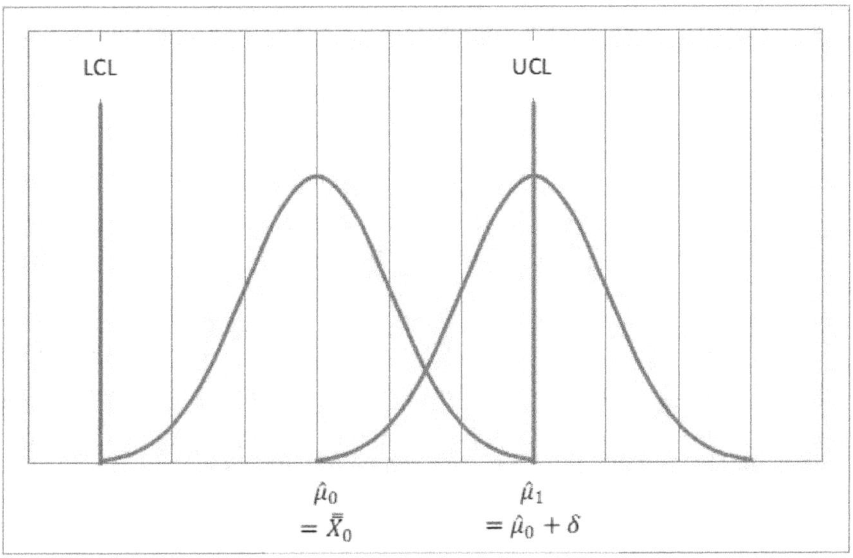

Figure 6.3 Original and new distribution when the process shifts by $\delta = 3\sigma_{\bar{X}}$

It should be obvious to the reader that the chance of catching the shift in Figure 6.3 with the next subgroup Average is 50%. The new bell curve is split exactly in two by the upper control limit. This means that the β error is 50%; there is a 50% chance that the next sample (subgroup) will not detect the presence of the special cause of variation. Although the β error is easy to quantify when the process shifts by exactly three standard shifts, other shifts won't be. The following will go through how to use the standard normal distribution (Table 12.3) to determine the β error. For Figure 6.3, the β error is the percentage of the shifted distribution that is between the LCL and UCL. Z-values can be determined for the LCL and UCL using Equation 2.2:

For UCL:

$$Z = \frac{X - \mu_X}{\sigma_X} = \frac{UCL_{\overline{X}} - \hat{\mu}_1}{\hat{\sigma}_{\overline{X}}} = \frac{(\hat{\mu}_0 + 3\hat{\sigma}_{\overline{X}}) - (\hat{\mu}_0 + \delta)}{\hat{\sigma}_{\overline{X}}} = \frac{\left(\hat{\mu}_0 + \frac{3\hat{\sigma}_X}{\sqrt{n}}\right) - (\hat{\mu}_0 + \delta)}{\frac{\hat{\sigma}_X}{\sqrt{n}}} = 3 - \frac{\delta}{\hat{\sigma}_X}\sqrt{n}$$

For Figure 6.3, $\frac{\delta}{\hat{\sigma}_X}\sqrt{n} = \frac{\delta}{\hat{\sigma}_{\overline{X}}} = 3$, so

$= 3 - 3 = \mathbf{0}$

For LCL:

$$Z = \frac{X - \mu_X}{\sigma_X} = \frac{LCL_{\overline{X}} - \hat{\mu}_1}{\hat{\sigma}_{\overline{X}}} = \frac{(\hat{\mu}_0 - 3\hat{\sigma}_{\overline{X}}) - (\hat{\mu}_0 + \delta)}{\hat{\sigma}_{\overline{X}}} = \frac{\left(\hat{\mu}_0 - \frac{3\hat{\sigma}_X}{\sqrt{n}}\right) - (\hat{\mu}_0 + \delta)}{\frac{\hat{\sigma}_X}{\sqrt{n}}}$$

$$= -3 - \frac{\delta}{\hat{\sigma}_X}\sqrt{n}$$

For Figure 6.3, $\frac{\delta}{\hat{\sigma}_X}\sqrt{n} = \frac{\delta}{\hat{\sigma}_{\overline{X}}} = 3$, so

$= -3 - 3 = \mathbf{-6}$

$\beta = P(LCL < \overline{X} < UCL) = P(0 < z < 0) + P(-6 < z < 0) = P(0 < z < 0) + P(0 < z < 6) = 0 + 0.5 =$ **0.5 or 50%**

It should be pointed out that the shift in Example 6.1 is $3\hat{\sigma}_{\overline{X}}$. In terms of standard deviations of the individual values, the shift was $1.34\hat{\sigma}_X = \frac{3\hat{\sigma}_{\overline{X}}}{\sqrt{5}}$. By substituting various sample sizes and shifts in the preceding example, the following Operating Characteristic (OC) curves for the Average chart can be obtained (Figure 6.4).

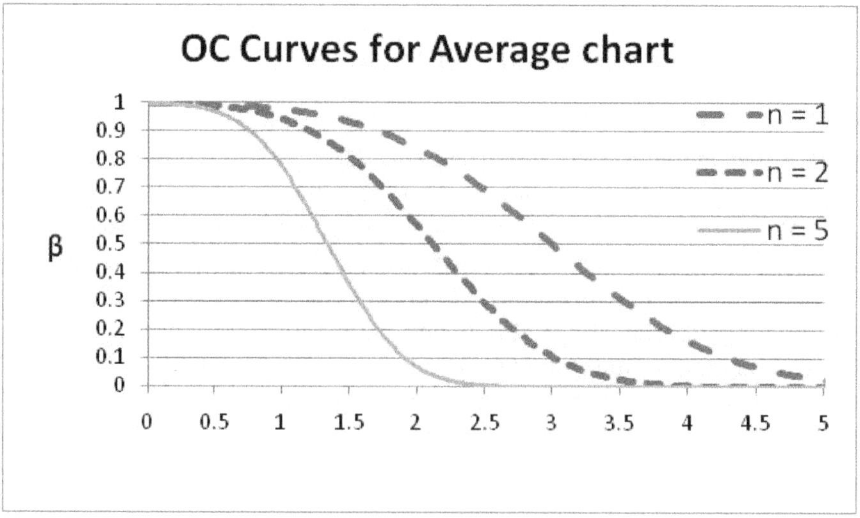

Figure 6.4 OC curve for the Average chart (β versus $\frac{\delta}{\sigma_X}$ for various values of n)

6.2 Statistical Power

As can be seen in Figure 6.4, and was discussed in Chapter 3, the larger the sample size, the smaller is the shift that can be detected (a small x marks the point calculated in Example 6.1). It is also obvious that the β error increases when trying to detect smaller shifts. Instead of reporting β error, it is more common to speak in terms of the power of a statistical test. The formula for Power is simply as shown in Equation 6.1.

$$\text{Power} = 1 - \beta \tag{6.1}$$

The power of a statistical test's relationship to β is analogous to the confidence coefficient's relationship to α. When determining the sample size for a statistical test, it can be useful to rearrange the OC curve as shown in Figure 6.5 (a small x marks the point calculated in Example 6.1).

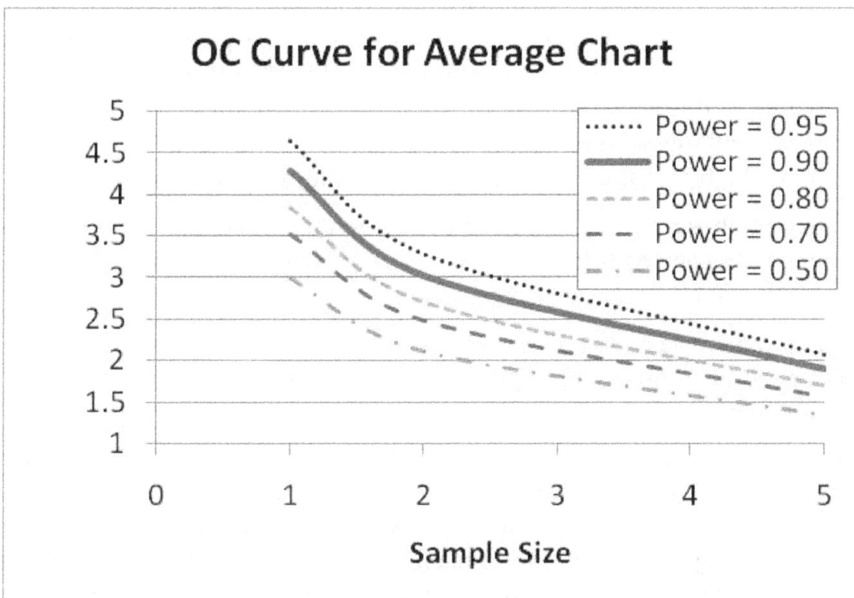

Figure 6.5 OC curve for the Average chart ($\frac{\delta}{\sigma_x}$ versus n for various values of Power)

Table 12.9 (see also Figure 12.1) in the Appendix has three different OC curves in the format of Figure 6.5, which will be used later in this chapter. For the hypothesis tests that will be reviewed in this chapter, the primary use of the OC curves is to determine what sample size is needed to detect a certain size difference, δ/σ_x, in whatever characteristic is being studied.

The reader should note here that a power of 50% for a sample size of five ($n = 5$) to detect the shift in Example 6.1 ($1.34\hat{\sigma}_X$ or $3\hat{\sigma}_{\overline{X}}$) isn't very good and it might therefore be thought that control charts aren't very useful. However, a control chart provides multiple chances to detect the shift, allowing the benefits of smaller sample sizes to be fully realized. From probability theory, the following is known (Equation 6.2):

$$\mathrm{P(AorB)} = \mathrm{P(A)} + \mathrm{P(B)} - \mathrm{P(AandB)} = \mathrm{P(A)} + \mathrm{P(B)} - \mathrm{P(A)P(B)} \tag{6.2}$$

So, if the probability of detecting the shift after one sample (subgroup) is 50%, then the probability of detecting it after two subgroups is:

P(A or B) = P(A) + P(B) – P(A)P(B) = 0.50 + 0.50 – 0.50 × 0.50 = 0.75 or 75%

Thus, the power is effectively increased from 50% to 75% after addition of the second subgroup. After three subgroups, the probability of detecting the shift is:

P(A or B or C) = P(A or B) + P(C) – P((A or B) P(C) = 0.75 + 0.50 – 0.75 × 0.50 = 0.875 or 87.5%

So, the power is 87.5% or, in other words, there is an 87.5% chance of detecting the shift from Example 6.1 after three subgroups have been taken. With hypothesis testing, however, it is generally desired to take one sample from the parts in question and decide based on that sample. Minimum powers of 70% to 95% are common. 90% will generally be used in this book. The cost of part inspection and testing is the reason that lower power will sometimes be accepted. Two more points will be made here on the power of the test before the steps of hypothesis testing are reviewed.

1. The power, and thus the sample size, is based on a minimum difference, δ, that the test is intended to detect. If the difference is larger than δ, then the test's power increases and the β error decreases.
2. If the measurement variation is large, the sample size will need to be larger since, for the same δ, the $\frac{\delta}{\sigma_X}$ is smaller because the measurement variation increases, σ_X. This is another reason that it is important to have consistent measurements.

There is a useful analogy between a statistical hypothesis test and a jury trial. In a trial, there are two competing hypotheses: *guilty* and *not guilty*. If the prosecuting attorney cannot provide sufficient evidence to prove the *guilty* hypothesis, the jury cannot reject the *not guilty* hypothesis. Most readers can think of a real or televised case where the defendant was probably guilty, yet the jury found him/her not guilty. This would be an example of a β error. Conversely, if the jury finds the defendant guilty and he/she is in fact innocent, that would be an example of an α error. What the jury considers "beyond reasonable doubt" is the α error level. For a statistical hypothesis test, the evidence playing this role is the data collected.

6.3 Steps of a Hypothesis Test

The following are the steps involved in performing a statistical hypothesis test:

1. State the null hypothesis, H_0
2. State the alternative hypothesis, H_1
3. Choose α
4. Choose the acceptable β and sample size, n
5. Determine the appropriate statistical test(s)
6. Collect the data and calculate the test statistics
7. Compare the test statistics to the acceptance/rejection values
8. Make a decision based on the data

Each step will be reviewed and then followed by several examples that show the calculations done by hand and using Excel and Minitab.

1. **State the null hypothesis, H_0.** The null hypothesis is analogous to the *not guilty* verdict in the jury trial; it cannot be rejected unless there is sufficient evidence to disprove it. For statistical hypothesis testing, it will always include the term *equals* in the statement and will be expressed in terms of the distributions rather than sample statistics; for example, one mean, μ, or standard deviation, σ, is stated to be equal to another.

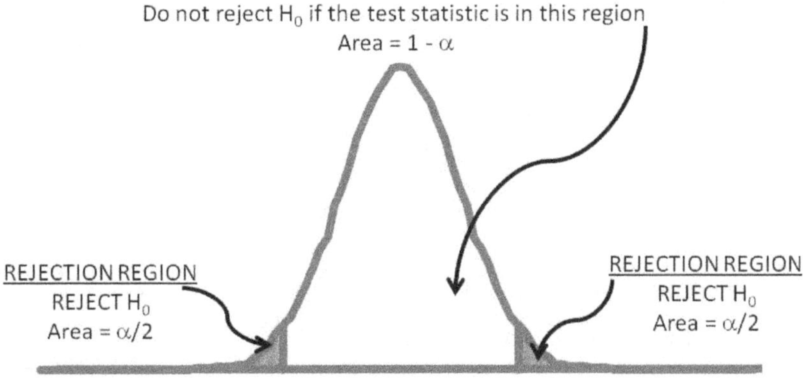

Figure 6.6 Two-tailed test to determine if two parameters are not equal. Note: normal distribution (Z-test) shown, but the same logic could be applied to t-, X^2-, and F-distributions

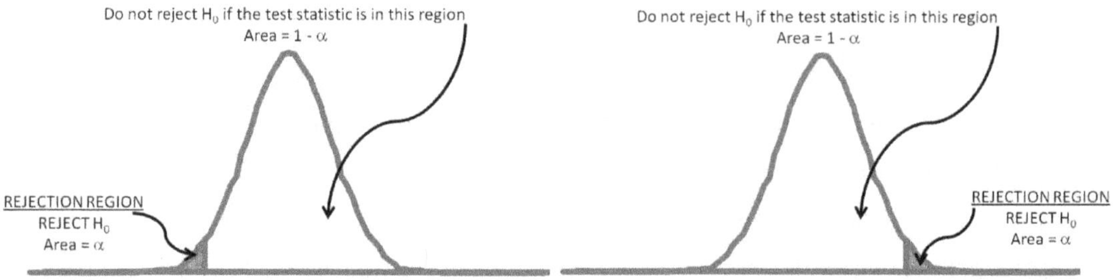

Figure 6.7 One-tailed tests. Left: testing whether one parameter is smaller. Right: testing whether one parameter is larger. Note: normal distribution (Z-test) shown, but the same logic could be applied to t-, X^2-, and F-distributions.

2. **State the alternative hypothesis, H_1.** The alternative hypothesis is the opposite of the null hypothesis. If it is a two-tailed test (see Figure 6.6), it will state that either a mean, μ, or a standard deviation, σ, is *not equal* to another. If it is a one-tailed test (see Figure 6.7), it will state that either a mean, μ, or a standard deviation, σ, is either *less than* (left tail) or *greater than* (right tail) another.
3. **Choose α.** A level of acceptable Type I error will typically be set at 0.05 or 5%.
4. **Choose the acceptable β and sample size, n.** As discussed earlier in this chapter, this will depend on the α level, the acceptable Type II or β error, and the size of the difference it is intended to detect. To do this, it is necessary to have some historical data or to run a baseline study to obtain an estimated standard deviation. The acceptable β error will typically be 0.10 or 10% (power of 90%).
5. **Determine the appropriate statistical test(s).** The following is a list of F-tests that will be reviewed in this book and a description of what each is testing.

Table 6.1 Descriptions of Several Hypothesis Tests

Test	Description
1-sample t-test	Compares a mean to a given value when the standard deviation is not known
2-sample t-test	Compares two means when the standard deviation is not known
X^2 test	Compares a standard deviation to a given value
F-test	Compares two standard deviations. This is commonly used to determine which 2-sample t-test statistic to use for the means

Notes:

1. The reference values used for comparison are usually based on historical data from control charts or other inspection/testing data.

2. There are several other tests that can be performed. The Z-test is basically the same as the t-test if the standard deviation is known or the sample size is large. The t-test for differences in related means is useful when comparing different testing or inspection methods. More on these tests can be found elsewhere (Montgomery, 1999).

6. **Collect the data and calculate the test statistics**. Here, it is very important that good sampling procedures be followed. The parts to be tested or measured need to represent the actual process that they would come from in a production setting. For example, if a new material is being tested and regrind is used in the material mix during production, it is important that the amount of regrind and the heat history of the regrind be correct for the samples.

   Table 6.2 shows the equations for the test statistics for each type of statistical test and the degrees of freedom.

Table 6.2 Hypothesis Test Equations for Calculating Test Statistics and Degrees of Freedom

Test	Test Statistic	Degrees of Freedom
1-sample t-test	$t = \frac{\overline{x} - \mu_0}{s/\sqrt{n}}$	$df = n - 1$
2-sample t-test	For equal standard deviations: $t = \frac{\overline{x}_1 - \overline{x}_2}{\sqrt{\frac{1}{n_1} + \frac{1}{n_2}}} \sqrt{\frac{n_1 + n_2 - 2}{(n_1 - 1){s_1}^2 + (n_2 - 1){s_2}^2}}$ For unequal standard deviations: $t = \frac{\overline{x}_1 - \overline{x}_2}{\sqrt{\frac{{s_1}^2}{n_1} + \frac{{s_2}^2}{n_2}}}$	For equal standard deviations: $df = n_1 + n_2 - 2$ For unequal standard deviations: $df = \frac{({s_1}^2/n_1 + {s_2}^2/n_2)^2}{\frac{({s_1}^2/n_1)^2}{n_1 - 1} + \frac{({s_2}^2/n_2)^2}{n_2 - 1}}$
X^2-test	$X^2 = \frac{(n-1)s^2}{{\sigma_0}^2}$	$df = n - 1$
F-test	$F = \frac{{s_1}^2}{{s_2}^2}$	$df_1 = n_1 - 1$ $df_2 = n_2 - 1$

7. **Compare the test statistics to the accept/reject values**. The tabulated value needs to be selected from Tables 12.4–12.7 to determine the rejection region (see Figure 6.6 and Figure 6.7).

Table 6.3 Rejection Regions for Hypothesis Tests and Excel Commands for Accept/Reject Values

Test	Alternative Hypothesis	Rejection Region (see Figure 6.6 and Figure 6.7)	Excel Commands
1-sample t-test or 2-sample t-test	$\mu_1 \neq \mu_0$ two tail	$t < -t_{\alpha/2,df}$ or $t > t_{\alpha/2,df}$	$t_{\alpha/2,\nu}$ = t.inv.2t(α,ν) $-t_{\alpha,\nu}$ = t.inv(α,ν)
	$\mu_1 < \mu_0$ left tail	$t < -t_{\alpha,df}$	
	$\mu_1 > \mu_0$ right tail	$t > t_{\alpha,df}$	
X^2-test	$\sigma_1 \neq \sigma_0$ two tail	$X^2 < X^2_{1-\alpha/2,df}$ or $X^2 > X^2_{\alpha/2,df}$	$X^2_{\alpha/2,\nu}$ = chisq.inv(α/2,ν) $X^2_{1-\alpha/2,\nu}$ = chisq.inv.rt(1-α/2,ν) $X^2_{\alpha,\nu}$ = chisq.inv(α,ν) $X^2_{1-\alpha,\nu}$ = chisq.inv.rt(1-α,ν)
	$\sigma_1 < \sigma_0$ left tail	$X^2 < X^2_{1-\alpha,df}$	
	$\sigma_1 > \sigma_0$ right tail	$X^2 > X^2_{\alpha,df}$	
F-test	$\sigma_1 \neq \sigma_2$ two tail	$F < 1/F_{\alpha/2,df_1,df_2}$ or $F > F_{\alpha/2,df_1,df_2}$	$F_{\alpha/2\nu1,\nu2}$ = f.inv.rt(α/2,ν₁,ν₂) $F_{\alpha,\nu1,\nu2}$ = f.inv.rt(α,ν₁,ν₂)
	$\sigma_1 < \sigma_2$ left tail	$F < 1/F_{\alpha/2,df_1,df_2}$	
	$\sigma_1 > \sigma_2$ right tail	$F > F_{\alpha,df_1,df_2}$	

8. **Make a decision based on the data**. If the test statistic is in the rejection region, then the null hypothesis is rejected in favor of the alternative hypothesis. If the test statistic is not in the rejection region, then the null hypothesis cannot be rejected.

These steps will now be illustrated with two common injection molding scenarios. The first involves trying out a new material variation that is reported to improve impact strength compared with the current material. The second examines whether parts obtained from two different suppliers differ in quality.

6.3.1 Example 6.2: Hypothesis of Improved Physical Properties

A CTQ characteristic of a molded part is the impact strength of a particular area of the part. Historically, the part has a mean impact strength of 1.50 foot-pounds per inch and a standard deviation of 0.230 foot-pounds per inch. A material supplier claims to have substitute material that will improve the impact strength of the part by at least 10%. To test the material, samples are to be molded and tested. For this part, the runner is 28% of the total shot size and the regrind is mixed back with the virgin material with a proportional loader at the press.

The steps for this hypothesis testing are as follows.

1. State the null hypothesis, H_0.

   The null hypothesis is $\mu_0 \leq 1.65$ ftlb/in since the material will not be considered unless the data shows that the null hypothesis is not true. This is the null hypothesis to use if the increase in impact strength needs to be 10% higher than the 1.50 ftlb/in historical level to be used.

2. State the alternative hypothesis, H_1.

   The alternative hypothesis is $\mu_1 > 1.65$ ftlb/in since the material will be considered if the data proves that the null hypothesis should be rejected in favor of this alternative hypothesis.

3. Choose α. The default of 0.05 will be used.

4. Choose the acceptable β and sample size, n.

   An acceptable β of 0.10 (power of 0.90) is common and will be used. Since the aim is to detect that an increase of 10% in the impact strength, $\delta = 0.10 \times 1.5$ ftlb/in = 0.15 ftlb/in. Therefore, $\frac{\delta}{\sigma_x} = \frac{0.15 \text{ ftlb/in}}{0.230 \text{ ftlb/in}} = 0.65$. From Figure 12.1 in the Appendix, a sample size of about 27 is needed. This assumes that the standard deviation with the new material will not be higher than it is with the current material.

5. Determine the appropriate statistical test(s).

   Since an estimated mean is to be compared to a given value (historical value), the 1-sample t-test will be used.

6. Collect the data and calculate the test statistics.

   Here, it is crucial that the parts be representative of what will be made in production. For this process, 28% of the shot is regrind and so it is important that the parts made contain 28% regrind. That does not mean that parts are molded with virgin, ground up and mixed back in at 28%. That would merely give parts with 28% first-generation regrind which, in most cases, would not be much different from using just virgin material. It is important that the regrind have the same thermal history that will be used in production; many materials do not start to significantly change in physical properties until at least the third time through the process (3rd generation regrind, R3). The steady state make of material should be 72% virgin, 20.2% R1(0.28 × 72), 5.6% R2(0.28 × 20.2) + 1.6% R3(0.28 × 5.6) + 0.4% R4 (0.28 × 1.6) + 0.2% R5 and higher generations. The process could either be run long enough to get to steady-state or the make-up could be arrived at by making the generations of regrind in the process and mixing off-line.

Once the correct material mix is obtained and the process reaches steady-state, the 27 samples are collected for testing. After the parts are tested, it is found for the 27 samples that the average is 1.75 $\frac{\text{ftlb}}{\text{in}}$ ($\overline{x}$ =1.75 ftlb/in) and the sample standard deviation is 0.19 $\frac{\text{ftlb}}{\text{in}}$ (s = 0.19 ftlb/in).

Using the equations from Table 6.2,

$$t = \frac{\overline{x} - \mu_0}{s/\sqrt{n}} = \frac{1.75\frac{\text{ftlb}}{\text{in}} - 1.65\frac{\text{ftlb}}{\text{in}}}{0.19\frac{\text{ftlb}}{\text{in}}/\sqrt{27}} = 2.7348$$

$$df = n - 1 = 27 - 1 = 26$$

7. Compare the test statistics to the acceptance/rejection values.

   The decision rule, in this case, is to reject H_0 if $t > t_{0.05,26}$.

   From Table 12.4 in the Appendix, $t_{0.05,\,26} = 1.7056$

8. Make a decision based on the data.

   Since is $t > t_{0.05,26}$ (2.7348 > 1.7056), the null hypothesis, H_0, can be rejected in favor of the alternative hypothesis, H_1, and it can be said that the impact strength of the new material is at least 10% greater than the original material.

6.3.2 Example 6.3: Comparing Two Processes

A molded part is purchased from two different suppliers. One supplier's parts are causing more problems in an assembly operation. This example will look at the hypothesis tests that can be used to test whether the parts are statistically the same. It should be noted that some other comparisons that these same tests can be used for include parts molded in the same mold on two different molding machines, with materials from different suppliers, or using different processing strategies.

It should also be pointed out that this example will do the calculations for just one CTQ dimension – the one that is believed to be causing the assembly problem. Typically, there would be several CTQ characteristics that would need to be examined. The calculations would be the same as shown here. Because of this fact, it is common to start by measuring 25 parts and checking what the β error is. If β is too high, more parts will need to be measured. Consequently, it is recommended to obtain or mold at least 100 parts from each process (more if destructive tests are needed) in case a sample size of 25 proves to be insufficient.

The following data is from a CTQ dimension measured on 25 random parts obtained from two different suppliers:

Supplier 1:	Average = 12.3288 inches
	Sample standard deviation = 0.0033 inches

Supplier 2:	Average = 12.3309 inches
	Sample standard deviation = 0.0026 inches

Tests will be done to compare both the standard deviations and the means of the parts from the two suppliers. The standard deviations will be compared first since that outcome – whether they are different or not – will determine which equations are used to compare the means (see Table 6.2).

6.4 Comparing Standard Deviations

The steps for this hypothesis testing are as follows.

1. State the null hypothesis, H_0.

H_0: $\sigma_1 = \sigma_2$	Where σ_1 = standard deviation from supplier 1, and σ_2 = standard deviation from supplier 2

2. State the alternative hypothesis, H_1.

   H_1: $\sigma_1 \neq \sigma_2$

3. Choose α. The default of 0.05 will be used.

4. Choose the acceptable β and sample size, n.

   An initial sample size of 25 was predetermined. The actual β will be checked after comparison of the means.

5. Determine the appropriate statistical test(s).

   Since two variances are being compared, the F-statistic will be calculated.

6. Collect the data and calculate the test statistics.

   $F = s_1^2/s_2^2 = (0.0033 \text{ inches})^2/(0.0026 \text{ inches})^2 = 1.61$

   $df_1 = n_1 - 1 = 25 - 1 = 24$

   $df_2 = n_2 - 1 = 25 - 1 = 24$

7. Compare the test statistics to the accept/reject values.

   Reject H_0 if $F > F_{0.05/2,24,24} = 2.27$ [from Table 12.7]

8. Make a decision based on the data.

   Since the F-statistic, 1.61, is not greater than 2.27, the null hypothesis, H_o, cannot be rejected and it is assumed that the variances are equal.

6.5 Comparing Means

1. State the null hypothesis, H_0.

H_0: $\mu_1 = \mu_2$	Where μ_1 = mean from supplier 1, and μ_2 = mean from supplier 2

2. State the alternative hypothesis, H_1.

   H_1: $\mu_1 \neq \mu_2$

3. Choose α. The default of 0.05 will be used.

4. Choose the acceptable β and sample size, n.

   An initial sample size of 25 was predetermined. The actual β will be checked after comparison of the means.

5. Determine the appropriate statistical test(s).

   Since two means are being compared and the variances are assumed to be equal, the 2-sample t-test for equal standard deviations will be used.

6. Collect the data and calculate the test statistics.

$$t = \frac{\overline{x}_1 - \overline{x}_2}{\sqrt{\frac{1}{n_1} + \frac{1}{n_2}}} \sqrt{\frac{n_1 + n_2 - 2}{(n_1 - 1)s_1{}^2 + (n_2 - 1)s_2{}^2}} = -2.5$$

$$df = n_1 + n_2 - 2 = 25 + 25 - 2 = 48$$

7. Compare the test statistics to the accept/reject values.

   Reject H_0 if $t < -t_{0.05/2,48}$ or $t > t_{0.05/2,48} = 2.0106$ (from Table 12.4).

8. Make a decision based on the data.

   Since the t-statistic, −2.5, is less than −2.0106, the null hypothesis, H_0, can be rejected in favor of the alternative hypothesis, H_1, indicating that means are not equal.

It is interesting to note that the same conclusion would have been reached by making a 95% confidence interval graph (as discussed in Chapter 2). The graph below shows some overlap between the two suppliers. However, since the standard deviations were statistically the same, the outcome must be that one interval overlaps the other mean in order to be able to conclude that the means are not statistically different.

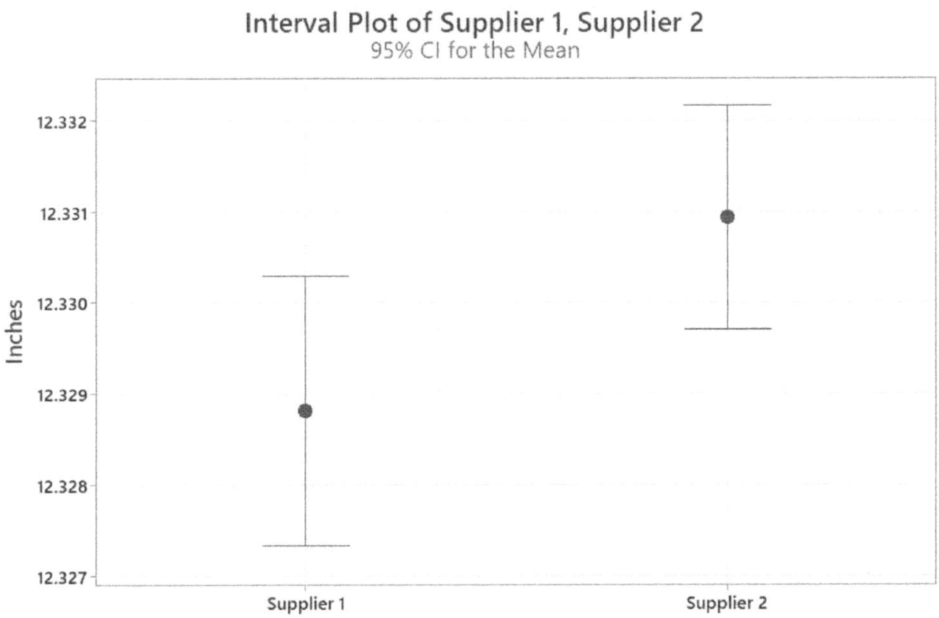

Figure 6.8 95% confidence intervals of the means for Example 6.3

6.6 Determine the β Error

The following OC curve (from Figure 12.1) is for $\alpha = 0.05$. In order to determine the Power and, thus, the β error, the δ/σ_x ratio is needed. Where that ratio intercepts the sample size of 25 line will indicate what the Power for this test was.

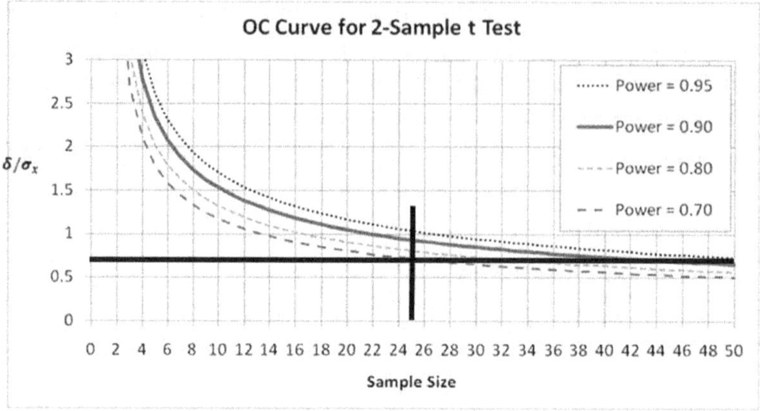

Figure 6.9 OC curve for Example 6.3

The difference, δ, for this test is the difference in the two estimated means:

$$\delta = |\overline{x}_1 - \overline{x}_2| = 12.3309 \text{ inches} - 12.3288 \text{ inches} = 0.0021 \text{ inches}$$

The standard deviation, σ_x, is the square root of the average of the two sample variances:

$$\sigma_X = \sqrt{\frac{(s_1{}^2 + s_2{}^2)}{2}} = \sqrt{\frac{(0.0033^2 \text{ inches}^2 + 0.0026^2 \text{ inches}^2)}{2}} = 0.0030 \text{ inches}$$

The δ/σ_x ratio is then:

$$\frac{\delta}{\sigma_X} = \frac{0.0021 \text{ inches}}{0.0030 \text{ inches}} = 0.70$$

The following curve shows that the Power is about 0.70 and, therefore, the β error is about 0.30 (1 – 0.70). If a Power of 0.90 were desired, the sample size would have to be increased to about 44. Since, in this case, the suppliers' parts were already shown to be not equal, a larger sample is probably not necessary. However, if the processes were shown to be equal with a 0.3 β error, it would be wise to redo the tests and calculations with a sample size of 42. The cost of further inspection would also enter this decision.

6.6.1 Example 6.4: Using Minitab

The following shows how to complete the previous example using Minitab. First, the standard deviations are compared.

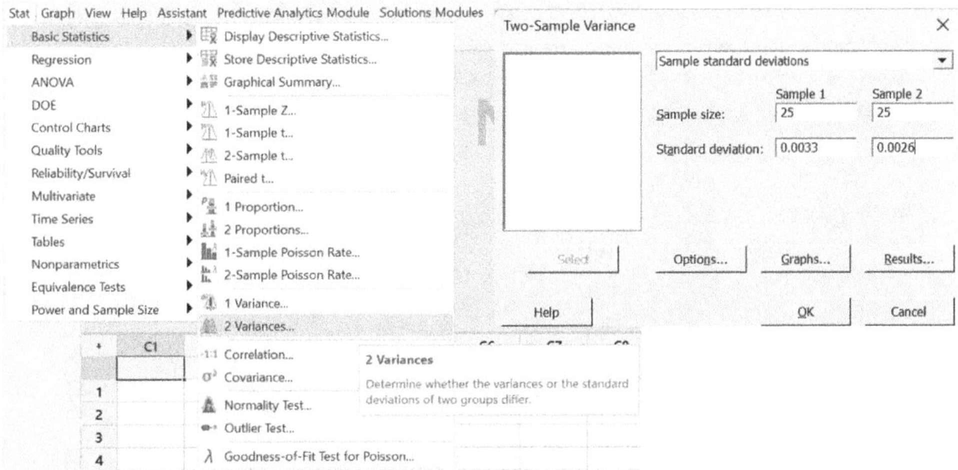

Figure 6.10 Sequence for hypothesis testing to compare standard deviations in Minitab

Test

Null hypothesis	H_0: $\sigma_1 / \sigma_2 = 1$
Alternative hypothesis	H_1: $\sigma_1 / \sigma_2 \neq 1$
Significance level	$\alpha = 0.05$

Method	Test Statistic	DF1	DF2	P-Value
F	1.61	24	24	0.250

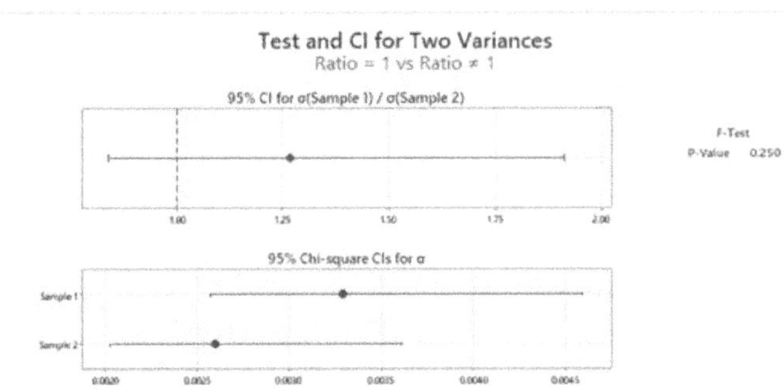

The F-statistic of 1.61 is the same as obtained earlier. The p-value indicates the minimum α value that would be required to reject H₀, based on this data. Since the p-value is greater than 0.05, H₀ cannot be rejected for α = 0.05.

The two graphs shown also clearly show that standard deviations cannot be said to be different because

1. the 95% confidence interval of the standard deviation ratio includes 1.0 and
2. both standard deviation 95% confidence intervals include the other sample standard deviation.

Figure 6.11 Results of F-test for Example 6.3

The test for comparing means can now be completed.

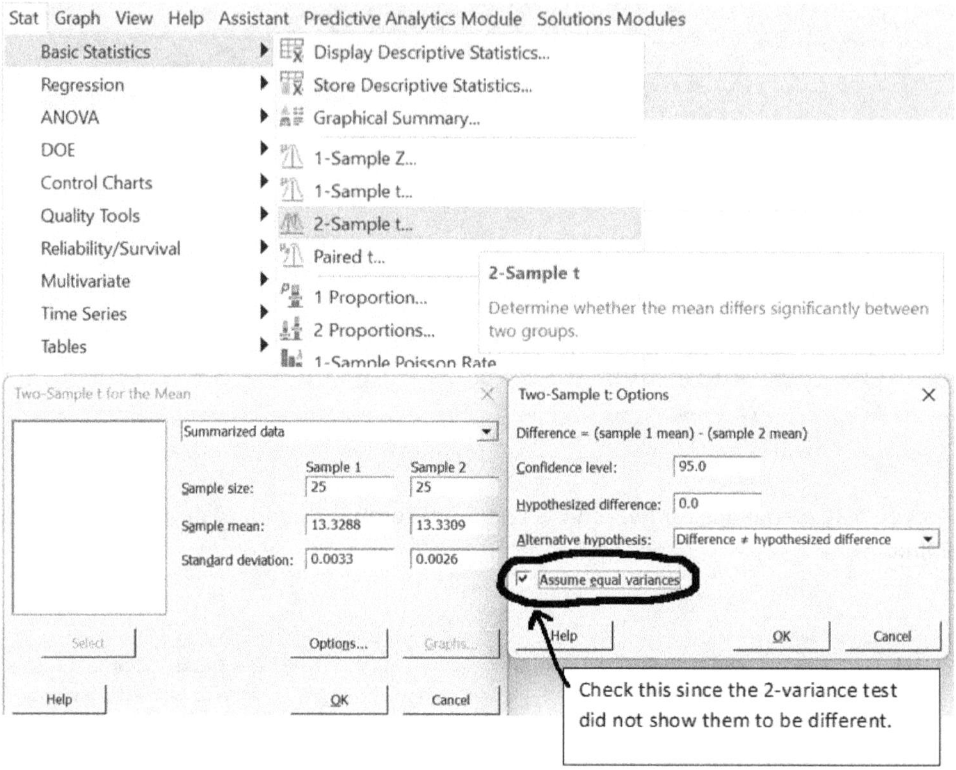

Figure 6.12 Sequence for hypothesis testing to compare means in Minitab

Two-Sample T-Test and CI

Descriptive Statistics

Sample	N	Mean	StDev	SE Mean
Sample 1	25	13.32880	0.00330	0.00066
Sample 2	25	13.33090	0.00260	0.00052

Estimation for Difference

Difference	Pooled StDev	95% CI for Difference
-0.002100	0.002971	(-0.003789, -0.000411)

Test

Null hypothesis H_0: $\mu_1 - \mu_2 = 0$

Alternative hypothesis H_1: $\mu_1 - \mu_2 \neq 0$

T-Value	DF	P-Value
-2.50	48	0.016

Figure 6.13 Results of t-test for Example 6.3

The t-statistic of −2.5 is the same as obtained earlier. The *p*-value indicates the minimum α value that would be required to reject H_0, based on this data. Since the *p*-value is less than 0.05, H_0 can be rejected in favor of H_1 for our maximum $\alpha = 0.05$.

The Power of the sample size chosen can also be found using Minitab instead of the OC curve. The steps are as follows.

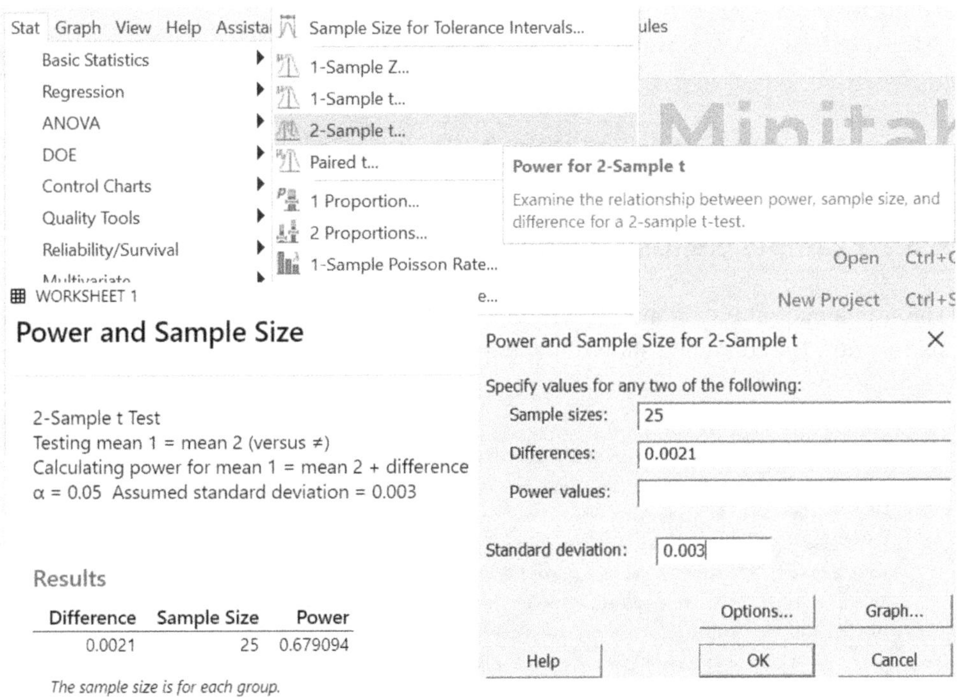

Figure 6.14 Determining Power in Minitab for Example 6.3

As can be seen, the calculated Power is 68%. Minitab could have been used at the outset if the minimum difference to be detected, 0.0021 inches, was known and an estimate of the standard deviation, 0.0030, was known from historical data.

As was seen on the OC curve, a sample size of 44 would have been needed for a Power of 90%.

The useful Table 12.9 in the Appendix was generated with Minitab; see also the OC graph for a 2-sample t-test in Figure 12.1.

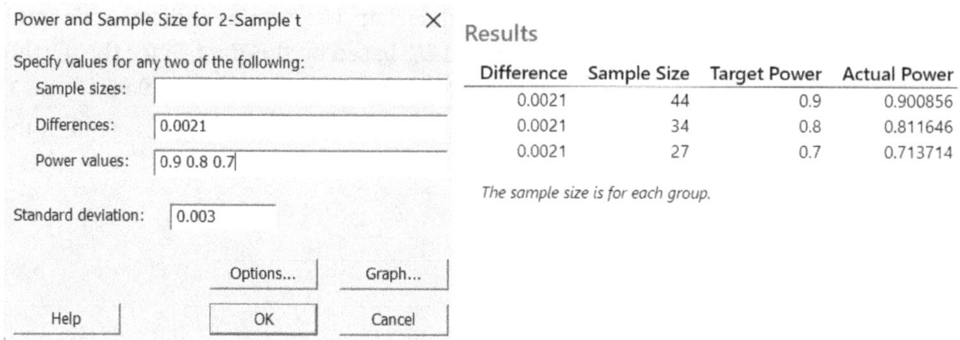

Figure 6.15 Determining Power in Minitab if delta and standard deviation are known

6.6.2 Example 6.5: Using Excel

The supplemental files available with this book include an Excel file called "Hypothesis Testing". The following shows how to use that file to complete Example 6.5.

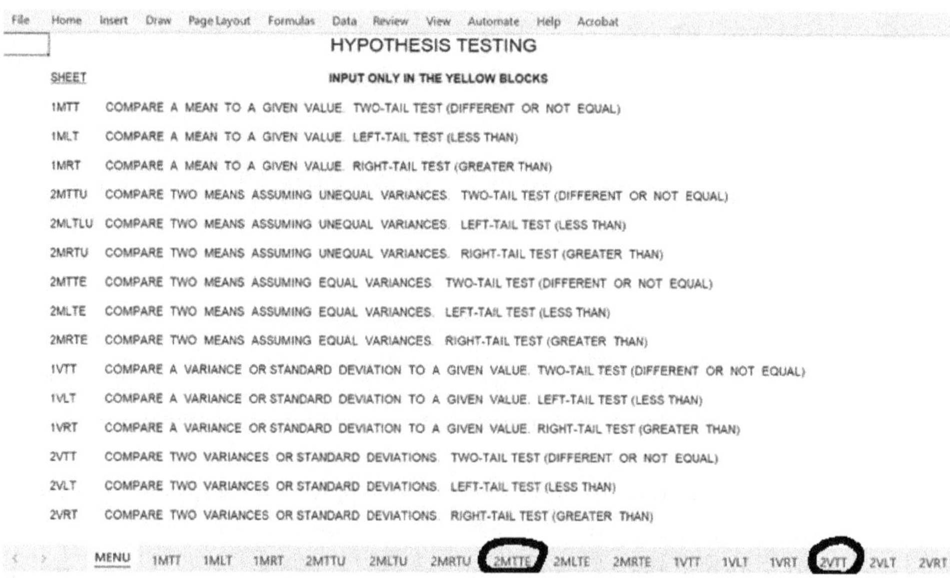

Figure 6.16 Menu for Excel spreadsheet on hypothesis testing

For this example, the 2VTT worksheet would first be used to compare the two variances (or standard deviations). After it is found that the variances must be assumed to be equal, the 2MMTTE worksheet would be used to compare the means. If the variances were shown to be unequal, the 2MTTU worksheet would be used to compare the means.

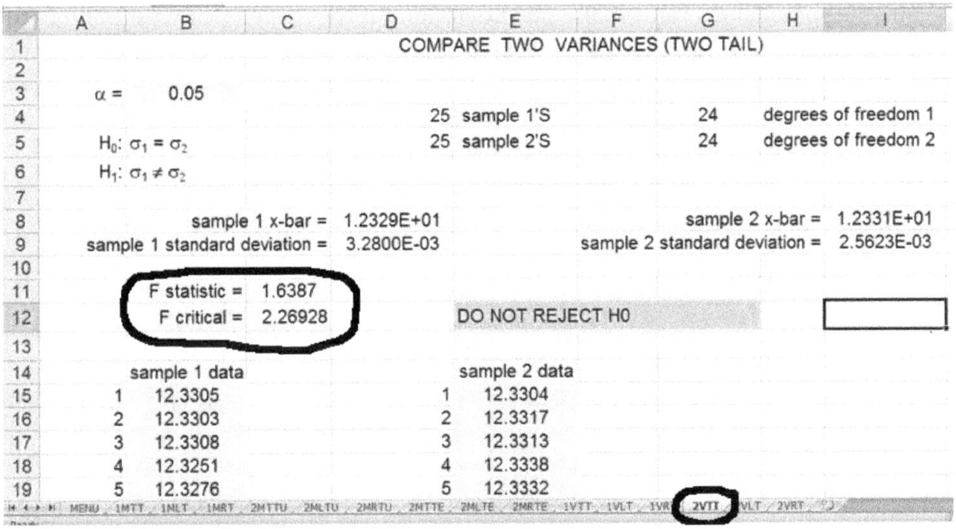

Figure 6.17 Excel worksheet for F-test for Example 6.3

The spreadsheet returns the same decision for the standard deviations that was found in the original example. The spreadsheet, as configured, can handle 70 data points for each sample. The reader may unprotect any sheet (Password "PSU") to add data points.

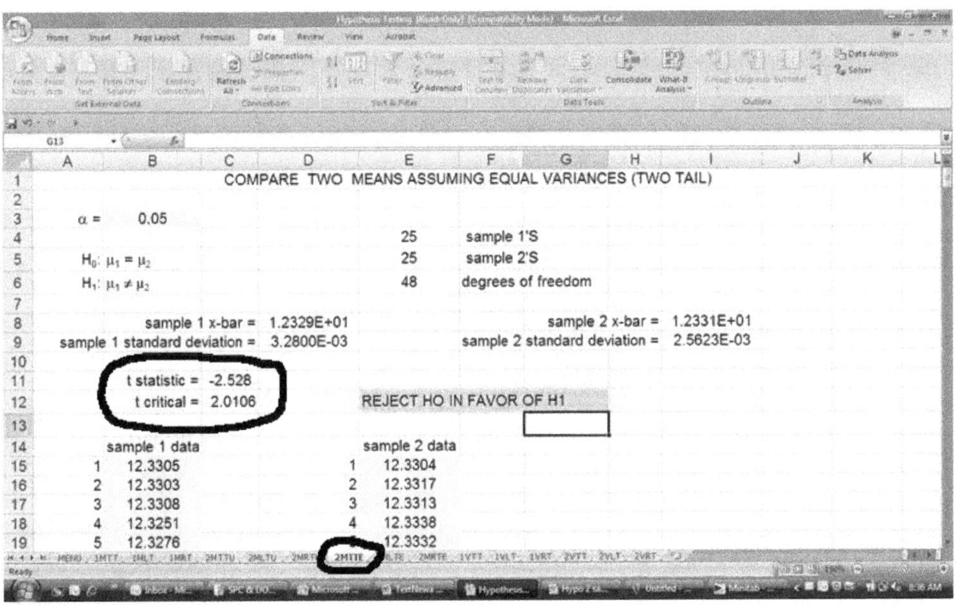

Figure 6.18 Excel worksheet for t-test for Example 6.3

Again, the spreadsheet returns the same decision for the means that was found in the original example. Table 12.9/Figure 12.1 would then be used to determine whether the sample size was sufficient, as was done in the original example.

7 Full Factorial Design of Experiments

As discussed in Chapter 1, DOE (Design of Experiments) is a method of directed experimentation in which systematic changes are made to the inputs or factors to make observations of the effect on the outputs or responses. A **full factorial or fully crossed experiment** is an experiment whose design consists of two or more factors, each with discrete possible values or levels, and whose runs or treatments take on all possible combinations of these levels across all such factors. A **fractional factorial or partially crossed experiment** is one in which only some (a fraction) of the possible combinations are run. This chapter will discuss full factorial experiments, Chapter 8 will discuss the extra considerations for fractional factorial experiments and other designs, and Chapter 9 will discuss strategy for carrying out an injection molding DOE.

To determine the number of runs that are required to complete all possible combinations of factors and levels, the following equation is used.

$$\text{No.of runs} = (\text{No.of levels})^{(\text{No.of factors})} \tag{7.1}$$

To do a full factorial design of 2 factors at 2 levels each would require 4 runs ($2^2 = 4$), 3 factors at 2 levels each would require 8 runs ($2^3 = 8$), 4 factors at two levels each would require 16 runs ($2^4 = 16$), etc. To study 2 factors at 3 levels each would require 9 runs ($3^2 = 9$), 3 factors at 3 levels each 27 runs ($3^3 = 27$), etc. It is easy to see that the number of runs required to do a full factorial experiment increases rapidly as you add factors and/or levels. It is also possible to mix factors and factor levels through multiplication. For example four factors at two levels each and two factors at three levels each would require 144 runs ($2^4 \times 3^2 = 144$).

7.1 2^2 Full Factorial

For the simplest case of 2 factors (A and B) at 2 levels (– and +) each, the matrix given in Table 7.1 would give every possible combination of factors and levels.

Table 7.1 2^2 Full Factorial DOE

Run	A	B	AB	Response
1	+1	+1	+1	$\overline{x}_1$
2	+1	–1	–1	$\overline{x}_2$
3	–1	+1	–1	$\overline{x}_3$
4	–1	–1	+1	$\overline{x}_4$

In Table 7.1, the four runs would consist of 1) both factors set at the high (+) level, 2) "A" high and "B" low, 3) "A" low and "B" high, and 4) both factors set at the low level. This "recipe" is indicated by columns A and B. Column AB is the interaction column and is not needed to carry out the experiment but will be used in the calculations of the results. Instead of just using + and –, +1 and –1 are used and will be referred to as contrast coefficients in later calculations. The response column represents the average of several samples run at each setting. Only one response column is shown here but, for most experiments, more than one response is generally measured for each run. At a minimum, each CTQ characteristic of the part is measured. Other characteristics that are not CTQ, but are easier to measure, may be included to see if they correlate with the CTQ characteristics. Part weight is an example of a characteristic that is rarely a CTQ characteristic but is easily measured and can sometimes be correlated with one or more CTQ characteristics.

To determine the estimated effect, $\hat{l}$, of each factor and the interaction between the two factors, Equation 7.2 to Equation 7.5 are used for each column:

$$\hat{l} = \text{Average of } (+\text{'s}) - \text{Average of } (-\text{'s}) = \sum_{i=1}^{k} c_i \overline{x}_i \Big/ \left(\sum_{i=1}^{k} |c_i|/2 \right) \tag{7.2}$$

Where:

c_i = contrast coefficient for run i,

$\overline{x}_i$ = average response for run i, and

k = number of runs in the experiment

So, for the DOE shown in Table 7.1:

$$\hat{l}_A = \frac{(\overline{x}_1 + \overline{x}_2)}{2} - \frac{(\overline{x}_3 + \overline{x}_4)}{2} = \frac{\overline{x}_1 + \overline{x}_2 - \overline{x}_3 - \overline{x}_4}{2} \tag{7.3}$$

$$\hat{l}_B = \frac{(\overline{x}_1 + \overline{x}_3)}{2} - \frac{(\overline{x}_2 + \overline{x}_4)}{2} = \frac{\overline{x}_1 + \overline{x}_3 - \overline{x}_2 - \overline{x}_4}{2} \tag{7.4}$$

$$\hat{l}_{AB} = \frac{(\overline{x}_1 + \overline{x}_4)}{2} - \frac{(\overline{x}_2 + \overline{x}_3)}{2} = \frac{\overline{x}_1 + \overline{x}_4 - \overline{x}_2 - \overline{x}_3}{2} \tag{7.5}$$

To illustrate what these estimated effects are, two examples will be looked at. The first will be a case where the interaction effect is not present and the second will be a case where there is an interaction between the two factors.

7.1.1 Example 7.1: Calculating Estimated Effects (No Interaction Present)

A full factorial experiment is carried out to investigate the effects of hold time and hold pressure on the part weight of an injection molded part. Five parts are collected and weighed for each run. Table 7.2 shows the results.

Table 7.2 DOE Settings and Results for Example 7.1

Run	A Hold Pressure, psi	B Hold Time, s	1, g	2, g	3, g	4, g	5, g	Response Average Weight, g
1	12,000 (+)	8 (+)	7.612	7.622	7.602	7.542	7.632	7.602
2	12,000 (+)	4 (−)	7.490	7.530	7.550	7.500	7.480	7.510
3	9,000 (−)	8 (+)	7.544	7.454	7.514	7.494	7.514	7.504
4	9,000 (−)	4 (−)	7.424	7.394	7.414	7.374	7.464	7.414

It can be helpful to look at Range and Average charts of the data before calculating the estimated effects. For these charts, there are 4 subgroups (each run is a subgroup) and the sample size is five.

In this case, the Range chart is in control, which shows that the within-subgroup variation is consistent. If the Range chart had been grossly out of control, it would generally have indicated a poor measurement or typo in the data. Occasionally, the Range chart will be slightly out of control, which can indicate that the runs that are above the UCL are not as consistent as the others. This is usually due to the parts having flash or being close to short shots (more on this in Chapter 9).

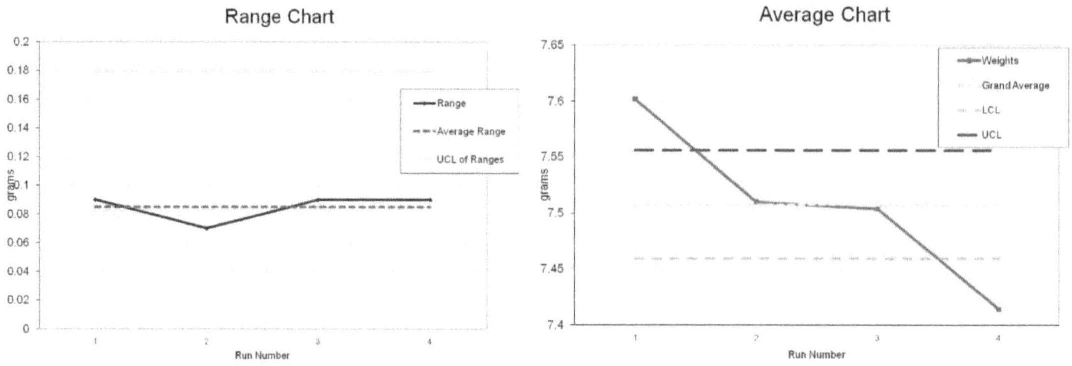

Figure 7.1 Range and Average charts for Example 7.1

In this example, the Average chart is very out of control, which shows that, given the within-run variation, there are detectable differences between runs. This indicates that at least one factor that was changed between runs did cause the response, the part weight, to change. It is important to note that the whole chart is out of control, not just runs 1 and 4. If this chart had been in control, then the between-subgroup variation would be of the same order of magnitude as the within-subgroup variation and the resulting estimated effects would likely be due to just experimental noise (process variation, measurement variation, etc.). If the Average chart is even slightly out of control, there is generally at least one signal (factor causing the response to change) present. It is even still possible that there are small signals present if the Average chart is in control. It is important to pay close attention to sample size to minimize the β error in such cases. Sample size considerations will be discussed in Chapter 9.

The estimated effects can be calculated using Equation 7.3, Equation 7.4, and Equation 7.5:

$$\hat{l}_A = \frac{\overline{x}_1 + \overline{x}_2 - \overline{x}_3 - \overline{x}_4}{2} = \frac{7.602\text{ g} + 7.510\text{ g} - 7.504\text{ g} - 7.414\text{ g}}{2} = 0.097\text{ g}$$

$$\hat{l}_B = \frac{\overline{x}_1 + \overline{x}_3 - \overline{x}_2 - \overline{x}_4}{2} = \frac{7.602\text{ g} + 7.504\text{ g} - 7.510\text{ g} - 7.414\text{ g}}{2} = 0.091\text{ g}$$

$$\hat{l}_{AB} = \frac{\overline{x}_1 + \overline{x}_4 - \overline{x}_2 - \overline{x}_3}{2} = \frac{7.602\text{ g} + 7.414\text{ g} - 7.510\text{ g} - 7.504\text{ g}}{2} = 0.001\text{ g}$$

In words, these effects can be stated as follows:

1. The estimated effect of factor A, hold pressure was 0.097 g. This means that the part weight averaged 0.097 g more when the process was run at a 12,000 psi hold pressure than when it was run at a 9,000 psi hold pressure.
2. The estimated effect of factor B, hold time, was 0.091 g. This means that the part weight averaged 0.091 g more when the process was run at an 8 second hold time than when it was run at a 4 second hold time.

3. The estimated effect of the interaction AB, hold pressure/hold time interaction, was 0.001 g. This means that the part weight change averaged 0.002 (2 × 0.001) g more when the other factor was at the high level, as compared to when it was at the low level. In other words, with an 8 second hold time, the hold pressure effect was 0.098 (7.602 – 7.504) g; with a 4 second hold time, the hold pressure effect was 0.096 (7.510 – 7.414) g. The interaction effect was (0.098 – 0.096)/2 = 0.001 g. Similarly, with a 12,000 psi hold pressure, the hold time effect was 0.092 (7.602 – 7.510) g; with a 9,000 psi hold pressure, the hold time effect was 0.090 (7.504 – 7.414) g. The interaction effect was (0.092 – 0.090)/2 = 0.001 g. It should also be noted here that this interaction effect is a very small number, the reason for which is most likely experimental noise (random process variation, measurement variation, etc.) and there really isn't a hold pressure/hold time interaction.

Graphically, the following charts (Figure 7.2 to Figure 7.4) are the three primary methods that are generally used to show the estimated effects.

1. Bar chart of the results

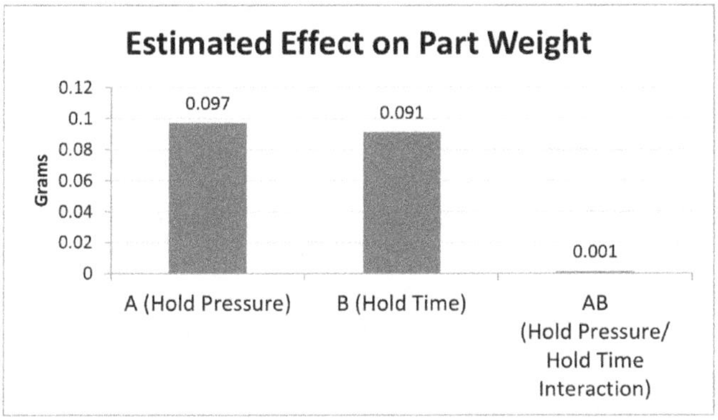

Figure 7.2 Bar chart of estimated effect for Example 7.1

2. Response plot

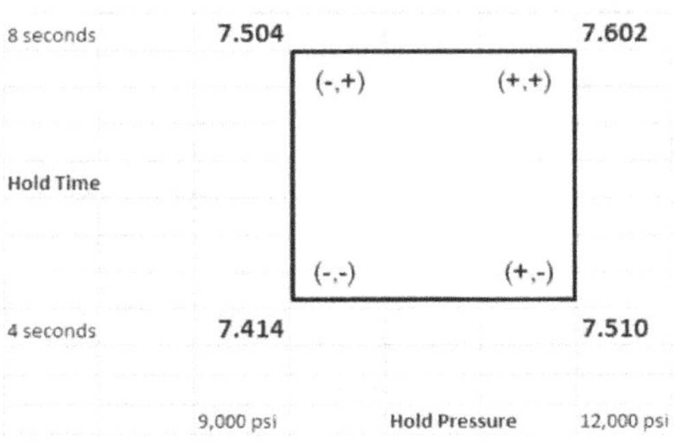

Figure 7.3 Response plot of hold pressure and hold time for Example 7.1

3. Interaction plot

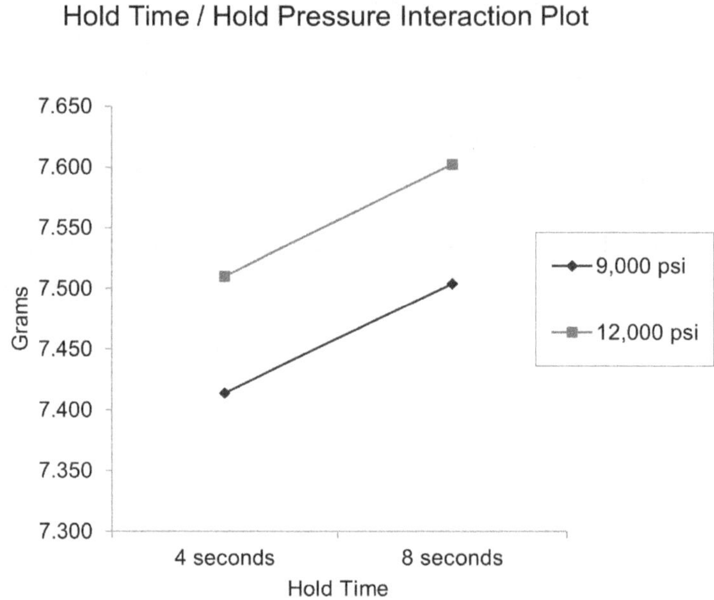

Figure 7.4 Interaction plot of hold pressure and hold time for Example 7.1

In this example, since there was little-to-no interaction, the two lines on the interaction plot, Figure 7.4, are approximately parallel. Had we known beforehand that there was no interaction (can be a risky hypothesis), then we could have gotten the same information with just 3 of the 4 runs. You could calculate any missing corner of the response plot, Figure 7.3, using the other 3 corners. This would be an example of the OFAT (one factor at a time) method that is commonly used when troubleshooting production problems. The next example will show what happens when there is a significant interaction.

7.1.2 Example 7.2: Calculating Estimated Effects (Interaction Present)

Using the same base process as was used for Example 7.1, a full factorial experiment is carried out to investigate the effects of hold time and mold temperature on the part weight of an injection molded part. In Example 7.1, the mold temperature was held constant at 85 °F; in this example, hold pressure is 12,000 psi for all runs. Again, five parts are collected and weighed for each run. Table 7.3 shows the results.

Table 7.3 DOE Settings and Results for Example 7.2

Run	A Mold Temperature, °F	B Hold Time, s	1, g	2, g	3, g	4, g	5, g	Response Average Weight, g
1	115 (+)	8 (+)	7.598	7.608	7.568	7.598	7.568	7.588
2	115 (+)	4 (−)	7.430	7.400	7.420	7.380	7.470	7.420
3	85 (−)	8 (+)	7.612	7.622	7.602	7.542	7.632	7.602
4	85 (−)	4 (−)	7.490	7.530	7.550	7.500	7.480	7.510

The Range and Average charts for these data are shown in Figure 7.5.

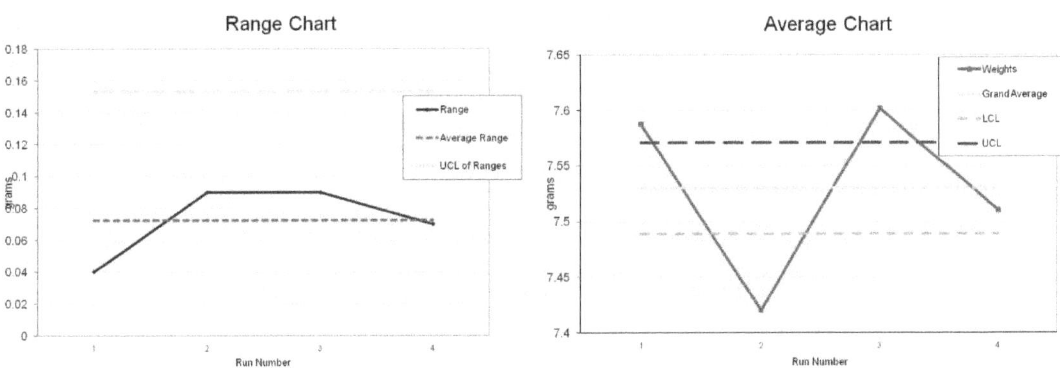

Figure 7.5 Range and Average charts for Example 7.2

Just as in Example 7.1, the Range chart is in control and the Average chart is very out of control. This is the desired result for DOEs.

The estimated effects can be calculated from Equation 7.3, Equation 7.4, and Equation 7.5:

$$\hat{l}_A = \frac{\overline{x}_1 + \overline{x}_2 - \overline{x}_3 - \overline{x}_4}{2} = \frac{7.588\text{ g} + 7.420\text{ g} - 7.602\text{ g} - 7.510\text{ g}}{2} = -0.052\text{ g}$$

$$\hat{l}_B = \frac{\overline{x}_1 + \overline{x}_3 - \overline{x}_2 - \overline{x}_4}{2} = \frac{7.588\text{ g} + 7.602\text{ g} - 7.420\text{ g} - 7.510\text{ g}}{2} = 0.130\text{ g}$$

$$\hat{l}_{AB} = \frac{\overline{x}_1 + \overline{x}_4 - \overline{x}_2 - \overline{x}_3}{2} = \frac{7.588\text{ g} + 7.510\text{ g} - 7.420\text{ g} - 7.602\text{ g}}{2} = 0.038\text{ g}$$

In words, these effects can be stated as follows:

1. The estimated effect of factor A, mold temperature was −0.052 g. This means that the part weight averaged 0.052 g more when the process was run at an 85 °F mold temperature than when it was run at a 115 °F mold temperature. Note that the negative estimated effect means that the low level of the factor gave larger measurements of the response.
2. The estimated effect of factor B, hold time, was 0.130 g. This means that the part weight averaged 0.130 g more when the process was run at an 8 second hold time than when it was run at a 4 second hold time.
3. The estimated effect of the interaction AB, mold temperature/hold time interaction, was 0.038 g. This means that the part weight change averaged 0.076 (2 × 0.038) g more when the other factor was at the high level, as compared to when it was at the low level. In other words, with an 8 second hold time, the mold temperature effect was 0.014 (7.602 – 7.588) g; with a 4 second hold time, the mold temperature effect was 0.090 (7.510 – 7.420) g. The interaction effect was (0.090 – 0.014)/2 = 0.038 g. Similarly, with a 115 °F mold temperature, the hold time effect was 0.168 (7.588 – 7.420) g; with an 85 °F mold temperature, the hold time effect was 0.092 (7.602 – 7.510) g. The interaction effect was (0.168 – 0.092)/2 = 0.038 g.

These estimated effects are shown graphically in Figure 7.6 to Figure 7.8.

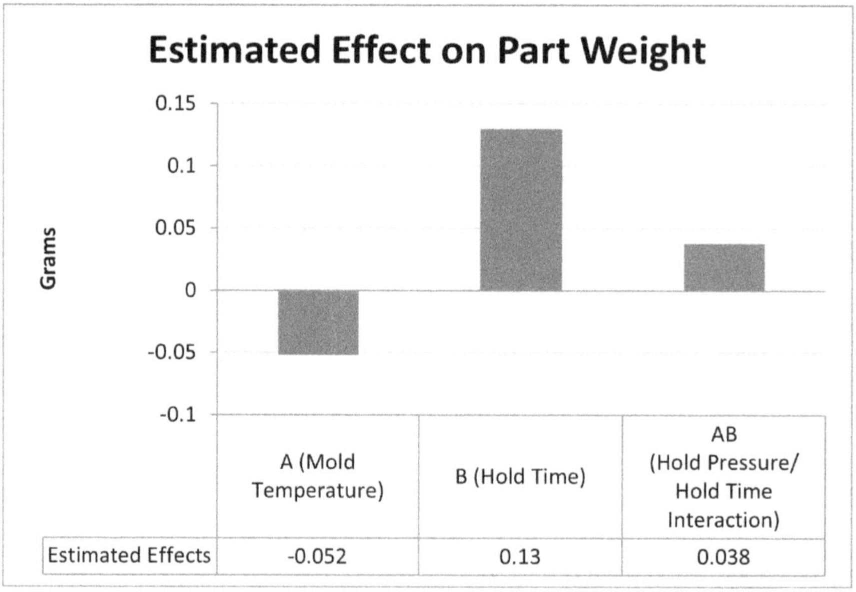

Figure 7.6 Bar chart of estimated effect for Example 7.2

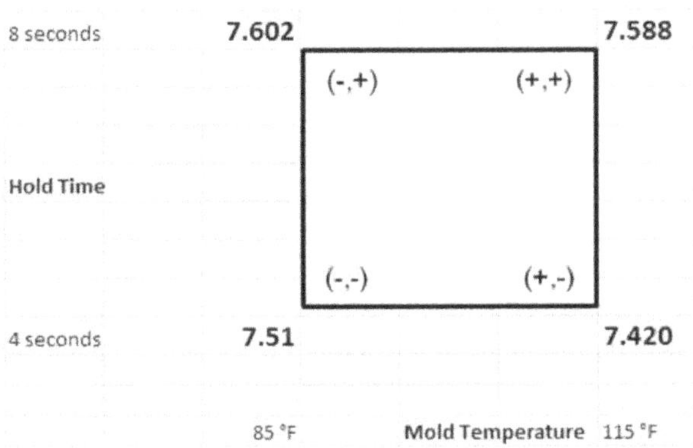

Figure 7.7
Response plot of mold temperature and hold time for Example 7.2

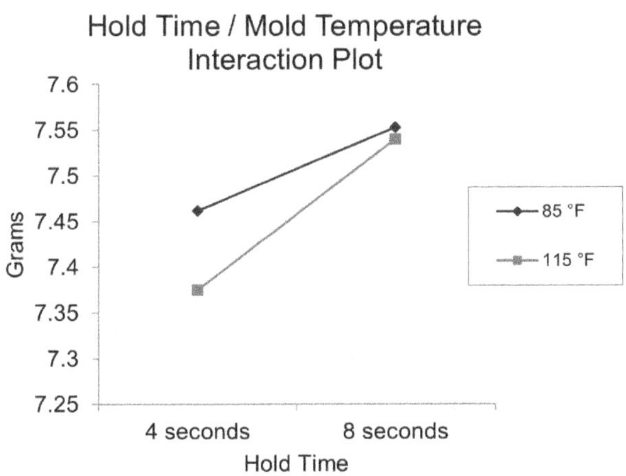

Figure 7.8
Interaction plot of mold temperature and hold time for Example 7.2

Unlike Example 7.1, this example does have a significant interaction term, which can be easily seen by the fact that the two lines on the Interaction plot are not close to being parallel. At the levels of mold temperature and hold time which were run, the effect of each factor depended on the level of the other. Hold time was a factor in both examples. In Example 7.1, the average effect of changing from 4 seconds to 8 seconds was a 0.091 g increase in part weight. In Example 7.2, the average effect of changing from 4 seconds to 8 seconds was a 0.130 g increase. The difference is explained by the interaction between hold time and mold temperature found in Example 7.2. If you look at just the runs at 85 °F from Example 7.2, the effect of hold time was 0.092 g, which is very close to the 0.091 g found when a constant mold temperature of 85 °F was used in Example 7.1. Again, looking at the runs at 115 °F, the effect of hold time was 0.168 g, so hold time had a greater effect on part weight at the high mold temperature than at the low mold temperature. The 0.013 g calculated estimated effect for hold time was the average of the 0.092 g and 0.168 g effects at the two different mold temperatures.

7.2 2^3 Full Factorial

Example 7.1 used a full factorial DOE to investigate the effect on part weight of hold time and hold pressure at two levels each. Example 7.2 used a full factorial DOE to investigate the effect on part weight of hold time and mold temperature on part weight. One full factorial experiment could be used to study all three factors at two levels each. Using Equation 7.1, the number of runs required will be 8 (2^3). Table 7.4 shows the matrix for all possible combinations of three factors (A, B, and C), as well as all possible interactions between those factors.

Table 7.4 2^3 Full Factorial DOE

Run	A	B	C	AB	AC	BC	ABC	Response
1	+1	+1	+1	+1	+1	+1	+1	$\overline{x}_1$
2	+1	+1	−1	+1	−1	−1	−1	$\overline{x}_2$
3	+1	−1	−1	−1	−1	+1	+1	$\overline{x}_3$
4	+1	−1	+1	−1	+1	−1	−1	$\overline{x}_4$
5	−1	−1	+1	+1	−1	−1	+1	$\overline{x}_5$
6	−1	−1	−1	+1	+1	+1	−1	$\overline{x}_6$
7	−1	+1	−1	−1	+1	−1	+1	$\overline{x}_7$
8	−1	+1	+1	−1	−1	+1	−1	$\overline{x}_8$

In addition to the A, B, and C columns which define the “recipe” for the experiment, there are three possible 2-factor interactions (AB, AC, and BC) and one possible 3-factor interaction (ABC). As discussed with Examples 7.1 and 7.2, the interaction columns will show whether the effect of any factor depends on the level of any other factor or, in the case of the ABC column, depends on the level of both other factors. To obtain the “+” and “–” pattern for each interaction column and thus determine whether the interaction exists, simply multiply the corresponding cells in the relevant columns. For example, for the AC interaction column, do the following (Table 7.5):

Table 7.5 Operation to Obtain an Interaction Column

A		C		AC
+1	×	+1	=	+1
+1		−1		−1
+1		−1		−1
+1		+1		+1
−1		+1		−1
−1		−1		+1
−1		−1		+1
−1		+1		−1

The same type of operation shown in Table 7.5 can be done for any 2-factor interaction column; for the 3-factor interaction column, all three columns are multiplied.

7.2.1 Example 7.3: Full Factorial for 3 Factors at 2 Levels

Using the same base process as was used for Examples 7.1 and 7.2, a full factorial experiment is carried out to investigate the effects of mold temperature, hold time and hold pressure on the part weight of an injection molded part. A look at Examples 7.1 and 7.2 shows that six of the eight runs have already been revealed. The two additional runs needed here are the high mold temperature and low hold pressure at each hold time. The two additional runs are necessary to determine whether there is a mold temperature/hold pressure interaction or an interaction between all three factors. Again, five parts are collected and weighed for each run. Table 7.6 shows the results.

Table 7.6 DOE Settings and Results for Example 7.3

Run	A Mold Temperature, °F	B Hold Time, s	C Hold Pressure, psi	1, g	2, g	3, g	4, g	5, g	Response Average Weight, g
1	115 (+)	8 (+)	12,000 (+)	7.598	7.608	7.568	7.598	7.568	7.588
2	115 (+)	8 (+)	9,000 (−)	7.502	7.512	7.472	7.502	7.472	7.492
3	115 (+)	4 (−)	9,000 (−)	7.370	7.280	7.340	7.320	7.340	7.330
4	115 (+)	4 (−)	12,000 (+)	7.430	7.400	7.420	7.380	7.470	7.420

Table 7.6 DOE Settings and Results for Example 7.3 *(continued)*

Run	A Mold Temperature, °F	B Hold Time, s	C Hold Pressure, psi	1, g	2, g	3, g	4, g	5, g	Response Average Weight, g
5	85 (−)	4 (−)	12,000 (+)	7.490	7.530	7.550	7.500	7.480	7.510
6	85 (−)	4 (−)	9,000 (−)	7.424	7.394	7.414	7.374	7.464	7.414
7	85 (−)	8 (+)	9,000 (−)	7.544	7.454	7.514	7.494	7.514	7.504
8	85 (−)	8 (+)	12,000 (+)	7.612	7.622	7.602	7.542	7.632	7.602

As done with the previous examples, the Range and Average charts will be examined first and are shown in Table 7.5.

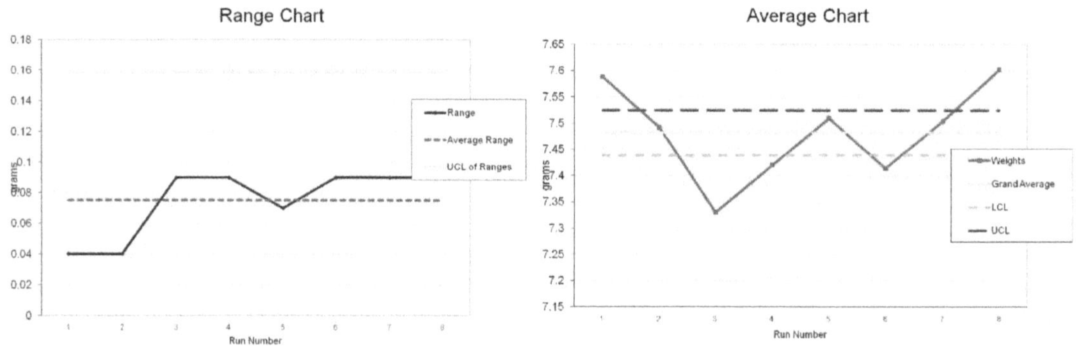

Figure 7.9 Range and Average charts for Example 7.3

The Range chart is in control and so the within-run variation was consistent in all eight runs. The Average chart is out of control and this shows that there are differences between the different runs. Both results are desirable when doing DOEs.

The estimated effects can be calculated using Equation 7.2 for each column.

$$\hat{l}_A = \frac{\overline{x}_1 + \overline{x}_2 + \overline{x}_3 + \overline{x}_4 - \overline{x}_5 - \overline{x}_6 - \overline{x}_7 - \overline{x}_8}{4} = -0.05\text{ g}$$

$$\hat{l}_B = \frac{\overline{x}_1 + \overline{x}_2 + \overline{x}_7 + \overline{x}_8 - \overline{x}_3 - \overline{x}_4 - \overline{x}_5 - \overline{x}_6}{4} = 0.128\text{ g}$$

$$\hat{l}_C = \frac{\overline{x}_1 + \overline{x}_4 + \overline{x}_5 + \overline{x}_8 - \overline{x}_2 - \overline{x}_3 - \overline{x}_6 - \overline{x}_7}{4} = 0.095\text{ g}$$

$$\hat{l}_{AB} = \frac{\overline{x}_1 + \overline{x}_2 + \overline{x}_5 + \overline{x}_6 - \overline{x}_3 - \overline{x}_4 - \overline{x}_7 - \overline{x}_8}{4} = 0.037\text{ g}$$

$$\hat{l}_{AC} = \frac{\overline{x}_1 + \overline{x}_4 + \overline{x}_6 + \overline{x}_7 - \overline{x}_2 - \overline{x}_3 - \overline{x}_5 - \overline{x}_8}{4} = -0.002 \text{ g}$$

$$\hat{l}_{BC} = \frac{\overline{x}_1 + \overline{x}_3 + \overline{x}_6 + \overline{x}_8 - \overline{x}_2 - \overline{x}_4 - \overline{x}_5 - \overline{x}_7}{4} = 0.002 \text{ g}$$

$$\hat{l}_{ABC} = \frac{\overline{x}_1 + \overline{x}_3 + \overline{x}_5 + \overline{x}_7 - \overline{x}_2 - \overline{x}_4 - \overline{x}_6 - \overline{x}_8}{4} = 0.001 \text{ g}$$

A bar chart of these results is shown in Figure 7.10.

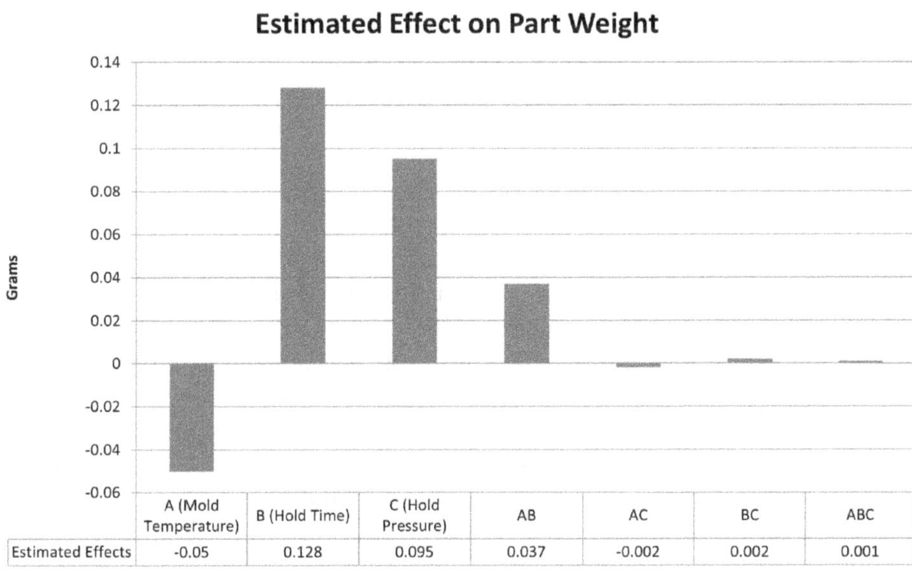

Figure 7.10 Bar chart of estimated effect for Example 7.3

Two things to note with these results are:

1. The average hold-time effect was the same as in Example 7.2 due to the mold temperature/hold time (AB) interaction. Example 7.3 had an average hold-time effect of 0.128 g and Example 7.2 had an average hold-time effect of 0.130 g, whereas Example 7.1 had an average hold-time effect of 0.91 g since the parts were run only at the low mold temperature.

2. No new significant effects on part weight were found compared with the combined results of Examples 7.1 and 7.2. The mold temperature/hold pressure (AC) interaction and the three-factor interaction were both close to zero.

To visually examine the interaction between mold temperature and hold time, the Response plot and Interaction plot could also be made. The only difference would be that each corner of the Response plot would have two runs and each point on the Interaction plot would represent the average of the two runs, as shown in Figure 7.8.

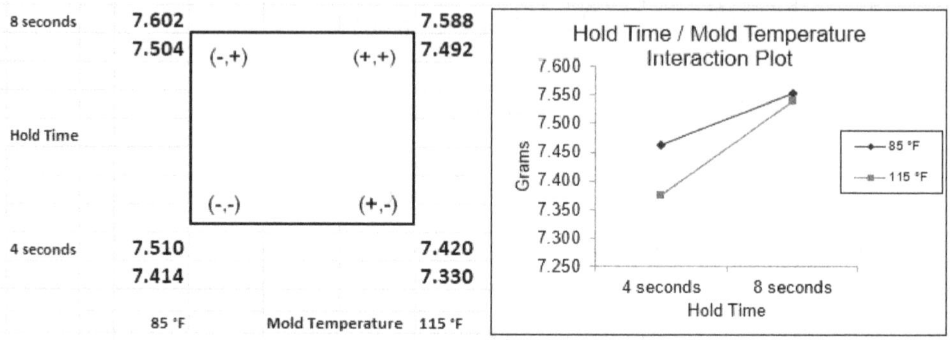

Figure 7.11 Response plot and Interaction plot for Example 7.3

7.3 2^4 Full Factorial

Table 7.7 shows the full factorial matrix for four factors at two levels each. Columns A, B, C, and D give the "recipe" for carrying out the DOE. There are also six possible 2-factor interactions, four possible 3-factor interactions, and one possible 4-factor interactions. The analysis is done the same as the 2^2 and 2^3 examples that have been reviewed; the fifteen estimated effects are calculated using Equation 7.2 for each column. For factors at two levels each, four factors are generally the most that are studied with full factorials with injection molding studies. With five or more factors, the fractional factorials discussed in Chapter 8 are generally used.

Table 7.7 2^4 Full Factorial DOE

Run	A	B	C	D	AB	AC	AD	BC	BD	CD	ABC	BCD	ACD	ABD	ABCD	Response
1	+1	+1	+1	+1	+1	+1	+1	+1	+1	+1	+1	+1	+1	+1	+1	$\bar{x}_1$
2	+1	+1	+1	−1	+1	+1	−1	+1	−1	−1	+1	−1	−1	−1	−1	$\bar{x}_2$
3	+1	+1	−1	−1	+1	−1	−1	−1	−1	+1	−1	+1	+1	−1	+1	$\bar{x}_3$
4	+1	+1	−1	+1	+1	−1	+1	−1	+1	−1	−1	−1	−1	+1	−1	$\bar{x}_4$
5	+1	−1	−1	+1	−1	−1	+1	+1	−1	−1	+1	+1	−1	−1	+1	$\bar{x}_5$
6	+1	−1	−1	−1	−1	−1	−1	+1	+1	+1	+1	−1	+1	+1	−1	$\bar{x}_6$
7	+1	−1	+1	−1	−1	+1	−1	−1	+1	−1	−1	+1	−1	+1	+1	$\bar{x}_7$
8	+1	−1	+1	+1	−1	+1	+1	−1	−1	+1	−1	−1	+1	−1	−1	$\bar{x}_8$

Run	A	B	C	D	AB	AC	AD	BC	BD	CD	ABC	BCD	ACD	ABD	ABCD	Response
9	−1	−1	+1	+1	+1	−1	−1	−1	−1	+1	+1	−1	−1	+1	+1	$\overline{x}_9$
10	−1	−1	+1	−1	+1	−1	+1	−1	+1	−1	+1	+1	+1	−1	−1	$\overline{x}_{10}$
11	−1	−1	−1	−1	+1	+1	+1	+1	+1	+1	−1	−1	−1	−1	+1	$\overline{x}_{11}$
12	−1	−1	−1	+1	+1	+1	−1	+1	−1	−1	−1	+1	+1	+1	−1	$\overline{x}_{12}$
13	−1	+1	−1	+1	−1	+1	−1	−1	+1	−1	+1	−1	+1	−1	+1	$\overline{x}_{13}$
14	−1	+1	−1	−1	−1	+1	+1	−1	−1	+1	+1	+1	−1	+1	−1	$\overline{x}_{14}$
15	−1	+1	+1	−1	−1	−1	+1	+1	−1	−1	−1	−1	+1	+1	+1	$\overline{x}_{15}$
16	−1	+1	+1	+1	−1	−1	−1	+1	+1	+1	−1	+1	−1	−1	−1	$\overline{x}_{16}$

7.4 3-Level Full Factorials

The estimated effects calculated with the two-level experiments are, by definition, based on just two points and are unable to determine whether there is any nonlinearity present in the factors effect on the response. Figure 7.12 illustrates how a third level can show nonlinear effects.

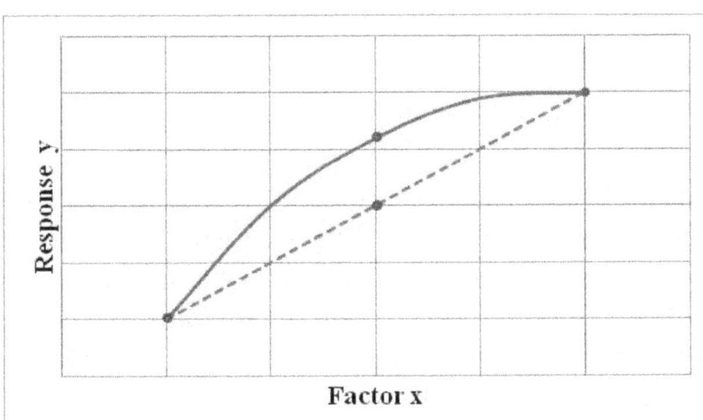

Figure 7.12
Linear (dashed line) versus non-linear (solid line) response

The dashed line in Figure 7.12 can be modeled by the familiar equation $y = mx + b$, where m is the slope and b is the intercept. The solid line is generally well represented by a quadratic polynomial $y = ax^2 + bx + c$. In order to quantify the curvature, a quadratic effect must be determined. This can be done by calculating the distance between the center of the solid curved line, B, and the dashed straight line, A, in Figure 7.12. If there is no curvature effect, the solid line will be right on top of the dashed line and the distance will be zero (A – B = 0). The solid line can be curved above the dashed line (A – B > 0) or it can be curved below the dashed line (A – B < 0). The arithmetic sign of the calculated quadratic effect will indicate whether the dashed line is above or below the solid line in Figure 7.12. The only 3-level full factorial DOE that will be shown in this book is the 3^2, a nine-run DOE with two factors at three levels each. Table 7.8 shows the matrix for this experiment. Looking for curvature with more than two factors will be a topic in Chapter 8.

Table 7.8 3^2 Full Factorial DOE

Run	A	B	A^2	B^2	AB	A^2B	AB^2	A^2B^2	Response
1	+1	+1	+1	+1	+1	+1	+1	+1	$\overline{x}_1$
2	+1	0	+1	–2	0	0	–2	–2	$\overline{x}_2$
3	+1	–1	+1	+1	–1	–1	+1	+1	$\overline{x}_3$
4	0	+1	–2	+1	0	–2	0	–2	$\overline{x}_4$
5	0	0	–2	–2	0	0	0	+4	$\overline{x}_5$
6	0	–1	–2	+1	0	+2	0	–2	$\overline{x}_6$
7	–1	+1	+1	+1	–1	+1	–1	+1	$\overline{x}_7$
8	–1	0	+1	–2	0	0	+2	–2	$\overline{x}_8$
9	–1	–1	+1	+1	+1	–1	–1	+1	$\overline{x}_9$

The “recipe” for the 3^2 full factorial DOE consists of columns “A” and “B” in Table 7.8. As in the two-level designs, +1 indicates when the high level of a factor should be set and –1 indicates when the low level should be set. The contrast coefficient for the mid-level is 0. Columns “A^2” and “B^2” are used to calculate the quadratic effects of each factor. It should be noted that the contrast coefficients for these two columns are not obtained by squaring the numbers in the “A” and “B” columns. Examination of these two columns shows that what they are doing is comparing, or contrasting, the average of the runs set at the mid-level to the average of all the other runs (set at the high or the low level). Since there are only half as many runs (3) when each factor is set at the mid-level compared to when set at high or low (3 + 3 = 6), a contrast coefficient of −2 is used for the runs at the mid-level and +1 for the other runs. This will allow Equation 7.2 to be used to calculate the estimated curvature effect; the average difference between the mid-level and the combined low-/high-level runs. It is important to note here that there needs to be an equal number of runs at each of the low and high levels for this to work. The remaining four columns in Table 7.8 are obtained by multiplying the other columns and will be used to determine if there are interactions between the factors (AB) or interactions involving the quadratic effects.

7.4.1 Example 7.4: Full Factorial for 2 Factors at 3 Levels

The following example is a continuation of Example 7.2 and will be expanded to include the mid-levels. The data for runs 1, 3, 7, and 9 come directly from Example 7.2; the other numbers are from runs in which at least one of the factors was set at the mid-level (100 °F mold, 6 second hold).

Table 7.9 DOE Settings and Results for Example 7.4

Run	A Mold Temperature, °F	B Hold Time, s	1, g	2, g	3, g	4, g	5, g	Response Average Weight, g
1	115 (+)	8 (+)	7.598	7.608	7.568	7.598	7.568	7.588
2	115 (+)	6 (0)	7.582	7.572	7.532	7.492	7.582	7.552
3	115 (+)	4 (−)	7.430	7.400	7.420	7.380	7.470	7.420
4	100 (0)	8 (+)	7.633	7.543	7.603	7.583	7.603	7.593
5	100 (0)	6 (0)	7.580	7.550	7.580	7.530	7.610	7.570
6	100 (0)	4 (−)	7.426	7.486	7.456	7.456	7.506	7.466
7	85 (−)	8 (+)	7.612	7.622	7.602	7.542	7.632	7.602
8	85 (−)	6 (0)	7.631	7.621	7.561	7.551	7.641	7.601
9	85 (−)	4 (−)	7.490	7.530	7.550	7.500	7.480	7.510

As done with the previous examples, the Range and Average charts will be examined first and are shown in Figure 7.13. Again, the Range chart is in control and the Average chart is out of control.

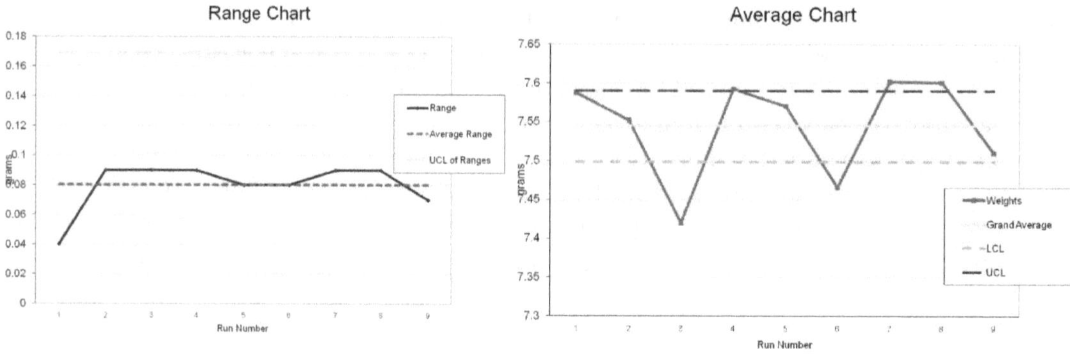

Figure 7.13 Range and Average charts for Example 7.4

The estimated effects can be calculated using Equation 7.2 for each column.

$$\hat{l}_A = \frac{\overline{x}_1 + \overline{x}_2 + \overline{x}_3 - \overline{x}_7 - \overline{x}_8 - \overline{x}_9}{3} = -0.051 \text{ g}$$

$$\hat{l}_B = \frac{\overline{x}_1 + \overline{x}_4 + \overline{x}_7 - \overline{x}_3 - \overline{x}_6 - \overline{x}_9}{3} = 0.129 \text{ g}$$

$$\hat{l}_{A2} = \frac{\overline{x}_1 + \overline{x}_2 + \overline{x}_3 - 2\overline{x}_4 - 2\overline{x}_5 - 2\overline{x}_6 + \overline{x}_7 + \overline{x}_8 + \overline{x}_9}{3} = 0.0025 \text{ g}$$

$$\hat{l}_{B2} = \frac{\overline{x}_1 - 2\overline{x}_2 + \overline{x}_3 + \overline{x}_4 - 2\overline{x}_5 + \overline{x}_6 + \overline{x}_7 - 2\overline{x}_8 + \overline{x}_9}{6} = -0.0445 \text{ g}$$

$$\hat{l}_{AB} = \frac{\overline{x}_1 - \overline{x}_3 - \overline{x}_7 + \overline{x}_9}{2} = 0.038 \text{ g}$$

$$\hat{l}_{A2B} = \frac{\overline{x}_1 - \overline{x}_3 - 2\overline{x}_4 + 2\overline{x}_6 + \overline{x}_7 - \overline{x}_9}{4} = 0.0015 \text{ g}$$

$$\hat{l}_{AB2} = \frac{\overline{x}_1 - 2\overline{x}_2 + \overline{x}_3 - \overline{x}_7 + 2\overline{x}_8 - \overline{x}_9}{4} = -0.0015 \text{ g}$$

$$\hat{l}_{A2B2} = \frac{\overline{x}_1 - 2\overline{x}_2 + \overline{x}_3 - 2\overline{x}_4 + 4\overline{x}_5 - 2\overline{x}_6 + \overline{x}_7 - 2\overline{x}_8 + \overline{x}_9}{8} = -0.003 \text{ g}$$

Looking at the estimated effects bar chart, Figure 7.14, in addition to the effects that were found in Example 7.2, a quadratic effect, B^2, is present for hold time.

Quadratic effect B^2 is shown graphically in Figure 7.15. As can be seen, the calculated quadratic effect, −0.0455 g, is the difference between the average of the combined high- and low-level runs and the average of the mid-level runs. Without knowing that a curvature effect was present, it would be assumed that the part weight for hold times between 4 seconds and 8 seconds would be found on the dotted line. The curvature of this graph should also be familiar to anyone who has set up an injection molding process and done a gate freeze study by weighing parts at different hold times.

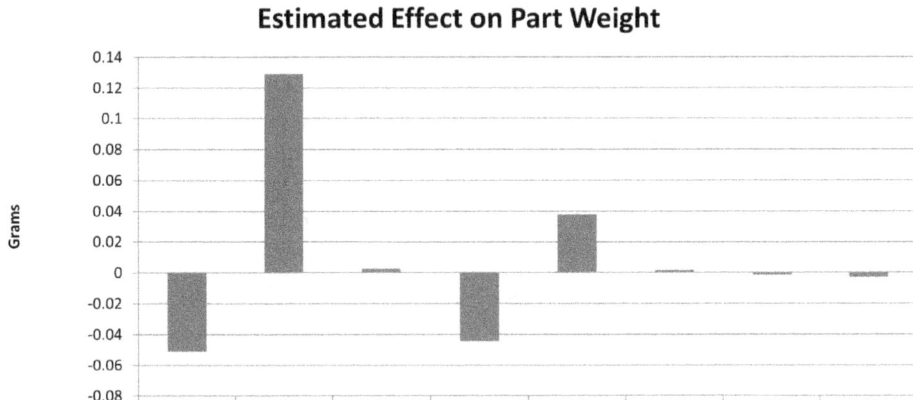

	A	B	A2	B2	AB	A2B	AB2	A2B2
Estimated Effects	-0.051	0.129	0.0025	-0.0445	0.038	0.0015	-0.0015	-0.003

Figure 7.14 Bar chart of estimated effects for Example 7.4

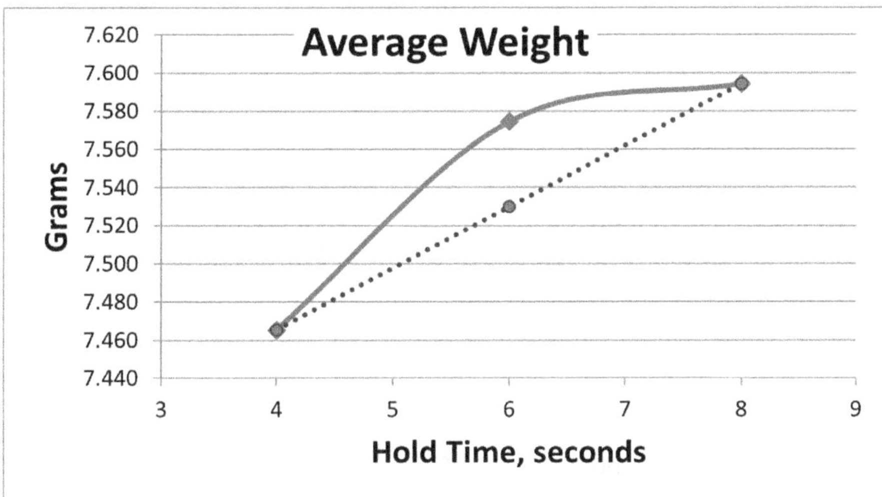

Figure 7.15 Hold-time quadratic effect for Example 7.4 (B^2 = −0.0445 g). What is the hold-time Effect? (B = ?)

To make the graph shown in Figure 7.15, it was necessary to calculate the average of the three runs at each level of hold time. The average at 4 seconds of hold was 7.465 g, the average at 6 seconds of hold was 7.574 g, and the average at 8 seconds of hold was 7.594 g. The calculation for the estimated effect of hold time, B, does not include the mid-level average. It is the difference between the high and low levels, which is 0.129 g (7.594 g–7.465 g).

7.5 Additional DOE Data Analysis

Up to this point, the only statistical methods that have been used to analyze the significance of the estimated effects are the Range and Average charts. A Confidence Interval chart is another method that can be used to show that the results obtained from the various runs were different and to catch typos (abnormally wide interval(s)). An example is shown in Figure 7.16. When Minitab is used for DOEs, the Confidence Interval graph is easier to generate than control charts and is easier for most to visualize when presenting the results.

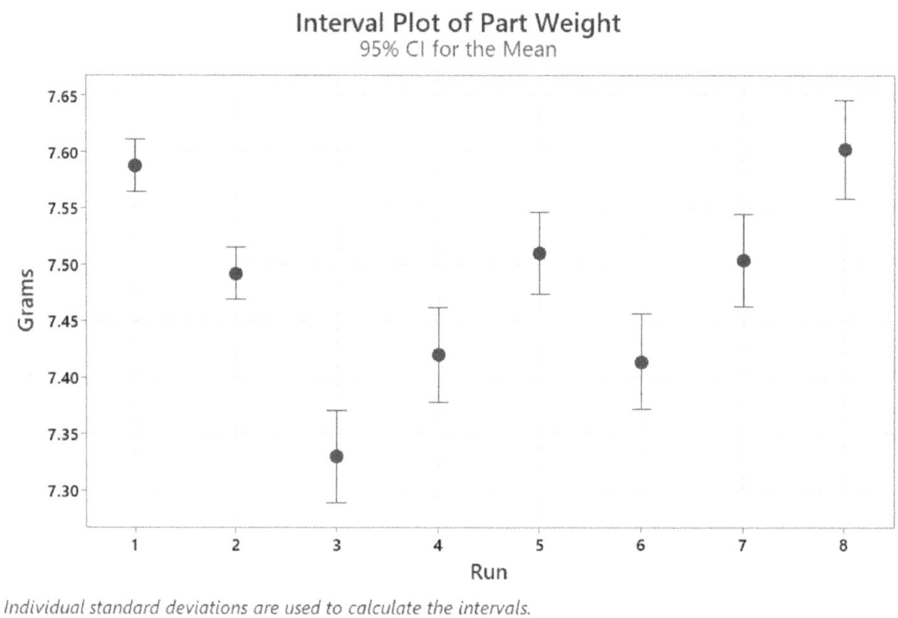

Figure 7.16 95% Confidence interval graph of the means for Example 7.3

Since the intervals for some of the runs do not intersect, this graph is also evidence that some of the factors that were changed influenced the part weight. It gives a picture of which runs produced the largest and smallest response – part weight in this example – but it isn't usually very clear when trying to determine which factors caused the changes. The estimated-effects bar chart is a good tool for seeing the relative size of the different effects. The next section will cover ANOVA, which is a method for determining whether a calculated estimated effect is statistically significant or just due to random variation. Then, an analysis using Minitab will be shown. Minitab and many other statistical software packages use regression analysis to determine which effects are significant.

7.6 Analysis of Variance (ANOVA)

ANOVA is a method of comparing variation between different runs with the average variation within the runs. More specifically, an F-test can be done for each estimated effect, which is the variation between the "+" levels and "–" levels, compared with the average variation within the runs. The hypotheses for this F-test are:

H_0: Variation between the "+" levels and "–" levels = Variation within the runs

H_1: Variation between the "+" levels and "–" levels > Variation within the runs

Unless enough evidence is present to prove the alternative hypothesis, H_1, the null hypothesis, H_0, must be accepted and the effect in question is assumed not to be significant.

For DOEs, the equations used for the F-ratio of each estimated effect, $\hat{l}_i$, are given in Equation 7.6 to Equation 7.8):

$$F_i = \frac{\text{Signal}}{\text{Noise}} = \frac{SS_i}{MSW} \tag{7.6}$$

Where:

SS_i = sum of squares for $\hat{l}_i$

MSW = mean square within

= average sample variance of all the runs

$$= n \times \left(\hat{l}_i \times \sum_{j=1}^{k} |c_j| \right)^2 \Big/ \left(4 \times \sum_{j=1}^{k} c_j^2 \right) \tag{7.7}$$

$$= \sum_{j=1}^{k} s_j^2 / k \tag{7.8}$$

Where:

c_j = contrast coefficient for run j

s_j^2 = sample variance for run j

n = number of replicates per run

k = number of runs in the experiment

For Examples 7.1–7.4, each experiment has had 5 replicates, i.e. the 5 measurements per run. A discussion of replicates and how they differ from repetitions will be presented in Chapter 9. It should also be pointed out that Equation 7.8 can only be used when there are replicates, since a sample variance cannot be calculated with just one number. Two methods of dealing with experiments with one replicate per run will be discussed later in this chapter.

Once the F-ratios are calculated, they are compared to the accept/reject value as was done in Chapter 6. From Figure 6.10, the rejection region for the F test in this case is for $F_i > F_{\alpha,df1,df2}$. As done throughout this book, an α of 0.05 will be used. Since the variation of each estimated effect is always based on the difference between the "+" and "–" contrast coefficients or 2 things, there is always only 1 degree of freedom in the numerator, d_{f1}. When determining the MSW, the sample variance for each run is averaged. Each run will have $n - 1$ degrees of freedom, so the total degrees of freedom are the number of runs times $(n - 1)$, as shown in Equation 7.9.

Degrees of freedom for MSW:

$$df_2 = k(n - 1) \tag{7.9}$$

Where:

k = number of runs

n = number of replicates

Example 7.5 will work through the calculations of the F-ratios for Examples 7.3 and 7.4.

7.6.1 Example 7.5: F-Tests for DOEs

The results for Example 7.3 are shown in Table 7.10. The sample variance for each run is needed for the calculation of the MSW and is obtained by squaring the sample standard deviation of the 5 replications.

Table 7.10 Results for Example 7.3 with Sample Variance

Run	1, g	2, g	3, g	4, g	5, g	Sample variance, g^2
1	7.598	7.608	7.568	7.598	7.568	0.00035
2	7.502	7.512	7.472	7.502	7.472	0.00035
3	7.370	7.280	7.340	7.320	7.340	0.00110
4	7.430	7.400	7.420	7.380	7.470	0.00115
5	7.490	7.530	7.550	7.500	7.480	0.00085
6	7.424	7.394	7.414	7.374	7.464	0.00115
7	7.544	7.454	7.514	7.494	7.514	0.00110
8	7.612	7.622	7.602	7.542	7.632	0.00125

Equation 7.8 is used to calculate the denominator of the F-ratio, the mean square within.

$$MSW = \sum_{j=1}^{k} s_j^2/k$$
$$= \frac{(0.00035 + 0.00035 + 0.00110 + 0.00115 + 0.00085 + 0.00115 + 0.00110 + 0.00125)\text{g}^2}{8}$$
$$= 0.0009125\ \text{g}^2$$

The number of degrees of freedom for the denominator is found using Equation 7.9.

$$df_2 = k(n-1) = 8 \times (5-1) = 32$$

The sums of squares is calculated for each column or effect using Equation 7.7.

$$SS_A = n \times \left(\hat{l}_A \times \sum_{j=1}^{k} |c_j| \right)^2 / \left(4 \times \sum_{j=1}^{k} c_j^2 \right)$$
$$= (5 \times (-0.05\ \text{g} \times (1+1+1+1+1+1+1+1))^2)/$$
$$(4 \times (1+1+1+1+1+1+1+1)) = 0.0250\ \text{g}^2$$

$$SS_B = (5 \times (0.128\ \text{g} \times (1+1+1+1+1+1+1+1))^2)/$$
$$(4 \times (1+1+1+1+1+1+1+1)) = 0.16384\ \text{g}^2$$

$$SS_C = (5 \times (0.095\ \text{g} \times (1+1+1+1+1+1+1+1))^2)/$$
$$(4 \times (1+1+1+1+1+1+1+1)) = 0.09025\ \text{g}^2$$

$$SS_{AB} = (5 \times (0.037\ \text{g} \times (1+1+1+1+1+1+1+1))^2)/$$
$$(4 \times (1+1+1+1+1+1+1+1)) = 0.01369\ \text{g}^2$$

$$SS_{AC} = (5 \times (-0.002\ \text{g} \times (1+1+1+1+1+1+1+1))^2)/$$
$$(4 \times (1+1+1+1+1+1+1+1)) = 0.00004\ \text{g}^2$$

$$SS_{BC} = (5 \times (0.002\ \text{g} \times (1+1+1+1+1+1+1+1))^2)/$$
$$(4 \times (1+1+1+1+1+1+1+1)) = 0.00004\ \text{g}^2$$

$$SS_{ABC} = (5 \times (-0.001\ \text{g} \times (1+1+1+1+1+1+1+1))^2)/$$
$$(4 \times (1+1+1+1+1+1+1+1)) = 0.00001\ \text{g}^2$$

The F-ratios can then be calculated using Equation 7.6.

$$F_A = \frac{\text{Signal}}{\text{Noise}} = \frac{SS_A}{MSW} = \frac{0.025\ \text{g}^2}{0.009125\ \text{g}^2} = 27.4$$

$$F_B = \frac{0.16384\ \text{g}^2}{0.0009125\ \text{g}^2} = 179.6$$

$$F_C = \frac{0.09025\ \text{g}^2}{0.0009125\ \text{g}^2} = 98.9$$

$$F_{AB} = \frac{0.01369\ \text{g}^2}{0.0009125\ \text{g}^2} = 15.00$$

$$F_{AC} = \frac{0.00004\ \text{g}^2}{0.0009125\ \text{g}^2} = 0.0438$$

$$F_{BC} = \frac{0.00004\ \text{g}^2}{0.0009125\ \text{g}^2} = 0.0438$$

$$F_{ABC} = \frac{0.00001\ \text{g}^2}{0.0009125\ \text{g}^2} = 0.0110$$

For the estimated effect to be statistically significant, the calculated F-ratio has to be greater than the acceptance/rejection value of $F_{0.05,1,32} = 4.15$ (see Table 12.6). This example has very clear results in that A, B, C, and AB are much greater than 4.15 and AC, BC, and ABC are much less than 4.15. The estimated effects for A, B, C, and AB are statistically significant since the null hypothesis is rejected for those tests, and we can say that the variation between the high and low levels is greater than the variation within the runs.

The same calculations are repeated here for Example 7.4, so that the reader can see an example with contrast coefficients that aren't all +1 and −1.

Table 7.11 Results for Example 7.4 with Sample Variance

Run	1, g	2, g	3, g	4, g	5, g	Sample Variance, g^2
1	7.598	7.608	7.568	7.598	7.568	0.00035
2	7.582	7.572	7.532	7.492	7.582	0.00155
3	7.430	7.400	7.420	7.380	7.470	0.00115
4	7.633	7.543	7.603	7.583	7.603	0.00110

Run	1, g	2, g	3, g	4, g	5, g	Sample Variance, g^2
5	7.580	7.550	7.580	7.530	7.610	0.00095
6	7.426	7.486	7.456	7.456	7.506	0.00095
7	7.612	7.622	7.602	7.542	7.632	0.00125
8	7.631	7.621	7.561	7.551	7.641	0.00175
9	7.490	7.530	7.550	7.500	7.480	0.00085

Equation 7.8 is used to calculate the denominator of the F-ratio, the mean square within.

$$MSW = \sum_{j=1}^{k} s_j^2/k$$

$$= \frac{(0.00035 + 0.00155 + 0.00115 + 0.00110 + 0.00095 + 0.00095 + 0.00125 + 0.00175 + 0.00085)\ \text{g}^2}{9}$$

$$= 0.0011\ \text{g}^2$$

The number of degrees of freedom for the denominator is found using Equation 7.9.

$$df_2 = k(n-1) = 9 \times (5-1) = 36$$

The sums of squares is calculated for each column or effect using Equation 7.7.

$$SS_A = n \times \left(\hat{l}_A \times \sum_{j=1}^{k} |c_j|\right)^2 \Bigg/ \left(4 \times \sum_{j=1}^{k} c_j^2\right)$$

$$= (5 \times (-0.051\ \text{g} \times (1+1+1+0+0+0+1+1+1))^2)/$$
$$(4 \times (1+1+1+0+0+0+1+1+1)) = 0.01951\ \text{g}^2$$

$$SS_B = (5 \times (0.129\ \text{g} \times (1+0+1+1+0+1+1+0+1))^2)/$$
$$(4 \times (1+0+1+1+0+1+1+0+1)) = 0.12481\ \text{g}^2$$

$$SS_{A2} = (5 \times (-0.0025\ \text{g} \times (1+1+1+2+2+2+1+1+1))^2)/$$
$$(4 \times (1+1+1+4+4+4+1+1+1)) = 0.00006\ \text{g}^2$$

$$SS_{B2} = (5 \times (0.0445\ \text{g} \times (1+2+1+1+2+1+1+2+1))^2)/$$
$$(4 \times (1+4+1+1+4+1+1+4+1)) = 0.01980\ \text{g}^2$$

$$SS_{AB} = (5 \times (0.038\,\text{g} \times (1+0+1+0+0+0+1+0+1))^2)/$$
$$(4 \times (1+0+1+0+0+0+1+0+1)) = 0.00722\,\text{g}^2$$

$$SS_{A2B} = (5 \times (-0.0015\,\text{g} \times (1+0+1+2+0+2+1+0+1))^2)/$$
$$(4 \times (1+0+1+4+0+4+1+0+1)) = 0.000015\,\text{g}^2$$

$$SS_{AB2} = (5 \times (0.0015\,\text{g} \times (1+2+1+0+0+0+1+2+1))^2)/$$
$$(4 \times (1+4+1+0+0+0+1+4+1)) = 0.000015\,\text{g}^2$$

$$SS_{A2B2} = (5 \times (-0.003\,\text{g} \times (1+2+1+2+4+2+1+2+1))^2)/$$
$$(4 \times (1+4+1+4+16+4+1+4+1)) = 0.00008\,\text{g}^2$$

The F-ratios can then be calculated using Equation 7.6.

$$F_A = \frac{\text{Signal}}{\text{Noise}} = \frac{SS_A}{MSW} = \frac{0.0195\,\text{g}^2}{0.0011\,\text{g}^2} = 17.7$$

$$F_B = \frac{0.12481\,\text{g}^2}{0.0011\,\text{g}^2} = 113.5$$

$$F_{A2} = \frac{0.00006\,\text{g}^2}{0.0011\,\text{g}^2} = 0.0568$$

$$F_{B2} = \frac{0.01980\,\text{g}^2}{0.0011\,\text{g}^2} = 18.00$$

$$F_{AB} = \frac{0.00722\,\text{g}^2}{0.0011\,\text{g}^2} = 6.56$$

$$F_{A2B} = \frac{0.000015\,\text{g}^2}{0.0011\,\text{g}^2} = 0.0136$$

$$F_{AB2} = \frac{0.000015\,\text{g}^2}{0.0011\,\text{g}^2} = 0.0136$$

$$F_{A2B2} = \frac{0.00008\,\text{g}^2}{0.0011\,\text{g}^2} = 0.0727$$

In order for the estimated effect to be statistically significant, the calculated F-ratio must be greater than the accept/reject value of $F_{0.05,1,36} = 4.11$ (see Table 12.6). The F-ratios greater than 4.11 are A, B, AB, and B^2.

7.6.2 Example 7.6: Using Minitab

As mentioned earlier, Minitab uses the regression method to determine whether estimated effects are statistically significant. Minitab can also generate the test matrix and calculate the estimated effects. This example will show how Example 7.3 can be done with Minitab.

The first step is to create a factorial design, as shown in the following.

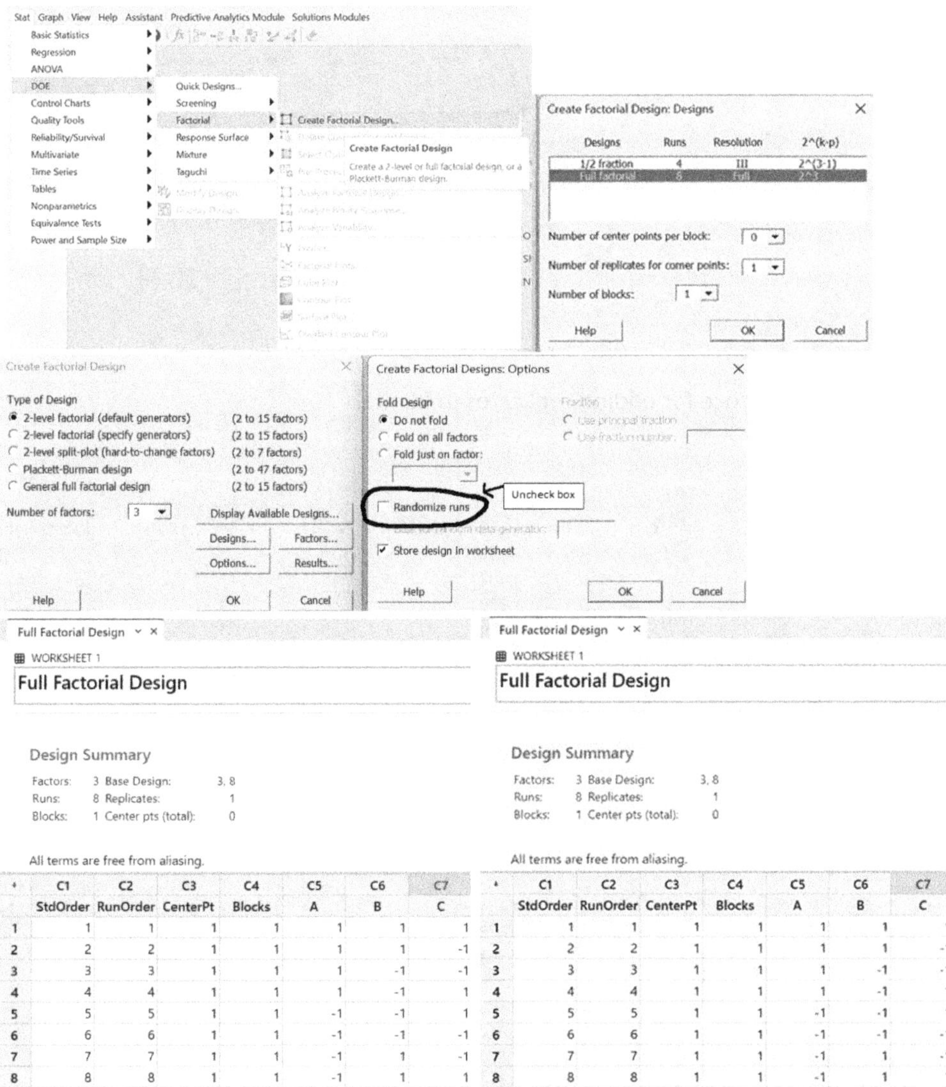

Figure 7.17 Sequence for creating a factorial design for Example 7.2 in Minitab

Columns C5, C6, and C7 (A, B, and C) give the recipe for the DOE. Note that the +1 and −1 pattern for the worksheet shown on the left is not the same as shown in Table 7.4, which was used for Example 7.3. The same 8 runs are there; they are just in different order. Chapter 9 will discuss why the runs were put in the order shown in Table 7.4. For this example, the pattern used in Figure 7.12 is typed into the matrix after the original matrix is deleted, or if the matrix is already in Excel it could have been copied to Minitab to yield the worksheet shown on the right.

Since there are five replicates, this matrix must now be repeated four more times.

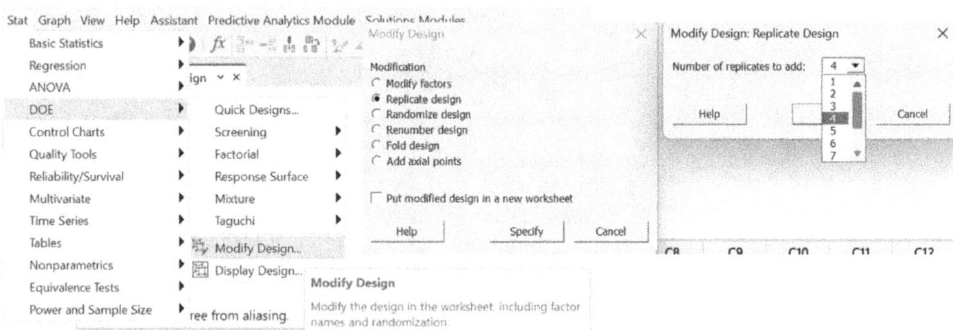

Figure 7.18 Sequence for adding replicates in Minitab

C4	C5	C6	C7	C8	C9
Blocks	A	B	C		Part Weight
1	1	1	1		7.598
1	1	1	-1		7.502
1	1	-1	-1		7.370
1	1	-1	1		7.430
1	-1	-1	1		7.490
1	-1	-1	-1		7.424
1	-1	1	-1		7.544
1	-1	1	1		7.612
2	1	1	1		7.608
2	1	1	-1		7.512
2	1	-1	-1		7.280
2	1	-1	1		7.400
2	-1	-1	1		7.530
2	-1	-1	-1		7.394
2	-1	1	-1		7.454
2	-1	1	1		7.622
3	1	1	1		7.568
3	1	1	-1		7.472
3	1	-1	-1		7.340
3	1	-1	1		7.420
3	-1	-1	1		7.550
3	-1	-1	-1		7.414
3	-1	1	-1		7.514
3	-1	1	1		7.602

C4	C5	C6	C7	C8	C9
Blocks	A	B	C		Part Weight
4	1	1	1		7.598
4	1	1	-1		7.502
4	1	-1	-1		7.320
4	1	-1	1		7.380
4	-1	-1	1		7.500
4	-1	-1	-1		7.374
4	-1	1	-1		7.494
4	-1	1	1		7.542
5	1	1	1		7.568
5	1	1	-1		7.472
5	1	-1	-1		7.340
5	1	-1	1		7.470
5	-1	-1	1		7.480
5	-1	-1	-1		7.464
5	-1	1	-1		7.514
5	-1	1	1		7.632

Figure 7.19 Experimental set-up and results for Example 7.6

There are now 40 rows to enter data into. The first replicate goes in the first 8 rows, the second replicate goes in rows 9 through 16, etc. The column labeled Part Weight in Figure 7.19 contains the experimental data.

After the data is entered, it can be analyzed in Minitab, as shown in Figure 7.20.

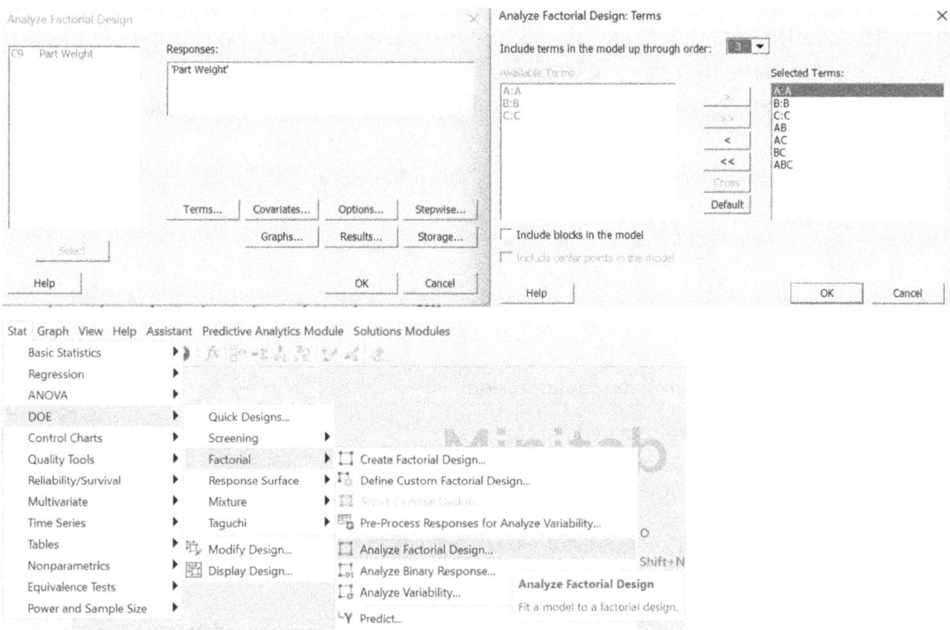

Figure 7.20 Sequence for analyzing a factorial experiment in Minitab

The results are then shown in the Session window in Minitab (Figure 7.21).

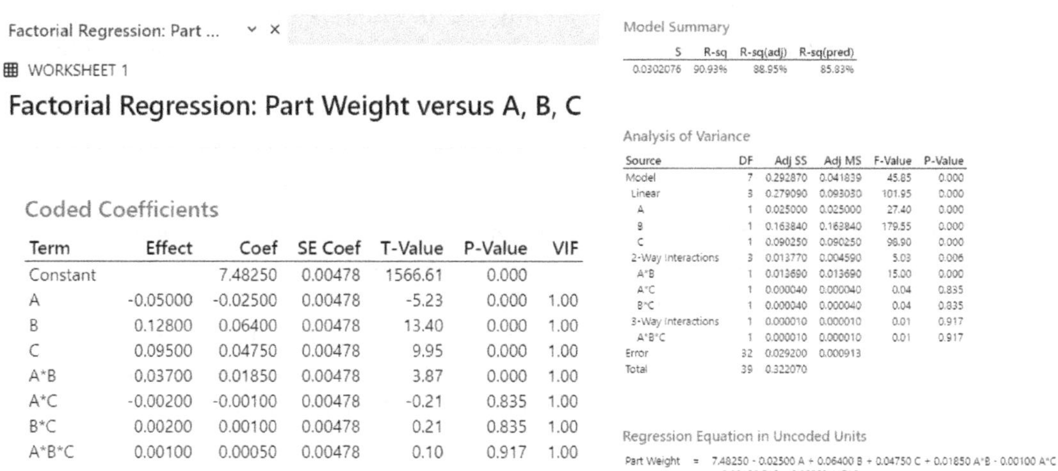

Factorial Regression: Part ...

WORKSHEET 1

Factorial Regression: Part Weight versus A, B, C

Coded Coefficients

Term	Effect	Coef	SE Coef	T-Value	P-Value	VIF
Constant		7.48250	0.00478	1566.61	0.000	
A	-0.05000	-0.02500	0.00478	-5.23	0.000	1.00
B	0.12800	0.06400	0.00478	13.40	0.000	1.00
C	0.09500	0.04750	0.00478	9.95	0.000	1.00
A*B	0.03700	0.01850	0.00478	3.87	0.000	1.00
A*C	-0.00200	-0.00100	0.00478	-0.21	0.835	1.00
B*C	0.00200	0.00100	0.00478	0.21	0.835	1.00
A*B*C	0.00100	0.00050	0.00478	0.10	0.917	1.00

Model Summary

S	R-sq	R-sq(adj)	R-sq(pred)
0.0302076	90.93%	88.95%	85.83%

Analysis of Variance

Source	DF	Adj SS	Adj MS	F-Value	P-Value
Model	7	0.292870	0.041839	45.85	0.000
Linear	3	0.279090	0.093030	101.95	0.000
A	1	0.025000	0.025000	27.40	0.000
B	1	0.163840	0.163840	179.55	0.000
C	1	0.090250	0.090250	98.90	0.000
2-Way Interactions	3	0.013770	0.004590	5.03	0.006
A*B	1	0.013690	0.013690	15.00	0.000
A*C	1	0.000040	0.000040	0.04	0.835
B*C	1	0.000040	0.000040	0.04	0.835
3-Way Interactions	1	0.000010	0.000010	0.01	0.917
A*B*C	1	0.000010	0.000010	0.01	0.917
Error	32	0.029200	0.000913		
Total	39	0.322070			

Regression Equation in Uncoded Units

Part Weight = 7.48250 - 0.02500 A + 0.06400 B + 0.04750 C + 0.01850 A*B - 0.00100 A*C + 0.00100 B*C + 0.00050 A*B*C

Figure 7.21 Initial regression analysis and ANOVA results for Example 7.6

As can be seen from the Minitab results in Figure 7.21, the same estimated effects calculated in Example 7.5 are shown in the column labeled "Effect" in the results of the regression analysis in "coded units". The same F-ratios from Example 7.5 can be seen in the F-Value column of the ANOVA. The regression equation shown is the result when the contrast coefficients (−1 and +1) are used for the levels instead of the actual engineering units. With a linear model and when doing the regression with coded units, the coefficients (Coef) are all the half effects or one half the estimated effects (it will be shown later that the coefficients are the full estimated effects for quadratic terms). Equation 7.10 shows the regression equation for dealing with coded units and linear models:

$$y = \text{Constant} + \sum_{i=1}^{j} \left(\frac{\hat{l}_i}{2}\right) x_i \tag{7.10}$$

Where:

$\hat{l}_i$ = estimated Effect i

j = number of terms in regression equation, and

x_i = actual coded value for estimated effect i

For this example, Equation 7.10 gives the following as shown in the Minitab results.

$$\text{Weight} = 7.4825 - 0.025A + 0.064B + 0.0475C + 0.0185AB - 0.001AC + 0.001BC + 0.005ABC$$

If predicting the weight of the parts for a given combination of settings, a number between −1 and 1 would be entered for each actual coded estimated effect. The easiest calculation is if all the settings are set at the midpoint (coded value = 0); the estimated weight would be 7.4825 g with this equation.

In Example 7.5, it was found that the F-ratios for the AC, BC, and ABC interactions were not statistically significant. So why are they in the equation? The answer is that they shouldn't be. The regression analysis does the following t-test for each term in the equation in coded units:

H_0: the coefficient $\left(\hat{l}_i/2\right) = 0$

H_1: the coefficient $\left(\hat{l}_i/2\right) \neq 0$

As discussed in Chapter 6, the p-value indicates the minimum α value that would be required to reject H_0, based on this data. If the p-value is greater than 0.05, H_0 cannot be rejected for an accepted α error of 0.05. Looking at Figure 7.18, AC (0.835), BC (0.835), and ABC (0.917) all have p-values much greater than 0.05. To eliminate these terms, a backwards stepwise regression is performed. This is done by eliminating the term with the highest p-value and continuing to do that until all the p-values are below 0.05 or whatever α error is specified. One way to do this in Minitab is to make the selection shown in Figure 7.22.

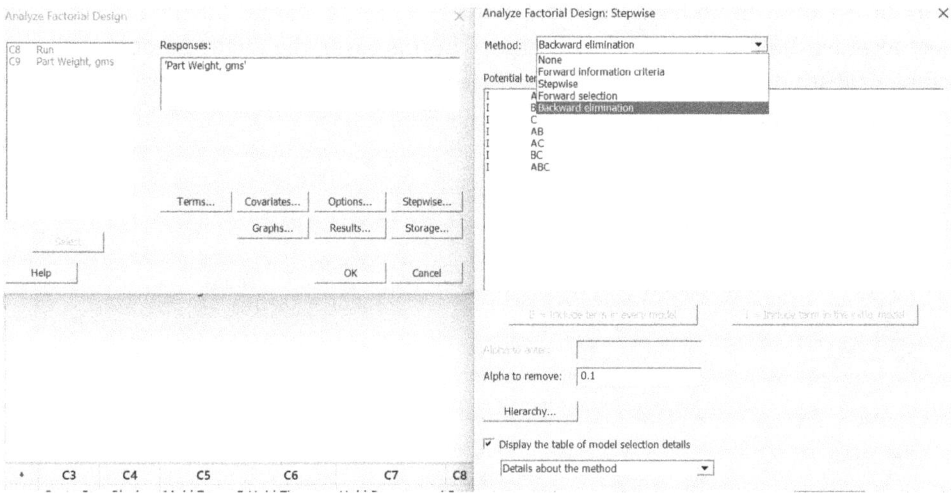

Figure 7.22 Sequence for eliminating insignificant terms in factorial DOE regression equation

The new regression results are then as follows (Figure 7.23).

WORKSHEET 1

Factorial Regression: Part Weight versus A, B, C

Coded Coefficients

Term	Effect	Coef	SE Coef	T-Value	P-Value	VIF
Constant		7.48250	0.00457	1635.88	0.000	
A	-0.05000	-0.02500	0.00457	-5.47	0.000	1.00
B	0.12800	0.06400	0.00457	13.99	0.000	1.00
C	0.09500	0.04750	0.00457	10.38	0.000	1.00
A*B	0.03700	0.01850	0.00457	4.04	0.000	1.00

Model Summary

S	R-sq	R-sq(adj)	R-sq(pred)
0.0289285	90.91%	89.87%	88.12%

Analysis of Variance

Source	DF	Adj SS	Adj MS	F-Value	P-Value
Model	4	0.29278	0.073195	87.46	0.000
Linear	3	0.27909	0.093030	111.17	0.000
A	1	0.02500	0.025000	29.87	0.000
B	1	0.16384	0.163840	195.78	0.000
C	1	0.09025	0.090250	107.84	0.000
2-Way Interactions	1	0.01369	0.013690	16.36	0.000
A*B	1	0.01369	0.013690	16.36	0.000
Error	35	0.02929	0.000837		
Total	39	0.32207			

Regression Equation in Uncoded Units

Part Weight = 7.48250 - 0.02500 A + 0.06400 B + 0.04750 C + 0.01850 A*B

Figure 7.23 Final regression results for Example 7.6

One other parameter to note in the regression results shown is R-Sq (R^2), which is the coefficient of determination. The coefficient of determination is about 90%, which means that the regression model explains 90% of the variation in the response (weight in this case). R-Sq(adj) or R-Sq(pred) are the better terms to look at since they don't go up when you add an insignificant term. The closer to 100% that these terms are, the better. R-Sq(adj) and R-Sq(pred) both improved slightly when the insignificant terms were eliminated in this example.

The final regression equation would then be $Weight = 7.4825 - 0.025A + 0.064B + 0.0475C + 0.0185AB$. In order to avoid converting actual values for the factors (mold temperature, hold time, and hold pressure in this case), the factor names and levels can be substituted for the contrast coefficients in Minitab and the regression can be carried out as shown in Figure 7.24.

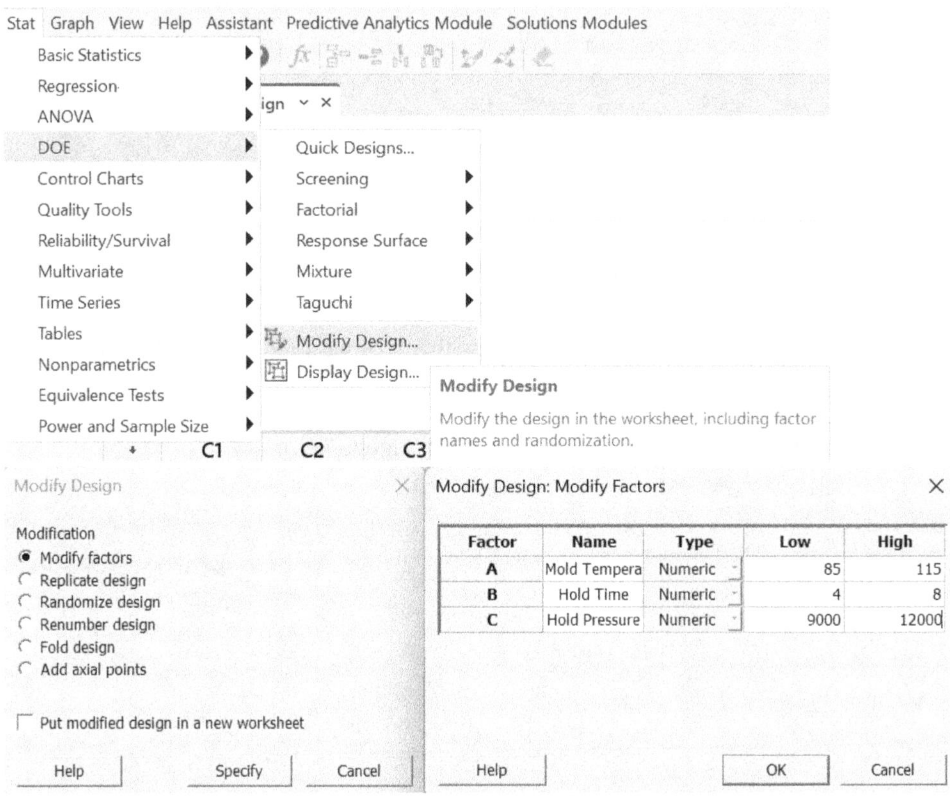

Figure 7.24 Sequence for adding actual factor names and levels to factorial DOE

When the DOE is again analyzed, the only difference is that an equation is given in uncoded units – again, this could have been done from the beginning.

Regression Equation in Uncoded Units

Part Weight = 7.495 - 0.005367 Mold Temperature - 0.0297 Hold Time + 0.00003167 Hold Pressure + 0.000617 Mold Temperature*Hold Time

Figure 7.25 Final regression results for Example 7.6 using uncoded units

The equation would give how many grams the model predicts a part would weigh if the mold temperature, hold time, and hold pressure settings were all entered into the equation in the same units used when generating the model. An important rule when applying regression equations is, however, that none of the settings can be outside the range of settings used when generating the model. Just to show one example of how the equation works, suppose the mold temperature is set at 110 °F, the hold time is set at 5 seconds, and the hold pressure is set at 11,000 psi. What would the predicted weight be?

$$\text{Weight} = 7.495 - 0.0005367(110^\circ\text{F}) - 0.0297(5\text{ s}) + (3.167 \times 10^{-5})(11{,}000\text{ psi}) + 0.0006167(110^\circ\text{F})(5\text{ s}) = 7.4435\text{ g}$$

It is not necessary to enter the numbers into the equation, as Minitab will calculate any combination of factor settings that you would like to test, as shown in the following (Figure 7.26).

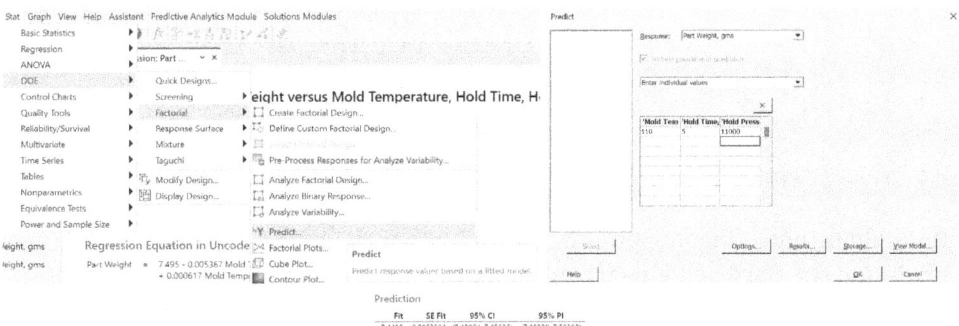

Figure 7.26 Sequence for predicting a response in Minitab for Example 7.6

Minitab also returns a 95% confidence interval (CI) and 95% prediction interval (PI) for the calculated value. The confidence interval applies to the average of a sample and the prediction interval applies to one measured part. If a sample of 5 (the number of replicates in our experiment) was taken with parts run at the conditions used in the equation, this model predicts that the average has a 95% chance of being between 7.43064 and 7.45636 g. Each individual part weight would have a 95% chance of being between 7.38338 and 7.50362 g. To test out how well an equation works, it is necessary to run various conditions and see how the predicted results compare with the actual results because, as George Box famously said, "all models are wrong, but some are useful."

If using Minitab, it is important to also know what kind of graphs are available to use when reporting the results. When running the analysis, the **normal plot of the effects** and the **Pareto effects plot** are two useful graphs that can be obtained. Figure 7.27 shows how to obtain them and what they look like.

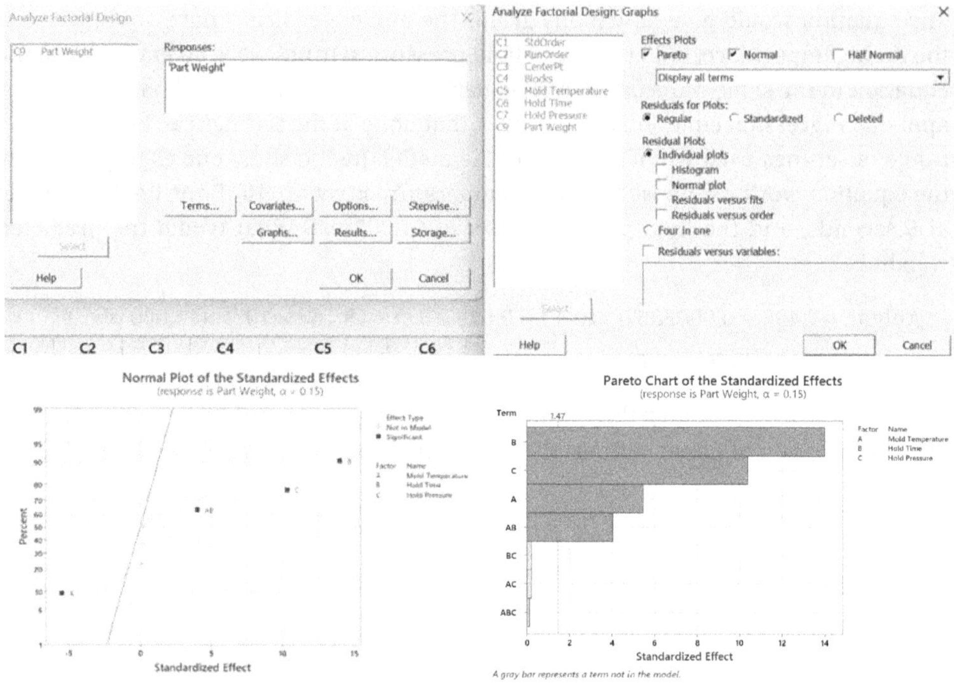

Figure 7.27 Normal plot and Pareto plot of the standardized effects for Example 7.6

It should be noted that the plots in Figure 7.27 are not of the estimated effects; they are of the standardized effects which are the t-statistic calculated for the t-test for each coefficient. For the Pareto chart, it is the absolute value of that t-statistic. For a normal plot, any points a certain distance from the line are deemed not to be due to noise or common cause variation and are said to be signals or statistically significant effects. The Pareto chart shows a line at 1.47; this is the accept/reject or t-critical line for the t-test. Anything above that line is deemed to be statistically significant.

Two other plots that can be obtained in Minitab after the analysis are the main effects plot and the interaction plot (Figure 7.28).

The one interaction plot (Hold Time/Mold Temperature) is the same as those shown in Figure 7.11. Minitab shows both ways that the interaction can be shown and the user can select both or one to explain the results when giving a presentation. In this case, the graph in Figure 7.11 and the upper right of the Minitab graph would be chosen. The main effects plot is a better visual indicator of the effect of the factors than the response plot shown in Figure 7.11. The steeper the slope, the more effect the change in factor levels had on the response. The difference between the high and low levels on the graph is the estimated effect of each factor.

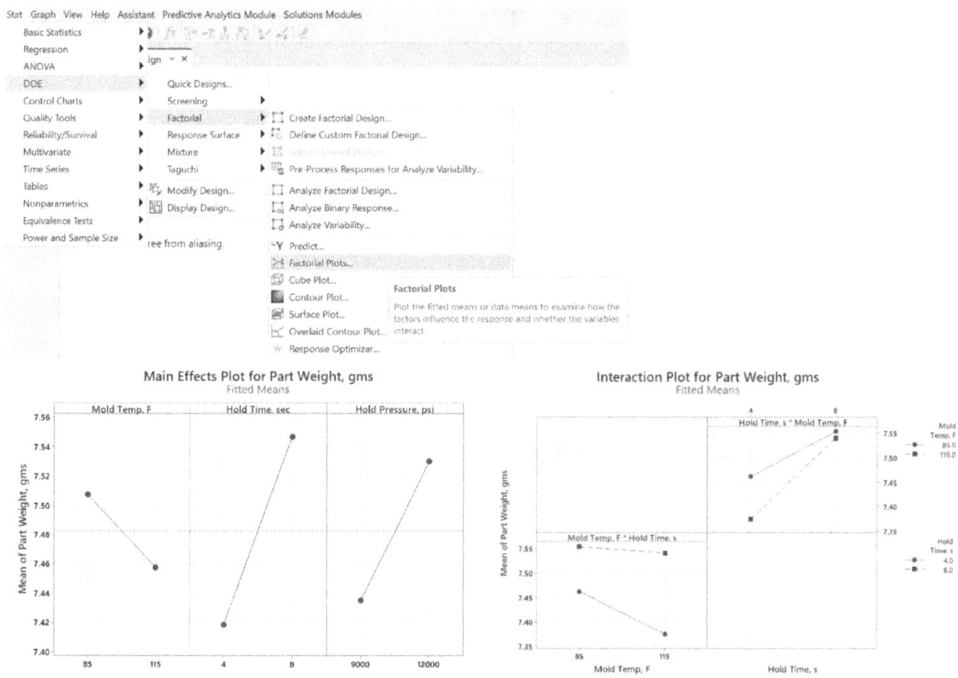

Figure 7.28 Sequence in Minitab for creating main effects and interaction plots for Example 7.6

One other graph that the author recommends generating first when conducting DOEs using Minitab is the interval graph shown in Figure 7.16. This graph is good for catching typos by showing runs with large intervals and for giving a good indication of the data spread. However, generating it requires either manipulating the data – as one would to make Average and Range charts – or adding an extra column. Adding the "Run" column as shown in Figure 7.29 is the easiest method.

C4 Blocks	C5 Mold Temp, F	C6 Hold Time, sec	C7 Hold Pressure, psi	C8 Run	C9 Part Weight, gms
1	115	8	12000	1	7.598
1	115	8	9000	2	7.502
1	115	4	9000	3	7.370
1	115	4	12000	4	7.430
1	85	4	12000	5	7.490
1	85	4	9000	6	7.424
1	85	8	9000	7	7.544
1	85	8	12000	8	7.612
2	115	8	12000	1	7.608
2	115	8	9000	2	7.512
2	115	4	9000	3	7.280
2	115	4	12000	4	7.400
2	85	4	12000	5	7.530
2	85	4	9000	6	7.394
2	85	8	9000	7	7.454
2	85	8	12000	8	7.622
3	115	8	12000	1	7.568
3	115	8	9000	2	7.472
3	115	4	9000	3	7.340
3	115	4	12000	4	7.420
3	85	4	12000	5	7.550
3	85	4	9000	6	7.414
3	85	8	9000	7	7.514
3	85	8	12000	8	7.602

C4 Blocks	C5 Mold Temp, F	C6 Hold Time, sec	C7 Hold Pressure, psi	C8 Run	C9 Part Weight, gms
4	115	8	12000	1	7.598
4	115	8	9000	2	7.502
4	115	4	9000	3	7.320
4	115	4	12000	4	7.300
4	85	4	12000	5	7.500
4	85	4	9000	6	7.374
4	85	8	9000	7	7.494
4	85	8	12000	8	7.542
5	115	8	12000	1	7.568
5	115	8	9000	2	7.472
5	115	4	9000	3	7.340
5	115	4	12000	4	7.470
5	85	4	12000	5	7.480
5	85	4	9000	6	7.464
5	85	8	9000	7	7.514
5	85	8	12000	8	7.632

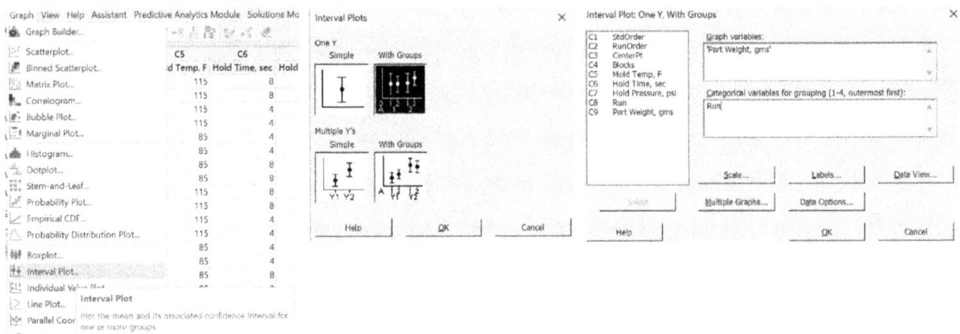

Figure 7.29 Sequence for making 95% confidence interval graph from Minitab DOE results

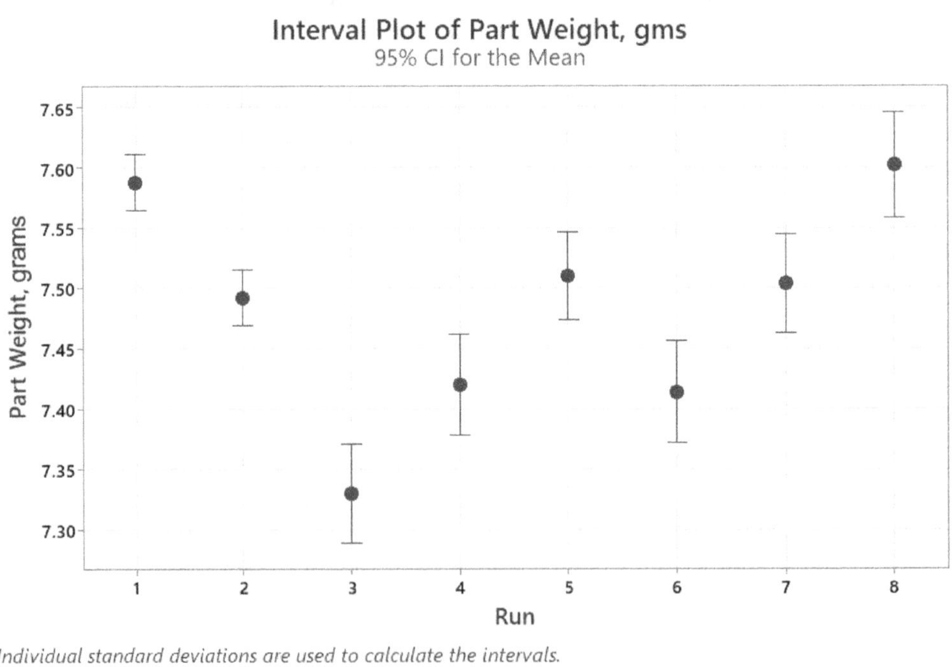

Figure 7.30 95% confidence interval graph made in Minitab for Example 7.6

7.7 Statistical Analysis of DOEs without Multiple Replicates

Only one replicate might be used if parts are expensive to make/measure or if the average of several measurements or repetitions is treated as one replicate. The distinction between replicates and repetitions and the implications for sample size determination will be discussed further in Chapter 9. The two primary methods used to determine the statistical significance of each estimated effect when only one replicate is used are,

1. Running several center points during the experiment to calculate the denominator of the F-ratio (experimental noise)
2. Using the small estimated effects to calculate the denominator of the F-ratio (experimental noise)

An example of each method follows. The first illustrates the eight-run experiment shown earlier in this chapter. However, it will be assumed that the average of the five readings from each run is the single replicate for that run. The second example will illustrate why this author does not recommend the second method listed above.

7.7.1 Example 7.7: Eight-Run DOE with Three Center-Point Runs

The following shows the 8-run experiment from Example 7.3, assuming that the average of each run is one replicate and with three extra center-point runs.

Table 7.12 DOE Settings and Results for Example 7.7

Run	A Mold Temperature, °F	B Hold Time, s	C Hold Pressure, psi	Response Weight, g
1	100 (0)	6 (0)	10,500 (0)	7.523
2	115 (+)	8 (+)	12,000 (+)	7.588
3	115 (+)	8 (+)	9,000 (−)	7.492
4	115 (+)	4 (−)	9,000 (−)	7.330
5	115 (+)	4 (−)	12,000 (+)	7.420
6	100 (0)	6 (0)	10,500 (0)	7.524
7	85 (−)	4 (−)	12,000 (+)	7.510
8	85 (−)	4 (−)	9,000 (−)	7.414

Table 7.12 DOE Settings and Results for Example 7.7 (*continued*)

Run	A Mold Temperature, °F	B Hold Time, s	C Hold Pressure, psi	Response Weight, g
9	85 (–)	8 (+)	9,000 (–)	7.504
10	85 (–)	8 (+)	12,000 (+)	7.602
11	100 (0)	6 (0)	10,500 (0)	7.531

As can be seen when comparing Table 7.12 to Table 7.6, one disadvantage of adding center-point runs will be the additional time that it takes to carry out the experiment since there are eleven runs versus the original eight, in addition to the extra time for the temperatures to stabilize.

The calculations of the estimated effects are the same as Example 7.3, with the addition of an estimated overall quadratic effect. Equation 7.2 can be used to compare the average of the three center-point runs against the eight original runs:

$$\hat{l}_{\text{quadratic}} = \frac{-8\overline{x}_1 + 3\overline{x}_2 + 3\overline{x}_3 + 3\overline{x}_4 + 3\overline{x}_5 - 8\overline{x}_6 + 3\overline{x}_7 + 3\overline{x}_8 + 3\overline{x}_9 + 3\overline{x}_{10} - 8\overline{x}_{11}}{24}$$

$$= -0.0435 \text{ g}$$

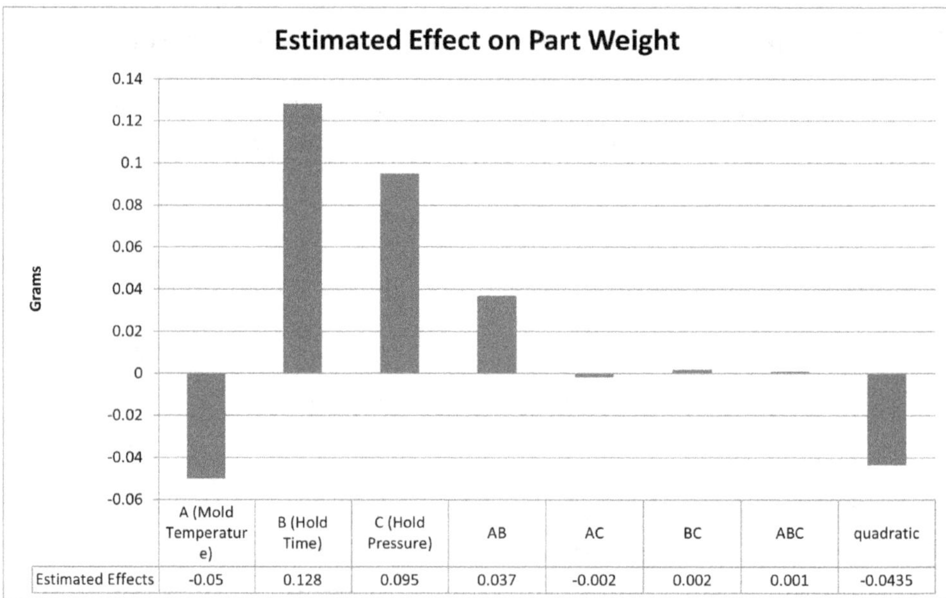

Figure 7.31 Bar chart of estimated effect for Example 7.7

Adding this estimated effect to the bar chart shown in Figure 7.10 yields the bar chart shown in Figure 7.32. Based on the size of the estimated effect (and Example 7.4), it is obvious that the quadratic effect is significant. However, based on just this experiment, it would not be known which factor or factors are nonlinear. Some experiments to separate the quadratic effects for three or more factors will be discussed in Chapter 8.

It should also be noted here that doing just one center run could give us information on the possible presence of quadratic effects. The reason for doing multiple center runs is to obtain an estimate of the experimental noise. In Example 7.5, the average sample variance (squared sample standard deviation) of the five replicates for each run was used for the MSW. With only one replicate from each run, the sample variance of the repeated center runs (7.523, 7.524, and 7.531) is used.

MSW = S^2 of 7.523, 7.524, and 7.531 = 0.0000190 g^2

The sums of squares and F-ratios are then calculated as they were in Example 7.5, with the following results:

Table 7.13 Sums of Squares and F-Ratios for Example 7.7

Effect	Sums of Squares, g^2	F-Ratio
A	5.00E-03	263.2
B	3.28E-02	1724.6
C	1.81E-02	950.0
AB	2.74E-03	144.1
AC	8.00E-06	0.4
BC	8.00E-06	0.4
ABC	2.00E-06	0.1
quadratic	4.13E-03	217.3

To be statistically significant, the calculated F-ratio must be greater than the accept/reject value of $F_{0.05,1,2}$ = 18.51. A, B, C, AB, and the quadratic term are all significant, as would be expected based on the earlier examples in this chapter.

Note that the MSW only has two degrees of freedom since it was based on just three replicates. This is a disadvantage of both this method and the following method as opposed to running replicates of the entire experiment. The fewer degrees of freedom make the estimate of the experimental noise less statistically sound.

To complete this example in Minitab, center points can be specified when specifying the design, as shown in Figure 7.32.

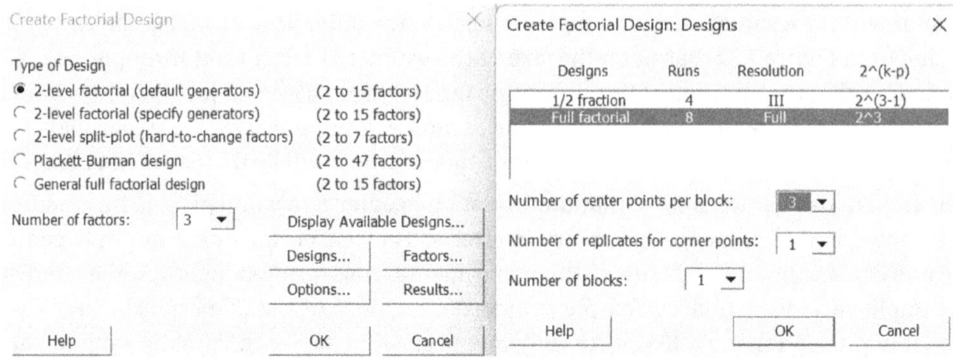

Figure 7.32 Sequence for specifying center points in Minitab

The Minitab worksheet would then appear or could be manually entered as follows. Note the placement of zeros and ones in the CenterPt column (Figure 7.33).

C1	C2	C3	C4	C5	C6	C7	C8	C9
StdOrder	RunOrder	CenterPt	Blocks	A	B	C		Weight, grams
1	1	0	1	0	0	0		7.523
2	2	1	1	1	1	1		7.588
3	3	1	1	1	1	-1		7.492
4	4	1	1	1	-1	-1		7.330
5	5	1	1	1	-1	1		7.420
6	6	0	1	0	0	0		7.524
7	7	1	1	-1	-1	1		7.510
8	8	1	1	-1	-1	-1		7.414
9	9	1	1	-1	1	-1		7.504
10	10	1	1	-1	1	1		7.602
11	11	0	1	0	0	0		7.531

Figure 7.33 Center-point design and results for Example 7.7

The regression results shown in Figure 7.34 have the same significant terms as Example 7.5 (see Figure 7.21), with the addition of showing that there is at least one significant quadratic effect (Ct Pt).

When only one replicate is used for each run of a response measurement and center points are not made, a special means of calculating whether an estimated effect is significant is sometimes used. It involves the use of the sums of squares that are found using Equation 7.7 when $n = 1$. A visual ranking of the sums of squares can be shown in what is called a scree plot. A scree plot for Example 7.3 is shown in Figure 7.35.

Factorial Regression: Weight, gms versus A, B, C, CenterPt

Coded Coefficients

Term	Effect	Coef	SE Coef	T-Value	P-Value	VIF
Constant		7.48250	0.00154	4855.29	0.000	
A	-0.05000	-0.02500	0.00154	-16.22	0.004	1.00
B	0.12800	0.06400	0.00154	41.53	0.001	1.00
C	0.09500	0.04750	0.00154	30.82	0.001	1.00
A*B	0.03700	0.01850	0.00154	12.00	0.007	1.00
A*C	-0.00200	-0.00100	0.00154	-0.65	0.583	1.00
B*C	0.00200	0.00100	0.00154	0.65	0.583	1.00
A*B*C	0.00100	0.00050	0.00154	0.32	0.776	1.00
Ct Pt		0.04350	0.00295	14.74	0.005	1.00

Analysis of Variance

Source	DF	Adj SS	Adj MS	F-Value	P-Value
Model	8	0.062703	0.007838	412.52	0.002
Linear	3	0.055818	0.018606	979.26	0.001
A	1	0.005000	0.005000	263.16	0.004
B	1	0.032768	0.032768	1724.63	0.001
C	1	0.018050	0.018050	950.00	0.001
2-Way Interactions	3	0.002754	0.000918	48.32	0.020
A*B	1	0.002738	0.002738	144.11	0.007
A*C	1	0.000008	0.000008	0.42	0.583
B*C	1	0.000008	0.000008	0.42	0.583
3-Way Interactions	1	0.000002	0.000002	0.11	0.776
A*B*C	1	0.000002	0.000002	0.11	0.776
Curvature	1	0.004129	0.004129	217.29	0.005
Error	2	0.000038	0.000019		
Total	10	0.062741			

Figure 7.34 Minitab regression results for Example 7.7

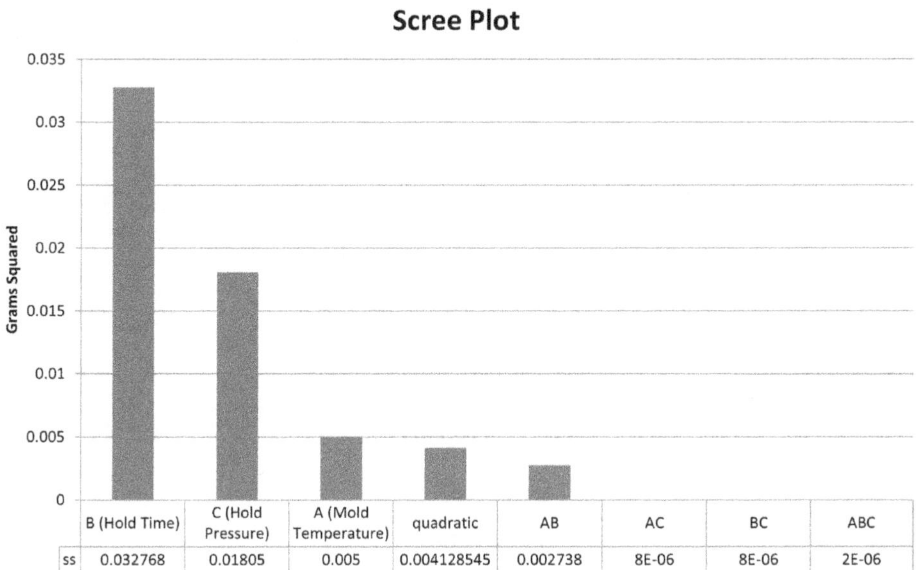

Figure 7.35 Scree plot for Example 7.3, assuming the run averages are one replicate

Scree is defined as the rubble at the bottom of a cliff. In Figure 7.22, the scree is represented by the sums of squares for AC, BC, and ABC. With this method, the scree is pooled or averaged together and is called the mean square error (MSE). It is substituted for the MSW in Equation 7.6 to yield the following equation (Equation 7.11) to determine the F-ratios when only replicate of each run is available.

$$F_i = \frac{\text{Signal}}{\text{Noise}} = \frac{SS_i}{MSE} = \frac{SS_i}{\text{Avg of Scree}} \tag{7.11}$$

The number of degrees of freedom is simply the number of SS that are determined to be scree, since each SS has one degree of freedom. The use of this equation and the problem associated with its use will be shown in the following example.

7.7.2 Example 7.8: DOEs with No Replicates

The problem with only one number for each run is that there is not a good estimate of the common cause variation. The pooling of the scree can very easily lead to erroneous conclusions. To illustrate this, 8 random numbers will be pulled from a random distribution with a mean of 10 and a standard deviation of 0.010. Since these numbers are all from one normal distribution, no statistically significant effects should be found. Either the Data Analysis feature (ToolPak add-in) in Excel or Minitab (Calc – Random data – normal) can be used to do this. The following is an example of 8 random numbers.

Table 7.14 Random Numbers for Example 7.8

Run	Response
1	10.0127
2	10.0080
3	9.9991
4	9.9970
5	9.9996
6	9.9947
7	10.0061
8	10.0043

If the same DOE matrix that was used in Example 7.3 is used here, the following scree plot is obtained.

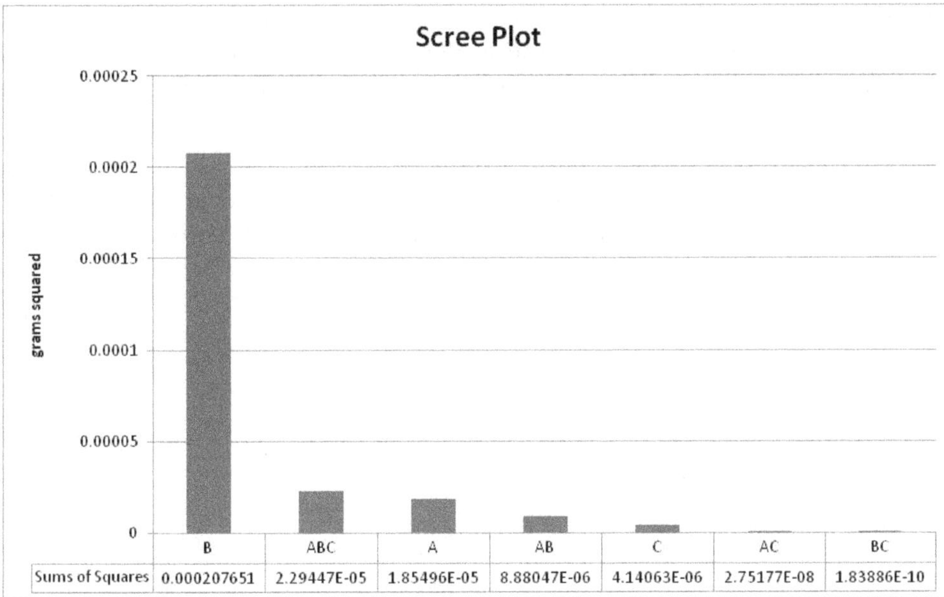

Figure 7.36 Scree plot for Example 7.8

Just looking at the chart, it appears that factor B had a significant effect. If the lowest three SS are considered scree, then the MSE can be calculated as follows.

$$\text{MSE} = \text{Avg of scree} = \frac{1.83886 \times 10^{-10} + 2.75177 \times 10^{-8} + 4.14063 \times 10^{-6}}{3}$$

$$= 1.3894 \times 10^{-6}$$

The F-ratios are calculated using Equation 7.11. The following is a table containing the SS and F-ratios (Table 7.15).

Table 7.15 Sums of Squares and F-Ratios for Example 7.8

Effect	SS	F-Ratio
A	1.85E-05	13.4
B	0.000208	149.4
C	4.14E-06	-
AB	8.88E-06	6.4
AC	2.75E-08	-
BC	1.84E-10	-
ABC	2.29E-05	16.5

The accept/reject or critical F-ratio is $F_{0.05,1,3} = 10.13$ (see Table 12.6). So, according to this analysis, B, ABC, and A are all statistically significant effects, since their F-ratios are greater than 10.13. This isn't the case and points out the problem of one replicate.

Chapter 9 will present more discussions on how to ensure that enough parts are measured to demonstrate an acceptable β error level, and will discuss the difference between repetitions and replicates. Chapter 8 will look at fractional factorial experiments and other experiments in which all the combinations of factors and levels do not need to be run.

8 Fractional Factorials and Response Surface Design of Experiments

Chapter 7 reviewed how to analyze full factorials: experiments in which all possible combinations of factors and levels are used. Even if using just two levels of each factor, it is easy to see how the number of runs that are needed will quickly get out of hand as the number of factors that are studied increases. Equation 7.1 and Table 8.1 illustrate this point.

Table 8.1 Runs Required for a Full Factorial DOE with Two Levels of Each Factor

Number of Factors	Number of Runs Required for a Full Factorial
2	4
3	8
4	16
5	32
6	64
7	128
8	256
9	512
10	1028

The time required to run and measure parts can quickly become a limiting factor when more than five factors are to be studied. This chapter will discuss how to use a fractional factorial DOE to get almost as much information without doing all the runs that would be required for a full factorial with factors at two levels each. For factors at three or more levels each, response surface designs (Box-Behnken and Central Composite Designs) will also be discussed.

8.1 Fractional Factorial DOEs

For a 2-level full factorial experiment, the number of estimated effects that are calculated is $k - 1$, where k is the number of runs. So, for a 4-run experiment, 3 estimated effects are calculated; for an 8 run, 7 estimated effects; for a 16 run, 15 estimated effects, etc. As was seen in the 2-level DOEs in Chapter 7, all the columns for calculating the estimated effects had a unique pattern of +1's and −1's as contrast coefficients. A fractional factorial simply assigns a factor to one of the interaction columns to use as the recipe when carrying out the experiment. For example, consider the 16-run experiment from Chapter 7 shown in Table 8.2 (Table 8.2 is the same as Figure 7.18, except that the response column has been left off). Columns A, B, C, and D are the four columns that are used as the recipe when doing a full factorial. The other columns represent interactions. To add a fifth factor and do a fractional factorial, which column should be chosen for the fifth factor?

Table 8.2 Matrix for a 16-Run Full Factorial DOE

Run	A	B	C	D	AB	AC	AD	BC	BD	CD	ABC	BCD	ACD	ABD	ABCD
1	+1	+1	+1	+1	+1	+1	+1	+1	+1	+1	+1	+1	+1	+1	+1
2	+1	+1	+1	−1	+1	+1	−1	+1	−1	−1	+1	−1	−1	−1	−1
3	+1	+1	−1	−1	+1	−1	−1	−1	−1	+1	−1	+1	+1	−1	+1
4	+1	+1	−1	+1	+1	−1	+1	−1	+1	−1	−1	−1	−1	+1	−1
5	+1	−1	−1	+1	−1	−1	+1	+1	−1	−1	+1	+1	−1	−1	+1
6	+1	−1	−1	−1	−1	−1	−1	+1	+1	+1	+1	−1	+1	+1	−1
7	+1	−1	+1	−1	−1	+1	−1	−1	+1	−1	−1	+1	−1	+1	+1
8	+1	−1	+1	+1	−1	+1	+1	−1	−1	+1	−1	−1	+1	−1	−1
9	−1	−1	+1	+1	+1	−1	−1	−1	−1	+1	+1	−1	−1	+1	+1
10	−1	−1	+1	−1	+1	−1	+1	−1	+1	−1	+1	+1	+1	−1	−1
11	−1	−1	−1	−1	+1	+1	+1	+1	+1	+1	−1	−1	−1	−1	+1
12	−1	−1	−1	+1	+1	+1	−1	+1	−1	−1	−1	+1	+1	+1	−1
13	−1	+1	−1	+1	−1	+1	−1	−1	+1	−1	+1	−1	+1	−1	+1

Run	A	B	C	D	AB	AC	AD	BC	BD	CD	ABC	BCD	ACD	ABD	ABCD
14	−1	+1	−1	−1	−1	+1	+1	−1	−1	+1	+1	+1	−1	+1	−1
15	−1	+1	+1	−1	−1	−1	+1	+1	−1	−1	−1	−1	+1	+1	+1
16	−1	+1	+1	+1	−1	−1	−1	+1	+1	+1	−1	+1	−1	−1	−1

The answer is that Column ABCD should be chosen. The reason is that a significant 4-factor interaction is extremely rare and, in fact, 3-factor interactions are very rare in injection molding. However, when another factor is added (factor O), interactions of that factor with the original factors are also possible. Table 8.3 shows that each column now represents more than one possible effect. The effects in each column are said to be **aliased** or **confounded** with each other.

Table 8.3 Matrix for a 16-Run Fractional Factorial DOE (2^{5-1}) with 5 Factors

Run	A BCDO	B ACDO	C ABDO	D ABCO	AB CDO	AC BDO	AD BCO	BC ADO	BD ACO	CD ABO	DO ABC	AO BCD	BO ACD	CO ABD	O ABCD
1	+1	+1	+1	+1	+1	+1	+1	+1	+1	+1	+1	+1	+1	+1	+1
2	+1	+1	+1	−1	+1	+1	−1	+1	−1	−1	+1	−1	−1	−1	−1
3	+1	+1	−1	−1	+1	−1	−1	−1	−1	+1	−1	+1	+1	−1	+1
4	+1	+1	−1	+1	+1	−1	+1	−1	+1	−1	−1	−1	−1	+1	−1
5	+1	−1	−1	+1	−1	−1	+1	+1	−1	−1	+1	+1	−1	−1	+1
6	+1	−1	−1	−1	−1	−1	−1	+1	+1	+1	+1	−1	+1	+1	−1
7	+1	−1	+1	−1	−1	+1	−1	−1	+1	−1	−1	+1	−1	+1	+1
8	+1	−1	+1	+1	−1	+1	+1	−1	−1	+1	−1	−1	+1	−1	−1
9	−1	−1	+1	+1	+1	−1	−1	−1	−1	+1	+1	−1	−1	+1	+1
10	−1	−1	+1	−1	+1	−1	+1	−1	+1	−1	+1	+1	+1	−1	−1
11	−1	−1	−1	−1	+1	+1	+1	+1	+1	+1	−1	−1	−1	−1	+1
12	−1	−1	−1	+1	+1	+1	−1	+1	−1	−1	−1	+1	+1	+1	−1
13	−1	+1	−1	+1	−1	+1	−1	−1	+1	−1	+1	−1	+1	−1	+1
14	−1	+1	−1	−1	−1	+1	+1	−1	−1	+1	+1	+1	−1	+1	−1
15	−1	+1	+1	−1	−1	−1	+1	+1	−1	−1	−1	−1	+1	+1	+1
16	−1	+1	+1	+1	−1	−1	−1	+1	+1	+1	−1	+1	−1	−1	−1

It should be noted that there is not a column which represents the 5-factor interaction ABCDO. Since column ABCD was used for the fifth factor, multiplying it by itself would give a column of all positive contrast coefficients. Since this column is not present, a significant 5-factor interaction would partially confound the other effects and could give erroneous results. However, this author has never seen an interaction of a higher order than a 3-factor and, as stated earlier, very rarely are 3-factor interactions found in injection molding. An example showing a 3-factor interaction will be discussed in Chapter 9. It is generally safe to assume that 3-factor and higher-order interactions are not present.

In the case of five factors and using the 16-run DOE as shown in Figure 8.2, the aliasing or confounding is not a big concern. If 3-factor and 4-factor interactions are negligible, then each column really represents either a main-factor effect or a 2-factor interaction effect. The confounding that is present is either a single main-factor effect aliased with a 4-factor interaction (I + IV confounding) or a 2-factor interaction aliased with a 3-factor interaction (II + III confounding). In either case, the **design resolution** is V (I + IV or II + III). The higher the design resolution, the easier the results are to interpret.

Adding a sixth factor to the 16-run matrix will make the confounding worse. Looking at Table 8.3, all the columns that are not used for main factors have at least one 2-factor interaction in the confounding pattern. Adding a factor to one of those columns would confound a main factor with a 2-factor interaction. This would give a design resolution of III (I + II). To keep from confounding a main factor with a 2-factor interaction, two of the original 3-factor interaction columns can be used instead of the ABCD column. If columns ABC and BCD are used for Factors K and L, the confounding pattern of main factors and 2-factor interactions is as shown in Table 8.4.

Table 8.4 Matrix for a 16-Run Fractional Factorial DOE (2^{6-2}) with 6 Factors

Run	A	B	C	D	AB CK	AC BK	AD KL	AK BC DL	BD CL	BL CD	K	L			AL DK
1	+1	+1	+1	+1	+1	+1	+1	+1	+1	+1	+1	+1	+1	+1	+1
2	+1	+1	+1	−1	+1	+1	−1	+1	−1	−1	+1	−1	−1	−1	−1
3	+1	+1	−1	−1	+1	−1	−1	−1	−1	+1	−1	+1	+1	−1	+1
4	+1	+1	−1	+1	+1	−1	+1	−1	+1	−1	−1	−1	−1	+1	−1
5	+1	−1	−1	+1	−1	−1	+1	+1	−1	−1	+1	+1	−1	−1	+1
6	+1	−1	−1	−1	−1	−1	−1	+1	+1	+1	+1	−1	+1	+1	−1
7	+1	−1	+1	−1	−1	+1	−1	−1	+1	−1	−1	+1	−1	+1	+1
8	+1	−1	+1	+1	−1	+1	+1	−1	−1	+1	−1	−1	+1	−1	−1

Run	A	B	C	D	AB CK	AC BK	AD KL	AK BC DL	BD CL	BL CD	K	L			AL DK
9	−1	−1	+1	+1	+1	−1	−1	−1	−1	+1	+1	−1	−1	+1	+1
10	−1	−1	+1	−1	+1	−1	+1	−1	+1	−1	+1	+1	+1	−1	−1
11	−1	−1	−1	−1	+1	+1	+1	+1	+1	+1	−1	−1	−1	−1	+1
12	−1	−1	−1	+1	+1	+1	−1	+1	−1	−1	−1	+1	+1	+1	−1
13	−1	+1	−1	+1	−1	+1	−1	−1	+1	−1	+1	−1	+1	−1	+1
14	−1	+1	−1	−1	−1	+1	+1	−1	−1	+1	+1	+1	−1	+1	−1
15	−1	+1	+1	−1	−1	−1	+1	+1	−1	−1	−1	−1	+1	+1	+1
16	−1	+1	+1	+1	−1	−1	−1	+1	+1	+1	−1	+1	−1	−1	−1

In Table 8.4, there are 2-factor interactions confounded with other 2-factor interactions, which gives a design resolution of IV (II + II). This means that the estimated effect calculated for a column with more than one 2-factor interaction represents the sum of the 2-factor interactions shown in that column. Without doing a follow-up experiment, one could only hypothesize as to which interaction or combination of interactions makes up the estimated effect for that column.

The matrix in Table 8.5 shows what the confounding pattern looks like when all 15 columns are used for main factors. Again, 3-factor and higher interactions are assumed to be negligible and are not shown.

Table 8.5 Matrix for a 16-Run Fractional Factorial DOE (2^{15-11}) with 15 Factors

Run	A BE CF DG HK IN JM LO	B AE CH DI FK GN JL MO	C AF BH DJ EK GM IL NO	D AG BI CJ EN FM HL KO	E AB CK DN FH GI JO LM	F AC BK DM EH GJ IO LN	G AD BN CM EI FJ HO KL	H AK BC DL EF GO IJ MN	I AN BD CL EG FO HJ KM	J AM BL CD EO FG HI KN	K AH BF CE DO GL IM JN	L AO BJ CI DH EM FN GK	M AJ BO CG DF EL HN IK	N AI BG CO DE FL HM JK	O AL BM CN DK EJ FI GH
1	+1	+1	+1	+1	+1	+1	+1	+1	+1	+1	+1	+1	+1	+1	+1
2	+1	+1	+1	−1	+1	+1	−1	+1	−1	−1	+1	−1	−1	−1	−1
3	+1	+1	−1	−1	+1	−1	−1	−1	−1	+1	−1	+1	+1	−1	+1
4	+1	+1	−1	+1	+1	−1	+1	−1	+1	−1	−1	−1	−1	+1	−1
5	+1	−1	−1	+1	−1	−1	+1	+1	−1	−1	+1	+1	−1	−1	+1

Table 8.5 Matrix for a 16-Run Fractional Factorial DOE (2^{15-11}) with 15 Factors (*continued*)

Run	A BE CF DG HK IN JM LO	B AE CH DI FK GN JL MO	C AF BH DJ EK GM IL NO	D AG BI CJ EN FM HL KO	E AB CK DN FH GI JO LM	F AC BK DM EH GJ IO LN	G AD BN CM EI FJ HO KL	H AK BC DL EF GO IJ MN	I AN BD CL EG FO HJ KM	J AM BL CD EO FG HI KN	K AH BF CE DO GL IM JN	L AO BJ CI DH EM FN GK	M AJ BO CG DF EL HN IK	N AI BG CO DE FL HM JK	O AL BM CN DK EJ FI GH
6	+1	−1	−1	−1	−1	−1	−1	+1	+1	+1	+1	−1	+1	+1	−1
7	+1	−1	+1	−1	−1	+1	−1	−1	+1	−1	−1	+1	−1	+1	+1
8	+1	−1	+1	+1	−1	+1	+1	−1	−1	+1	−1	−1	+1	−1	−1
9	−1	−1	+1	+1	+1	−1	−1	−1	−1	+1	+1	−1	−1	+1	+1
10	−1	−1	+1	−1	+1	−1	+1	−1	+1	−1	+1	+1	+1	−1	−1
11	−1	−1	−1	−1	+1	+1	+1	+1	+1	+1	−1	−1	−1	−1	+1
12	−1	−1	−1	+1	+1	+1	−1	+1	−1	−1	−1	+1	+1	+1	−1
13	−1	+1	−1	+1	−1	+1	−1	−1	+1	−1	+1	−1	+1	−1	+1
14	−1	+1	−1	−1	−1	+1	+1	−1	−1	+1	+1	+1	−1	+1	−1
15	−1	+1	+1	−1	−1	−1	+1	+1	−1	−1	−1	−1	+1	+1	+1
16	−1	+1	+1	+1	−1	−1	−1	+1	+1	+1	−1	+1	−1	−1	−1

Because of the possibility of interactions, it is not normally recommended that 15 factors be used with this 16-run experiment. The matrix in Table 8.5 is also shown in Table 12.15 of the Appendix, along with Table 8.6, which includes a list of the recommended factors to use and the design resolution when using different numbers of factors. To get the confounding pattern with different numbers of factors, just cross off any factors not used and the interactions that those factors could have been part of (the reader should try to duplicate Table 8.2 by doing this). Tables 12.11 and 12.13 in the Appendix show the same type of matrix for the 4- and 8-run DOEs.

Another thing to note in Table 8.4 is that, even with fifteen factors, this is still a design resolution of III. If someone tried to add a sixteenth factor, one of the columns would have to be reused, thus aliasing two main-factor effects. This would be an example of a resolution II design, which should never be used. The only time this is generally seen is when the experimenters mistakenly change two factors using the same column for the recipe and end up not being able to separate the two aliased effects at the conclusion of the experiment.

Table 8.6 Recommended Factors, Design Resolution and Nomenclature for the 16-Run DOE in Table 8.5

Number of Factors	Recommended Factors	$2^{(k-p)}$	Fraction	Design Resolution
4	A B C D	2^4	Full factorial	Full factorial
5	A B C D O	2^{5-1}	½	V
6	A B C D K L	2^{6-2}	¼	IV
7	A B C D K L M	2^{7-3}	1/8	IV
8	A B C D K L M N	2^{8-4}	1/16	IV
9	A B C D E K L M N	2^{9-5}	1/32	III
10	A B C D E F K L M N	2^{10-6}	1/64	III
11	A B C D E F G H I K N	2^{11-7}	1/128	III
12	A B C D E F G H I J K N	2^{12-8}	1/256	III
13	A B C D E F G H I J K M N	2^{13-9}	1/512	III
14	A B C D E F G H I J K L M N	2^{14-10}	1/1028	III
15	A B C D E F G H I J K L M N O	2^{15-11}	1/2048	III

Example 8.1 will demonstrate how confounding can make interpreting the results of a DOE more difficult. The results from Example 7.3 will be used to show what would happen if only 4 of the runs were originally done (a 2^{3-1} design). A method to separate the effects, called a foldover or reflected design, will then be reviewed in Example 8.2.

8.1.1 Example 8.1: Fractional Factorial for 3 Factors at 2 Levels

The following matrix from Table 12.11 in the Appendix shows the recipe used for the three factors.

Run	A BC	B AC	C AB
1	+1	+1	+1
2	+1	−1	−1
3	−1	+1	−1
4	−1	−1	+1

Table 8.7 shows the factor levels, the run number from Example 7.3 (in parentheses), and the results.

Table 8.7 Resolution III Design Fractional Factorial for 3 Factors at 2 Levels Each

Run	A Mold Temperature, °F	B Hold Time, sec	C Hold Pressure, psi	1 g	2 g	3 g	4 g	5 g	Response Average Weight, g
1(1)	115 (+)	8 (+)	12,000 (+)	7.598	7.608	7.568	7.598	7.568	7.588
2(3)	115 (+)	4 (−)	9,000 (−)	7.370	7.280	7.340	7.320	7.340	7.330
3(7)	85 (−)	8 (+)	9,000 (−)	7.544	7.454	7.514	7.494	7.514	7.504
4(5)	85 (−)	4 (−)	12,000 (+)	7.490	7.530	7.550	7.500	7.480	7.510

Calculating the estimated effects yields:

$$\hat{l}_A + \hat{l}_{BC} = \frac{7.588\ \text{g} + 7.330\ \text{g} - 7.504\ \text{g} - 7.510\ \text{g}}{2} = -0.048\ \text{g}$$

$$\hat{l}_B + \hat{l}_{Ac} = \frac{7.588\ \text{g} - 7.330\ \text{g} + 7.504\ \text{g} - 7.510\ \text{g}}{2} = 0.126\ \text{g}$$

$$\hat{l}_C + \hat{l}_{AB} = \frac{7.588\ \text{g} - 7.330\ \text{g} - 7.504\ \text{g} + 7.510\ \text{g}}{2} = 0.132\ \text{g}$$

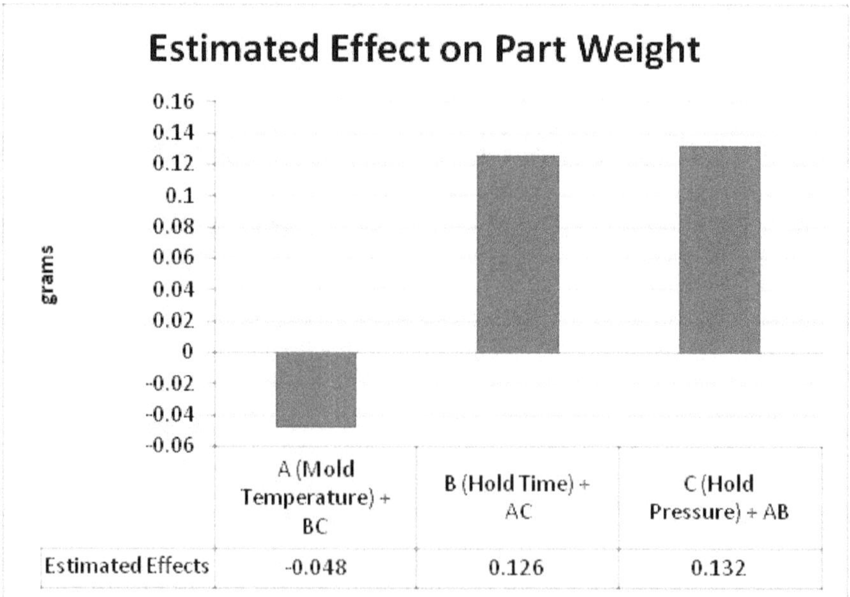

Figure 8.1 Estimated effects for Example 8.1

Note that each estimated effect is the sum of the two aliases for each column. If these were the only 4 runs that had been done, it would not be known whether the numbers calculated were due to the main-factor effect, the two-factor interaction or the combination of both. To separate out the two effects, it is necessary to do a foldover or reflected design. This involves reversing the signs of the contrast coefficients for some or all of the columns. The matrix used in Example 8.1 and the full reflection or complete foldover are shown in Table 8.8.

Table 8.8 Original Design (Left) and Complete Foldover (Full Reflection) (Right)

Run	A BC	B AC	C AB		Run	A –BC	B –AC	C –AB
1	+1	+1	+1		1	−1	−1	−1
2	+1	−1	−1		2	−1	+1	+1
3	−1	+1	−1		3	+1	−1	+1
4	−1	−1	+1		4	+1	+1	−1

What this will do is reverse the sign of the two-factor interactions, as shown in Table 8.8. Why this occurs is shown in Table 8.9. This is like the operation that was shown in Figure 7.13 except, in this case, there are no columns in the foldover design that match the interaction column. There is a column, however, that is the exact opposite or negative of the interaction.

Table 8.9 Operation to Obtain an Interaction Column for the Foldover Design in Table 8.8

A		B		AB
−1	×	−1	=	+1
−1		+1		−1
+1		−1		−1
+1		+1		+1

The following example will show how the foldover design separates the aliased effects. It should be noted that, since the original example is a ½ fraction, doing the foldover will gives us a full factorial. This is not always the case, as will be seen in Example 8.4.

8.1.2 Example 8.2: Using a Foldover Design

Using the foldover design shown in Table 8.8, the following Table 8.10 shows the factor levels, the run number from Example 7.3 (in parentheses), and the results.

Table 8.10 Foldover Design of DOE Used in Example 8.1

Run	A Mold Temperature, °F	B Hold Time, s	C Hold Pressure, psi	1, g	2, g	3, g	4, g	5, g	Response Average Weight, g
1(6)	85 (−)	4 (−)	9,000 (−)	7.424	7.394	7.414	7.374	7.464	7.414
2(8)	85 (−)	8 (+)	12,000 (+)	7.612	7.622	7.602	7.542	7.632	7.602
3(4)	115 (+)	4 (−)	12,000 (+)	7.430	7.400	7.420	7.380	7.470	7.420
4(2)	115 (+)	8 (+)	9,000 (−)	7.502	7.512	7.472	7.502	7.472	7.492

Calculating the estimated effects gives:

$$\hat{l}_A - \hat{l}_{BC} = \frac{-7.414\text{ g} - 7.602\text{ g} + 7.420\text{ g} + 7.492\text{ g}}{2} = -0.052\text{ g}$$

$$\hat{l}_B - \hat{l}_{Ac} = \frac{-7.414\text{ g} + 7.602\text{ g} - 7.420\text{ g} + 7.492\text{ g}}{2} = 0.130\text{ g}$$

$$\hat{l}_C - \hat{l}_{AB} = \frac{-7.414\text{ g} + 7.602\text{ g} + 7.420\text{ g} - 7.492\text{ g}}{2} = 0.058\text{ g}$$

Comparing Figure 8.2 and Figure 8.1, the only bar that looks significantly different is the one with C (hold pressure) and AB (mold temperature/hold time interaction) aliased. The fact that the other two bars are almost the same in both figures indicates that the interactions are small, probably non-existent. The results of this foldover DOE and the original experiment can be combined to calculate each estimated effect. The solution is that for two equations with two unknowns. Probably the easiest way to do this, in this case, is to first add the equations together to solve for the main-factor effect and, then, to use substitution to solve for the interaction. This method is shown below for all three sets of equations.

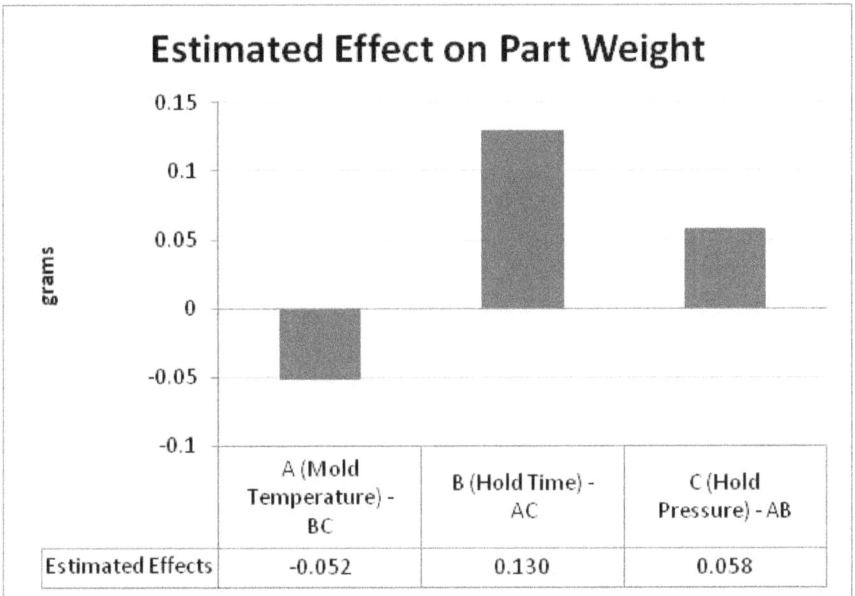

Figure 8.2 Estimated effects for Example 8.1

To separate the A and BC effects:

$$\hat{l}_A + \hat{l}_{BC} = -0.048 \text{ g}$$

$$\hat{l}_A - \hat{l}_{BC} = -0.052 \text{ g}$$

Adding these two equations together gives:

$$2\hat{l}_A = -0.100 \text{ g}$$

$$\hat{l}_A = -0.050 \text{ g}$$

Substituting this into one of the original equations gives:

$$-0.050 + \hat{l}_{BC} = -0.048 \text{ g}$$

$$\hat{l}_{BC} = 0.002 \text{ g}$$

To separate the B and AC effects:

$$\hat{l}_B + \hat{l}_{Ac} = 0.126 \text{ g}$$

$$\hat{l}_B - \hat{l}_{Ac} = 0.130 \text{ g}$$

Adding these two equations together gives:

$$2\hat{l}_B = 0.256\ \text{g}$$

$$\hat{l}_B = 0.128\ \text{g}$$

Substituting this into one of the original equations gives:

$$0.128 + \hat{l}_{Ac} = 0.126\ \text{g}$$

$$\hat{l}_{Ac} = -0.002\ \text{g}$$

To separate the C and AB effects:

$$\hat{l}_C + \hat{l}_{AB} = 0.132\ \text{g}$$

$$\hat{l}_C - \hat{l}_{AB} = 0.058\ \text{g}$$

Adding these two equations together gives:

$$2\hat{l}_C = 0.190\ \text{g}$$

$$\hat{l}_C = 0.095\ \text{g}$$

Substituting this into one of the original equations gives:

$$0.095 + \hat{l}_{AB} = 0.132\ \text{g}$$

$$\hat{l}_{AB} = 0.037\ \text{g}$$

Comparison of these results with those from Example 7.3 shows that the exact same numbers are obtained for the main-factor effects and the 2-factor interaction effects.

With the 4-run fractional factorial, it was only possible for one 2-factor interaction to be aliased with each main factor. For all the other two-level fractional factorials, with a resolution III design, some of the main-factor effects will be aliased with more than one 2-factor interaction. The following section will show an example of how to handle such a case.

8.1.3 Example 8.3: 8-Run, Resolution III Fractional Factorial DOE with 6 Factors

Using Table 12.13 from the Appendix, the following 8-run experiment was carried out to study the effect of six factors on the warpage of a particular part feature (Table 8.11).

Table 8.11 DOE Matrix for Example 8.3

Run	A Mold Temp, °F	B Melt Temp, °F	C Hold Press, psi	D Inject Speed, in/s	E Hold Time, s	F Screw Speed, RPM
1	180(+)	510(+)	550(+)	3.5(+)	9(+)	250(+)
2	180(+)	510(+)	450(−)	3.5(+)	6(−)	125(−)
3	180(+)	480(−)	450(−)	2.0(−)	6(−)	250(+)
4	180(+)	480(−)	550(+)	2.0(−)	9(+)	125(−)
5	140(−)	480(−)	550(+)	3.5(+)	6(−)	125(−)
6	140(−)	480(−)	450(−)	3.5(+)	9(+)	250(+)
7	140(−)	510(+)	450(−)	2.0(−)	9(+)	125(−)
8	140(−)	510(+)	550(+)	2.0(−)	6(−)	250(+)

The warpage measurements, five parts per run, were as follows (Table 8.12).

Table 8.12 DOE Response Measurements for Example 8.4

Run	Inches	Inches	Inches	Inches	Inches	Response Average Warpage, inches
1	0.0354	0.0116	0.0114	0.0402	0.0726	0.0343
2	0.1326	0.1610	0.1510	0.1412	0.1917	0.1555
3	0.1796	0.1731	0.1556	0.1759	0.1037	0.1618
4	0.1148	0.1305	0.1084	0.0950	0.1607	0.1219
5	0.1229	0.1197	0.1214	0.1313	0.1113	0.1213
6	0.1388	0.1834	0.1746	0.1730	0.1392	0.1576
7	0.1434	0.1637	0.1650	0.1249	0.1644	0.1523
8	0.0192	0.0552	0.0409	0.0219	0.0064	0.0287

The Range chart is in control and the Average chart is out of control – this is the desired outcome for both.

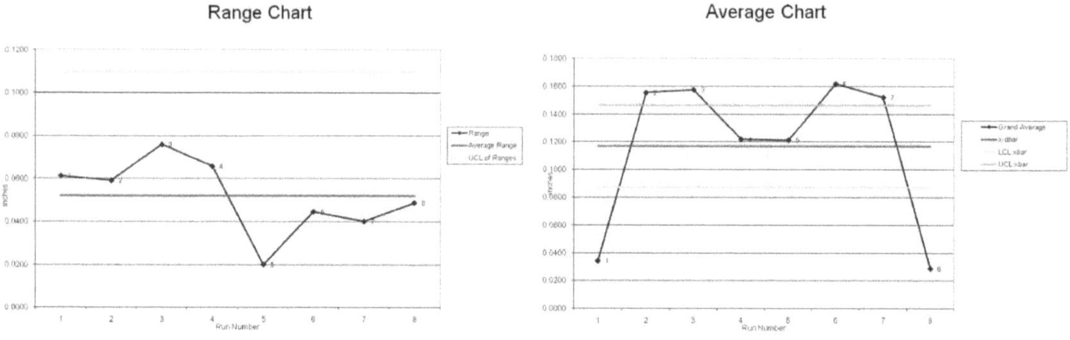

Figure 8.3 Average and Range charts for Example 8.3

Since there was no factor G in this experiment, the confounding pattern shown in Table 12.13 can be reduced to the following (Table 8.13).

Table 8.13 Confounding Table for Example 8.3

Run	A BD CE	B AD CF	C AE BF	D AB EF	E AC DF	F BC DE	AF BE CD
1	+1	+1	+1	+1	+1	+1	+1
2	+1	+1	−1	+1	−1	−1	−1
3	+1	−1	−1	−1	−1	+1	+1
4	+1	−1	+1	−1	+1	−1	−1
5	−1	−1	+1	+1	−1	−1	+1
6	−1	−1	−1	+1	+1	+1	−1
7	−1	+1	−1	−1	+1	−1	+1
8	−1	+1	+1	−1	−1	+1	−1

In addition to eliminating factor G from Table 12.13, all the 2-factor interactions containing G are also eliminated. As can be seen, each main-factor effect is aliased with two 2-factor interactions and the column where factor G was now has three 2-factor interactions aliased with each other.

The following estimated effects are calculated:

$\hat{l}_A + \hat{l}_{BD} + \hat{l}_{CE} = 0.0013$ inches

$\hat{l}_B + \hat{l}_{AD} + \hat{l}_{CF} = -0.0480$ inches

$\hat{l}_C + \hat{l}_{AE} + \hat{l}_{BF} = -0.0803$ inches

$\hat{l}_D + \hat{l}_{AB} + \hat{l}_{EF} = 0.0031$ inches

$\hat{l}_E + \hat{l}_{AC} + \hat{l}_{DF} = 0.0018$ inches

$\hat{l}_F + \hat{l}_{BC} + \hat{l}_{DE} = -0.0422$ inches

$\hat{l}_{AF} + \hat{l}_{BE} + \hat{l}_{CD} = -0.0006$ inches

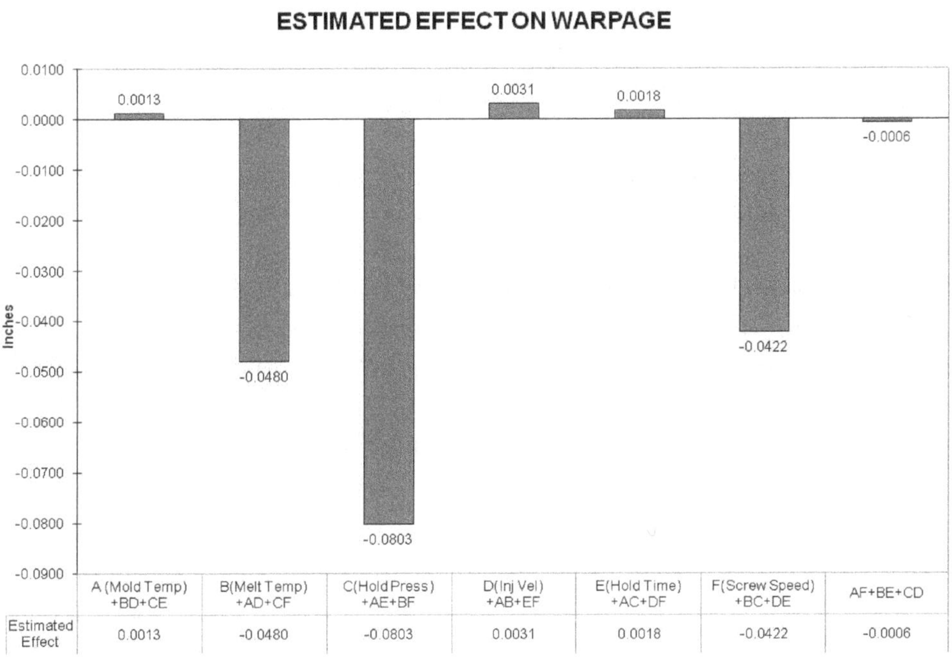

Figure 8.4 Original estimated effects for Example 8.3

As can be seen, each estimated effect represents three aliases. If it was assumed that 2-factor interactions were non-existent, then it would be said that the three big effects were caused by melt temperature, hold pressure and screw speed. As shall be seen, after analyzing the foldover design results, that would be a bad assumption.

A foldover design must now be carried out to determine whether the effects are due to main-factor effects, 2-factor interactions or a combination of both. The matrix and confounding pattern for a complete foldover are shown in Table 8.14.

Table 8.14 Confounding Table and Complete Foldover DOE for Example 8.3

Run	A –BD –CE	B –AD –CF	C –AE –BF	D –AB –EF	E –AC –DF	F –BC –DE	–AF –BE –CD
1	−1	−1	−1	−1	−1	−1	−1
2	−1	−1	+1	−1	+1	+1	+1
3	−1	+1	+1	+1	+1	−1	−1
4	−1	+1	−1	+1	−1	+1	+1
5	+1	+1	−1	−1	+1	+1	−1
6	+1	+1	+1	−1	−1	−1	+1
7	+1	−1	+1	+1	−1	+1	−1
8	+1	−1	−1	+1	+1	−1	+1

The eight new runs are shown in Table 8.15. After this DOE, sixteen of the sixty-four possible combinations of factors and levels will have been run.

Table 8.15 Foldover DOE Factors and Factor Levels for Runs

Run	A Mold Temp, °F	B Melt Temp, °F	C Hold Press, psi	D Inject Speed, in/s	E Hold Time, s	F Screw Speed, RPM
1	140	480	450	2.0	6	125
2	140	480	550	2.0	9	250
3	140	510	550	3.5	9	125
4	140	510	450	3.5	6	250
5	180	510	450	2.0	9	250
6	180	510	550	2.0	6	125
7	180	480	550	3.5	6	250
8	180	480	450	3.5	9	125

The warpage measurements, five parts per run, were as shown in Table 8.16.

Table 8.16 Foldover DOE Measurements

Run	Inches	Inches	Inches	Inches	Inches	Response Average Warpage, Inches
1	0.1753	0.1641	0.1595	0.1734	0.1928	0.1730
2	0.1386	0.1272	0.1503	0.1118	0.1322	0.1320
3	0.0029	0.0176	0.0141	0.0390	0.0339	0.0215
4	0.1657	0.1587	0.1665	0.1699	0.1617	0.1645
5	0.1405	0.1538	0.1540	0.1433	0.1510	0.1485
6	0.0255	0.0411	0.0340	0.0321	0.0703	0.0406
7	0.1053	0.1046	0.1188	0.1214	0.1373	0.1175
8	0.1642	0.1770	0.1555	0.1854	0.1609	0.1686

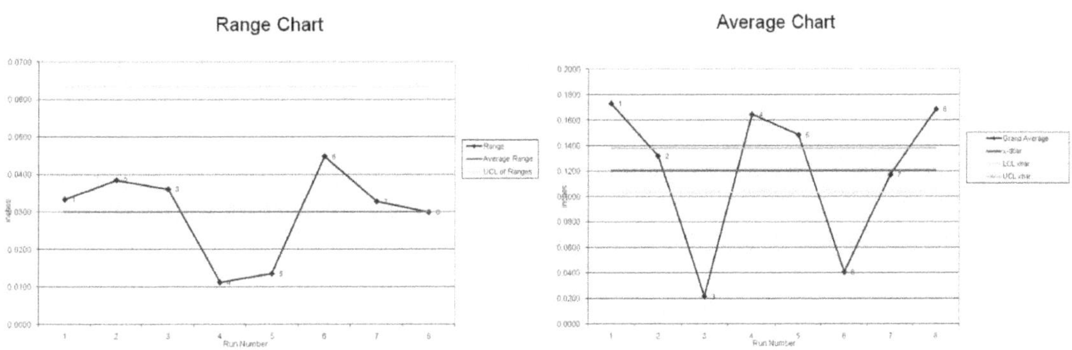

Figure 8.5 Average and Range charts for Example 8.3

Again, the Range chart is in control and the Average chart is out of control – which is the desired result.

The following estimated effects are calculated.

$\hat{l}_A - \hat{l}_{BD} - \hat{l}_{CE} = -0.0040$ inches

$\hat{l}_B - \hat{l}_{AD} - \hat{l}_{CF} = -0.0540$ inches

$\hat{l}_C - \hat{l}_{AE} - \hat{l}_{BF} = -0.0858$ inches

$\hat{l}_D - \hat{l}_{AB} - \hat{l}_{EF} = -0.0055$ inches

$\hat{l}_E - \hat{l}_{AC} - \hat{l}_{DF} = -0.0062$ inches

$\hat{l}_F - \hat{l}_{BC} - \hat{l}_{DE} = 0.0397$ inches

$-\hat{l}_{AF} - \hat{l}_{BE} - \hat{l}_{CD} = 0.0113$ inches

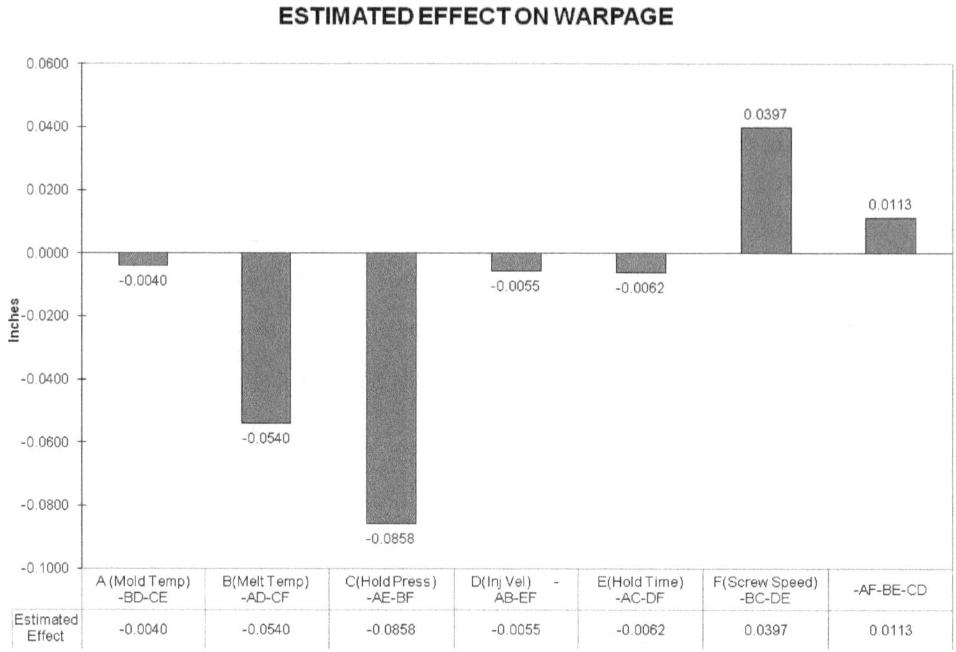

Figure 8.6 Estimated effects from foldover of the original design for Example 8.3

These results, along with the results from the original DOE can be used to solve for the estimated effects of the main factors (add the two equations) and the combined estimated effects of the 2-factor interactions (substitute the main factor estimated effect back into one of the equations). Table 8.17 shows the results.

Table 8.17 Estimated Effects for Example 8.3

Main Effect	Est. Effect, inches	Interactions	Est. Effect, inches
$\hat{l}_A$	-0.0013	$\hat{l}_{BD} + \hat{l}_{CE}$	0.0027
$\hat{\mathbf{l}}_{\mathbf{B}}$	**-0.0510**	$\hat{l}_{AD} + \hat{l}_{CF}$	0.0030
$\hat{\mathbf{l}}_{\mathbf{C}}$	**-0.0831**	$\hat{l}_{AE} + \hat{l}_{BF}$	0.0027
$\hat{l}_D$	-0.0012	$\hat{l}_{AB} + \hat{l}_{EF}$	0.0043
$\hat{l}_E$	-0.0022	$\hat{l}_{AC} + \hat{l}_{DF}$	0.0040
$\hat{l}_F$	-0.0012	$\hat{\mathbf{l}}_{\mathbf{BC}} + \hat{\mathbf{l}}_{\mathbf{DE}}$	**-0.0409**
		$\hat{l}_{AF} + \hat{l}_{BE} + \hat{l}_{CD}$	-0.0060

A few things to note about these results:

Two of the three large effects were due to main factors: melt temperature (B) and hold pressure (C). Compare the bars in Figure 8.4 and Figure 8.6 (same side).

The third large effect is due to the melt temperature/hold pressure (BC) interaction and/or the injection velocity/hold time (DE) interaction. Screw speed (F) had little to no effect. Compare the bars in Figure 8.4 and Figure 8.6 (flips side).

The 2-factor interaction effect in column G (made up of three 2-factor interactions) was obtained by subtracting the two equations. The fact that the effect did not just change algebraic signs from one experiment to the next is most likely due to measurement variation or, possibly, a small 3-factor interaction.

Since melt temperature and hold pressure were the only two large main-factor effects, it is safe to assume that the interaction in column F is due to the interaction between those two factors. Response plots of the combined results of both DOEs help to sort out interactions.

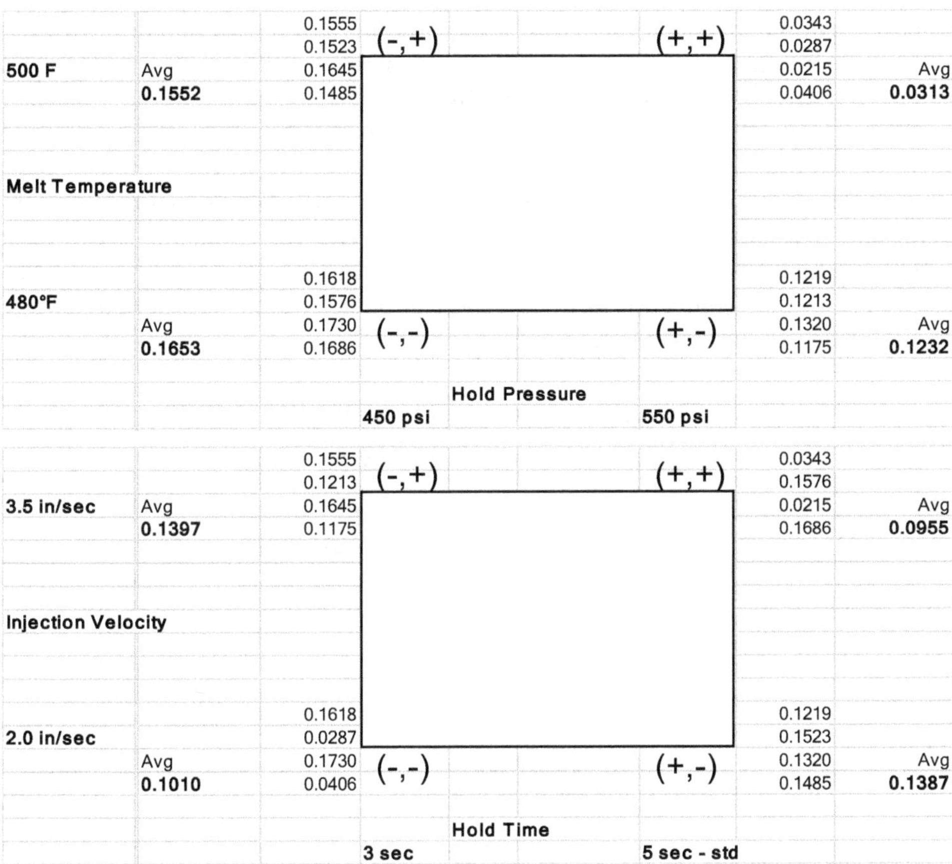

Figure 8.7 Response plots of possible 2-factor interactions for Example 8.3

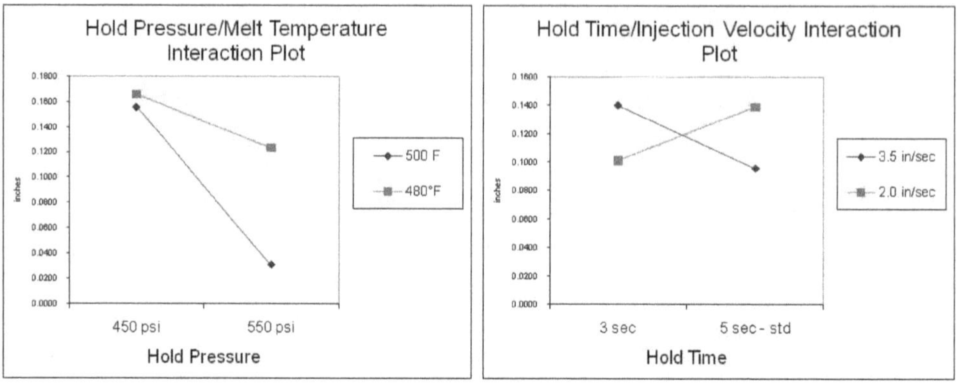

Figure 8.8 Factorial plots of possible 2-factor interactions for Example 8.3

Examination of the response and interaction plots confirms the hypothesis that the melt temperature/hold pressure interaction is the one that is occurring. The effect of hold pressure on warpage depends on the level of the melt temperature and vice-versa.

The choice of interaction is very clear in this example, but there are times when two or more interactions can both seem plausible. If there was still a question as to which interaction was occurring in this example, further experimentation would be needed. In this case, a simple full factorial with hold pressure and melt temperature at two levels would probably be all that would be necessary. In cases where more factors are discovered to be significant, doing another set of 8 runs could sort things out. A partial reflection or a foldover on just one column would give three equations with three unknowns for each column and make the solution for each interaction possible. Table 8.18 shows the contrast coefficients and confounding pattern if only folding on factor B from the original design.

Table 8.18 Confounding Table after the Fold of Factor B in the Original DOE

Run	A –BD CE	B AD CF	C AE –BF	D –AB EF	E AC DF	F –BC DE	AF –BE CD
1	+1	–1	+1	+1	+1	+1	+1
2	+1	–1	–1	+1	–1	–1	–1
3	+1	+1	–1	–1	–1	+1	+1
4	+1	+1	+1	–1	+1	–1	–1
5	–1	+1	+1	+1	–1	–1	+1
6	–1	+1	–1	+1	+1	+1	–1
7	–1	–1	–1	–1	+1	–1	+1
8	–1	–1	+1	–1	–1	+1	–1

Had the DOE defined in Table 8.18 been carried out for Example 8.3, the following set of three equations would allow the solution of each estimated effect.

Original experiment:

$$\hat{l}_F + \hat{l}_{BC} + \hat{l}_{DE} = -0.0422 \text{ inches}$$

Complete foldover:

$$\hat{l}_F - \hat{l}_{BC} - \hat{l}_{DE} = 0.0397 \text{ inches}$$

Fold on B only:

$$\hat{l}_F - \hat{l}_{BC} + \hat{l}_{DE} = ? \text{ inches}$$

8.1.4 Example 8.4: Foldover Designs in Minitab

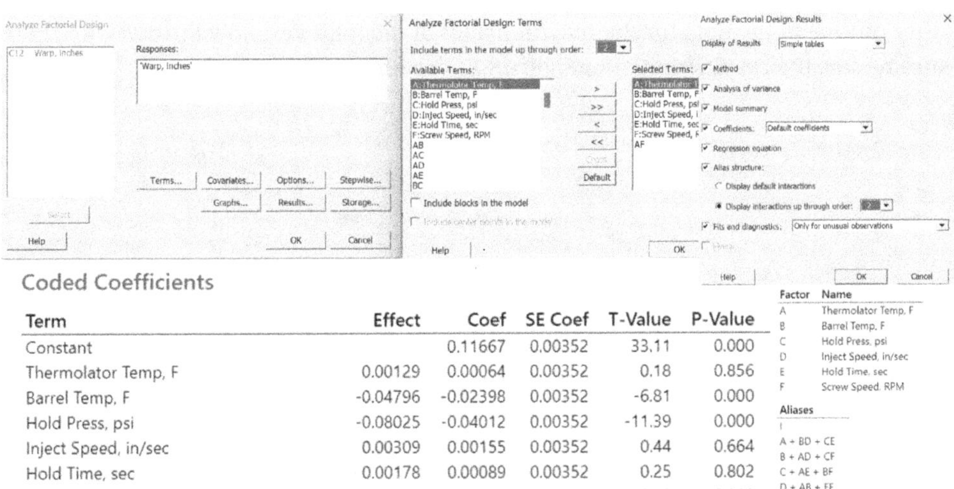

Coded Coefficients

Term	Effect	Coef	SE Coef	T-Value	P-Value
Constant		0.11667	0.00352	33.11	0.000
Thermolator Temp, F	0.00129	0.00064	0.00352	0.18	0.856
Barrel Temp, F	-0.04796	-0.02398	0.00352	-6.81	0.000
Hold Press, psi	-0.08025	-0.04012	0.00352	-11.39	0.000
Inject Speed, in/sec	0.00309	0.00155	0.00352	0.44	0.664
Hold Time, sec	0.00178	0.00089	0.00352	0.25	0.802
Screw Speed, RPM	-0.04215	-0.02108	0.00352	-5.98	0.000
Thermolator Temp, F*Screw Speed, RPM	-0.00062	-0.00031	0.00352	-0.09	0.930

Factor	Name
A	Thermolator Temp, F
B	Barrel Temp, F
C	Hold Press, psi
D	Inject Speed, in/sec
E	Hold Time, sec
F	Screw Speed, RPM

Aliases

I
A + BD + CE
B + AD + CF
C + AE + BF
D + AB + EF
E + AC + DF
F + BC + DE
AF + BE + CD

Figure 8.9 Original DOE from Example 8.3 analyzed in Minitab

Figure 8.9 shows the original DOE from Example 8.3 analyzed in Minitab. The estimated effects and the alias structure are the same as shown earlier. Figure 8.10 shows how to set up the foldover design in Minitab.

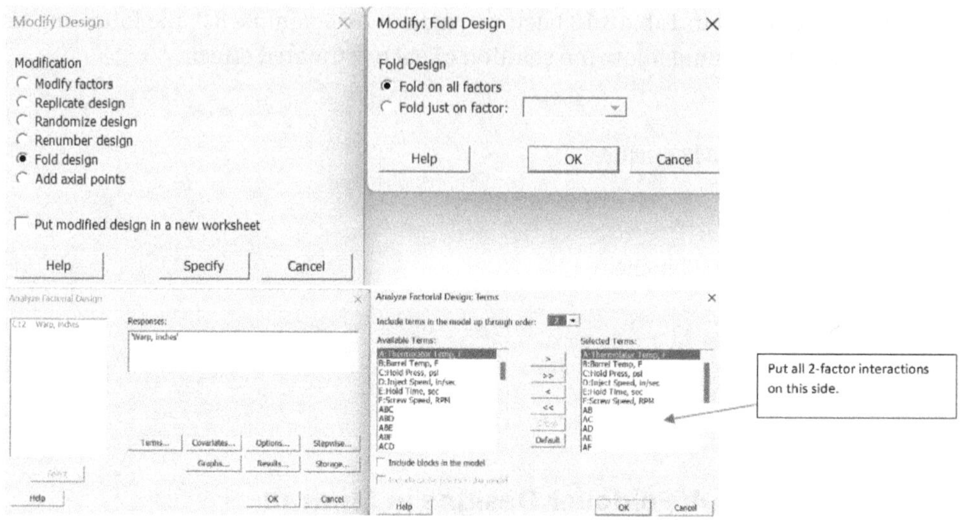

Figure 8.10 Sequence for setting up foldover DOE in Minitab

After putting in the data for the five replicates of the foldover (now 80 total warp measurements), the results shown in Figure 8.11 are obtained.

Coded Coefficients

Term	Effect	Coef	SE Coef	T-Value	P-Value
Constant		0.11872	0.00204	58.26	0.000
Thermolator Temp, F	-0.00133	-0.00067	0.00204	-0.33	0.745
Barrel Temp, F	-0.05098	-0.02549	0.00204	-12.51	0.000
Hold Press, psi	-0.08300	-0.04150	0.00204	-20.36	0.000
Inject Speed, in/sec	-0.00120	-0.00060	0.00204	-0.30	0.768
Hold Time, sec	-0.00223	-0.00112	0.00204	-0.55	0.585
Screw Speed, RPM	-0.00123	-0.00061	0.00204	-0.30	0.765
Thermolator Temp, F*Barrel Temp, F	0.00430	0.00215	0.00204	1.05	0.296
Thermolator Temp, F*Hold Press, psi	0.00401	0.00201	0.00204	0.99	0.328
Thermolator Temp, F*Inject Speed, in/sec	0.00302	0.00151	0.00204	0.74	0.461
Thermolator Temp, F*Hold Time, sec	0.00275	0.00138	0.00204	0.67	0.502
Thermolator Temp, F*Screw Speed, RPM	-0.00596	-0.00298	0.00204	-1.46	0.148
Barrel Temp, F*Hold Press, psi	-0.04093	-0.02046	0.00204	-10.04	0.000
Barrel Temp, F*Inject Speed, in/sec	0.00262	0.00131	0.00204	0.64	0.523

Factor	Name
A	Thermolator Temp, F
B	Barrel Temp, F
C	Hold Press, psi
D	Inject Speed, in/sec
E	Hold Time, sec
F	Screw Speed, RPM

Aliases
I
A
B
C
D
E
F
AB + EF
AC + DF
AD + CF
AE + BF
AF + BE + CD
BC + DE

Figure 8.11 Original and foldover DOEs from Example 8.3 analyzed in Minitab

There are two things that should be noted about the results shown in Figure 8.16.

The estimated effects for the main factors and 2-factor interactions are the same as those found in Example 8.3. This is the first of the two things.

Minitab always lists only one effect or interaction, the first that is listed in the alias structure. This is the second of the two things. It does show that the melt temp/hold press (BC) interaction is the one that is significant. However, note that the results do not list the inject speed/hold time (DE) interaction. Therefore, it is important to always check what other aliases could be responsible for each effect before making any conclusions. Note that if the alias structure was not checked in Figure 8.9, it would have been concluded that screw speed had a significant effect on warpage.

It should also be noted that the elimination of terms in the regression equation and the generation of graphs that were done for the full factorials in Chapter 7 can also be done with the fractional factorials.

8.2 Response Surface Designs

When investigating 3-levels or higher for three or more factors, the number of runs is usually not practical for injection molding experiments. For four factors, eighty-one runs (3^4) would be required to do a full factorial with 3 levels for each factor. Two popular designs are Box-Behnken (BB) and Box-Wilson Central Composite Designs (CCD). They both allow estimation of main effects, two-factor interactions and quadratic effects. As can be seen in Figure 8.12, these designs also require a number of runs that is not too practical if more than five factors are included.

Create Response Surface Design: Display Available Designs

Available Response Surface Designs

Design		Continuous Factors								
		2	3	4	5	6	7	8	9	10
Central composite full	unblocked	13	20	31	52	90	152			
	blocked	14	20	30	54	90	160			
Central composite half	unblocked				32	53	88	154		
	blocked				33	54	90	160		
Central composite quarter	unblocked							90	156	
	blocked							90	160	
Central composite eighth	unblocked									158
	blocked									160
Box-Behnken	unblocked		15	27	46	54	62		130	170
	blocked			27	46	54	62		130	170

Help OK

Figure 8.12 Minitab overview of the number of runs for various response surface designs

For 3 factors, a cube serves as a good graphical example of the differences in these designs. Example 7.3 was a full factorial experiment with three factors at 2 levels each. The 8 runs required for this experiment were the corner points on the cube, as shown in Figure 8.13.

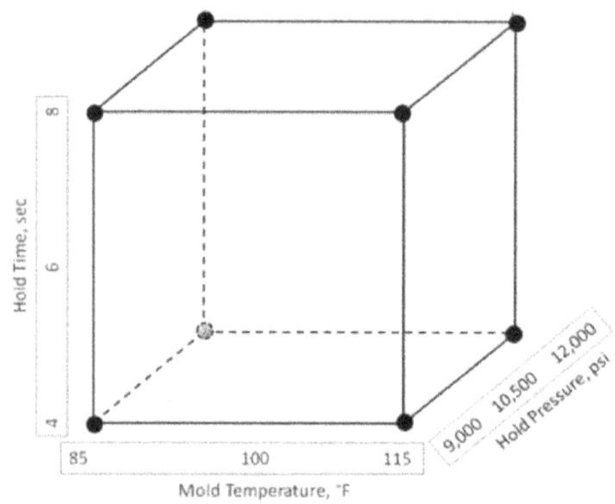

Run	MT, °F	HT, sec	HP, psi
1	115 (+)	8 (+)	12,000 (+)
2	115 (+)	8 (+)	9,000 (-)
3	115 (+)	4 (-)	9,000 (-)
4	115 (+)	4 (-)	12,000 (+)
5	85 (-)	4 (-)	12,000 (+)
6	85 (-)	4 (-)	9,000 (-)
7	85 (-)	8 (+)	9,000 (-)
8	85 (-)	8 (+)	12,000 (+)

Figure 8.13 Graphical representation of an 8-run full factorial DOE (Example 7.1)

Example 7.7 was the same experiment as 7.3, with the addition of three center-point runs. The center-point runs were done to give a measure of experimental noise when only one replication was collected for each run. In addition, the center point was used to check for the possibility of non-linearity in the effect of any of the factors on the response (part weight in this case). The position of the center point(s) in the cube is the exact center, as shown in Figure 8.19.

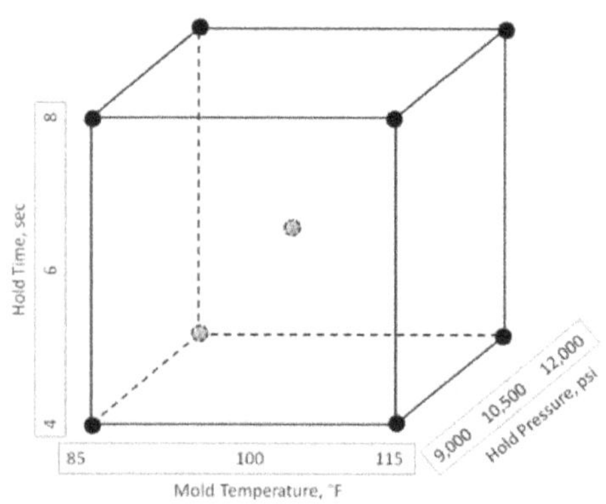

Run	MT, °F	HT, sec	HP, psi
2	115 (+)	8 (+)	12,000 (+)
3	115 (+)	8 (+)	9,000 (-)
4	115 (+)	4 (-)	9,000 (-)
5	115 (+)	4 (-)	12,000 (+)
7	85 (-)	4 (-)	12,000 (+)
8	85 (-)	4 (-)	9,000 (-)
9	85 (-)	8 (+)	9,000 (-)
10	85 (-)	8 (+)	12,000 (+)
1, 6, & 11	100 (0)	6 (0)	10,500 (0)

Figure 8.14 Graphical representation of an 8-run full factorial DOE with center-point runs (Example 7.7)

An advantage of the CCD designs over the BB designs is that some of the runs can be from a factorial design that was run with center points. If center-point runs show non-linearity, then the remainder of the CCD can be carried out to determine which factors demonstrated a quadratic effect. The CCD shown in Figure 8.15 shows the additional 6 points that would be needed to complete the experiment (more center-point runs would also be made). It should be noted that the additional 6 points in Figure 8.15 are shown in the center of each of the 6 faces of the cube. This is what is called a face-centered CCD. This is not the default in Minitab. The default would have the points outside the cube (at a distance equal to the number of runs in the factorial portion of the experiment raised to the ¼ power, $8^{1/4}$ or 1.682 in this example). With injection molding DOEs there is usually a risk of some of these parts being outside the region of operability (may result in short shots or flash). To decrease the risk and to reduce the complexity of the experiment, this author almost always recommends face-centered CCDs.

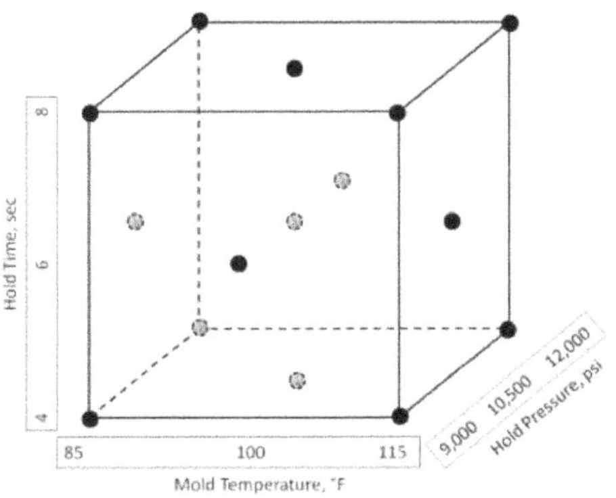

Run	MT, °F	HT, sec	HP, psi
2	115 (+)	8 (+)	12,000 (+)
3	115 (+)	8 (+)	9,000 (-)
4	115 (+)	4 (-)	9,000 (-)
5	115 (+)	4 (-)	12,000 (+)
7	85 (-)	4 (-)	12,000 (+)
8	85 (-)	4 (-)	9,000 (-)
9	85 (-)	8 (+)	9,000 (-)
10	85 (-)	8 (+)	12,000 (+)
12	100 (0)	6 (0)	9,000 (-)
13	100 (0)	6 (0)	12,000 (+)
15	100 (0)	4 (-)	10,500 (0)
16	100 (0)	8 (+)	10,500 (0)
18	85 (-)	6 (0)	10,500 (0)
20	115 (+)	6 (0)	10,500 (0)
1, 6, 11, 14, 17, & 19	100 (0)	6 (0)	10,500 (0)

Figure 8.15 Graphical representation of face-centered CCD runs (Example 8.5)

As can be seen in Figure 8.16, the BB designs do not contain any of the corner points. These designs would more likely be used when the idea is to do a three-level design from the beginning, not to add onto a factorial design like with the CCD, as they can give the same information with fewer runs in some cases (see Figure 8.12).

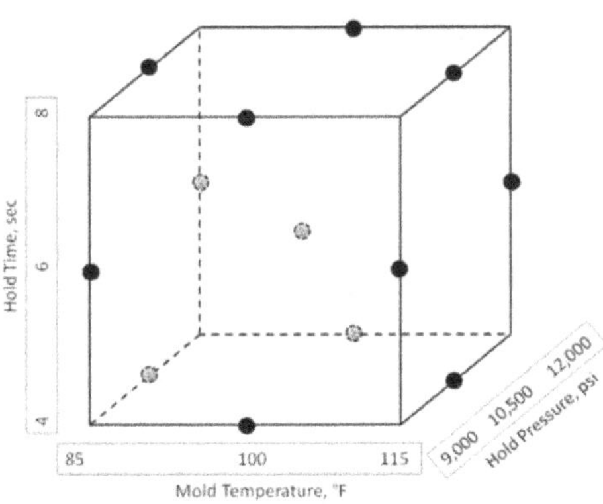

Run	MT, °F	HT, sec	HP, psi
1	85 (-)	4 (-)	10,500 (0)
2	85 (-)	8 (+)	10,500 (0)
3	85 (-)	6 (0)	9,000 (-)
4	85 (-)	6 (0)	12,000 (+)
6	100 (0)	4 (-)	9,000 (-)
7	100 (0)	4 (-)	12,000 (+)
9	100 (0)	8 (+)	9,000 (-)
10	100 (0)	8 (+)	12,000 (+)
12	115 (+)	4 (-)	10,500 (0)
13	115 (+)	8 (+)	10,500 (0)
14	115 (+)	6 (0)	9,000 (-)
15	115 (+)	6 (0)	12,000 (+)
5, 8, & 11	100 (0)	6 (0)	10,500 (0)

Figure 8.16 Graphical representation of BB design runs

The following example will review the use of a CCD by continuing the part-weight example from Chapter 7. This design will be a continuation of Example 7.7, with 11 of its 20 runs coming from that example. An additional 9 runs will allow determination of the individual quadratic effects. The BB designs and analysis are very similar to the CCD, so a separate example will not be shown here.

8.2.1 Example 8.5: CCD for 3 Factors at 3 Levels in Excel and Minitab

The CCD design, results and contrast coefficients used to calculate the estimated effects are shown in Table 8.19. Note that each quadratic effect estimate is based on comparing the average of the center-point runs with the average of the two runs when the factor in question was at the high and low levels.

Table 8.19 CCD Contrast Coefficients and Results for Example 8.5

Run	A	B	C	AB	AC	BC	A^2	B^2	C^2	Weight, g
1	0	0	0	0	0	0	−1	−1	−1	7.523
2	+1	+1	+1	+1	+1	+1	0	0	0	7.588
3	+1	+1	−1	+1	−1	−1	0	0	0	7.492
4	+1	−1	−1	−1	−1	+1	0	0	0	7.330
5	+1	−1	+1	−1	+1	−1	0	0	0	7.420
6	0	0	0	0	0	0	−1	−1	−1	7.524
7	−1	−1	+1	+1	−1	−1	0	0	0	7.510
8	−1	−1	−1	+1	+1	+1	0	0	0	7.414
9	−1	+1	−1	−1	+1	−1	0	0	0	7.504
10	−1	+1	+1	−1	−1	+1	0	0	0	7.602
11	0	0	0	0	0	0	−1	−1	−1	7.531
12	0	0	−1	0	0	0	0	0	+3	7.476
13	0	0	+1	0	0	0	0	0	+3	7.570
14	0	0	0	0	0	0	−1	−1	−1	7.534
15	0	−1	0	0	0	0	0	+3	0	7.421
16	0	+1	0	0	0	0	0	+3	0	7.545
17	0	0	0	0	0	0	−1	−1	−1	7.531

Run	A	B	C	AB	AC	BC	A^2	B^2	C^2	Weight, g
18	−1	0	0	0	0	0	+3	0	0	7.558
19	0	0	0	0	0	0	−1	−1	−1	7.523
20	+1	0	0	0	0	0	+3	0	0	7.495

The estimated effects, sums of squares, and F-ratios can be calculated as was done in Chapter 7, with the results shown in Table 8.20:

Table 8.20 Estimated Effects, Sums of Squares and F-Ratios for Example 8.5

Effect	Estimated Effect, g	SS, g^2	F-Ratio
A	−0.0526	6.92E-03	289.8
B	0.1272	4.04E-02	1694.8
C	0.0948	2.25E-02	941.4
AB	0.0370	2.74E-03	114.7
AC	−0.0020	8.00E-06	0.3
BC	0.0020	8.00E-06	0.3
A2	−0.0012	2.04E-06	0.1
B^2	−0.0447	2.99E-03	125.4
C^2	−0.0047	3.27E-05	1.4

In this case, the sample variance, s^2, of the six center-point replicates serves as the denominator in the F-ratio equation. Since there are six replicates, there are five degrees of freedom and the accept/reject value is $F_{0.05,1,5} = 6.61$. It can be clearly seen that the only significant quadratic effect is B^2, hold time, with an F-ratio of 125.4. The other two possible quadratic effects are shown here to be insignificant. The other significant effects are also the same as was found in Chapter 7. Again, a good way to show this quadratic effect graphically is as shown in Figure 7.24. For this example, the low hold-time value is from run 15, the high hold-time value is from run 16, and the midpoint is the average of the six center-point runs. Figure 8.17 is a graph of these values and is nearly identical in shape to Figure 7.24. However, the values for each of the three points are lower in Figure 8.17, due to the settings of the other two factors.

Using Minitab, the following screen shots show the sequence for setting up and carrying out this experiment. It should be noted that this shows setting up the experiment from scratch. How to add the axial points to the original factorial with center points (Example 7.7) will be shown in Example 8.6.

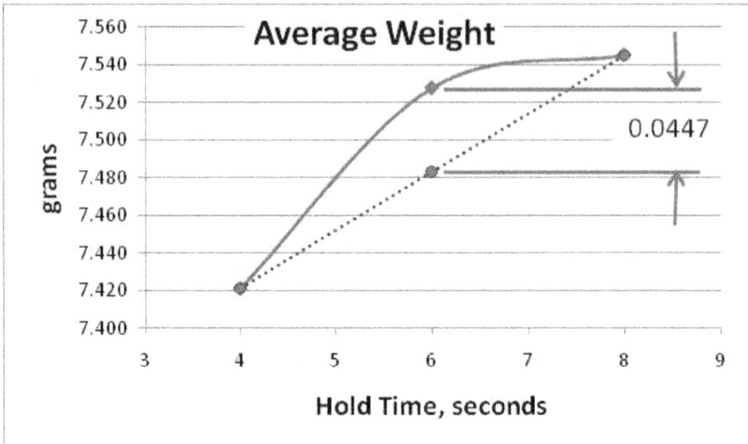

Figure 8.17 Hold-time quadratic effect for Example 8.5

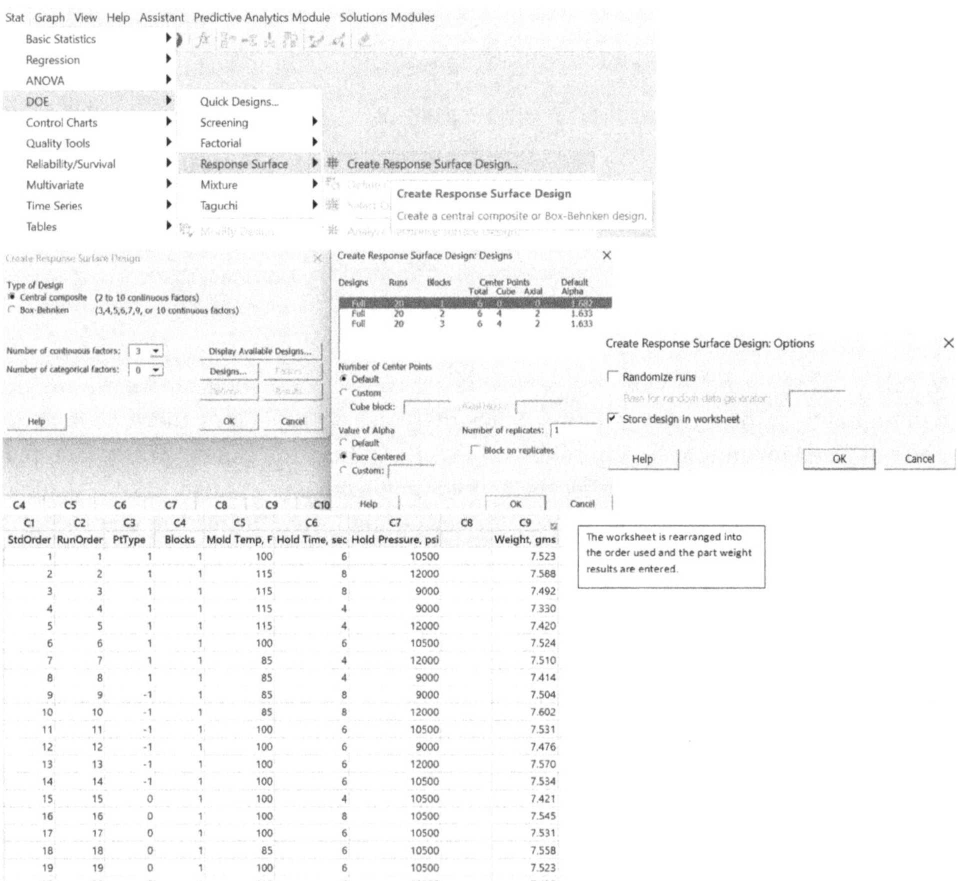

Figure 8.18 Sequence for setting up response surface DOE in Minitab for Example 8.5

An analysis was done as shown in Figure 8.19:

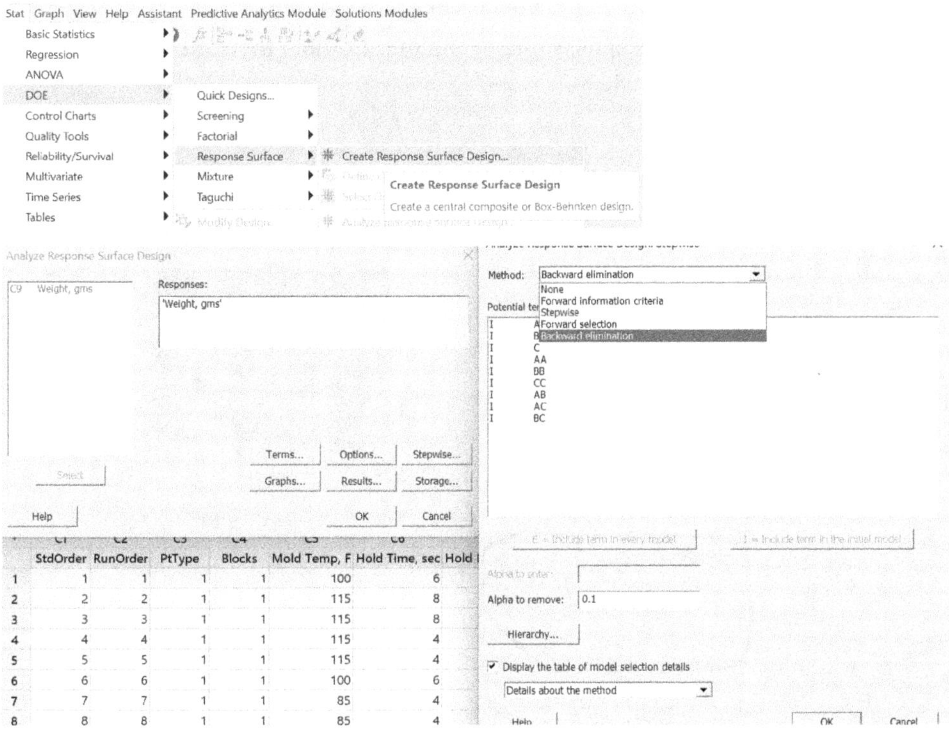

Figure 8.19 Sequence for analyzing response surface DOE in Minitab for Example 8.5

Since the stepwise regression was selected, the results show only the significant terms, and the regression equation is in uncoded units (Figure 8.20).

Coded Coefficients

Term	Coef	SE Coef	T-Value	P-Value
Constant	7.52650	0.00132	5691.82	0.000
Mold Temp, F	-0.02630	0.00132	-19.89	0.000
Hold Time, sec	0.06360	0.00132	48.10	0.000
Hold Pressure, psi	0.04740	0.00132	35.85	0.000
Hold Time, sec*Hold Time, sec	-0.04390	0.00187	-23.48	0.000
Mold Temp, F*Hold Time, sec	0.01850	0.00148	12.51	0.000

Regression Equation in Uncoded Units

Weight, gms = 7.1541 - 0.005453 Mold Temp, F + 0.10183 Hold Time, sec
+ 0.000031600 Hold Pressure, psi - 0.010975 Hold Time, sec*Hold Time, sec
+ 0.0006167 Mold Temp, F*Hold Time, sec

Figure 8.20 Example 8.5 regression results

Note that the response surface analysis does not give an "effect" column. As noted earlier, the estimated effect of the main factors (mold temp, hold time, and hold pressure) is just double the "Coef" in the Coded Coefficients table.

For the response surface designs, some of the output plots that can be generated in Minitab are shown in Figure 8.21.

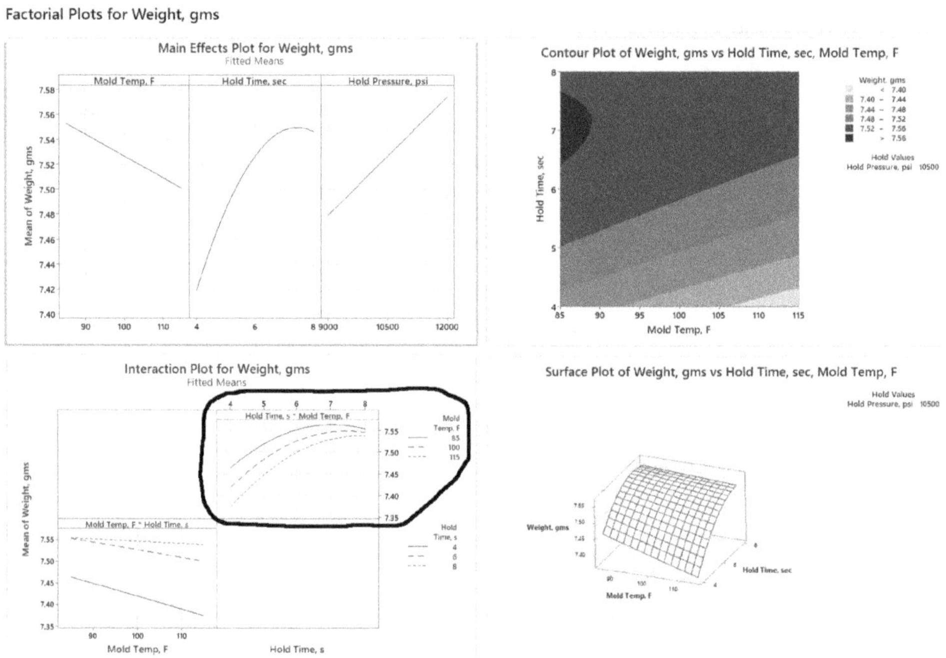

Figure 8.21 Factorial, surface and contour plots for mold temperature and hold time for Example 8.5

In this case, this author thinks that the circled interaction plot in Figure 8.21 shows both the curvature in hold time and interaction between hold time and mold temp more clearly than do either the contour plot or the surface plot. The factorial plots and the interaction plot can be used to explain the effects on part weight well.

The following example shows how the response surface design could have been created by adding axial points after the factorial experiment with center points, Example 7.7, revealed the presence of a curvature effect.

8.2.2 Example 8.6: Adding Axial Points in Minitab

Add the six axial points and three more center points as shown in Figure 8.22.

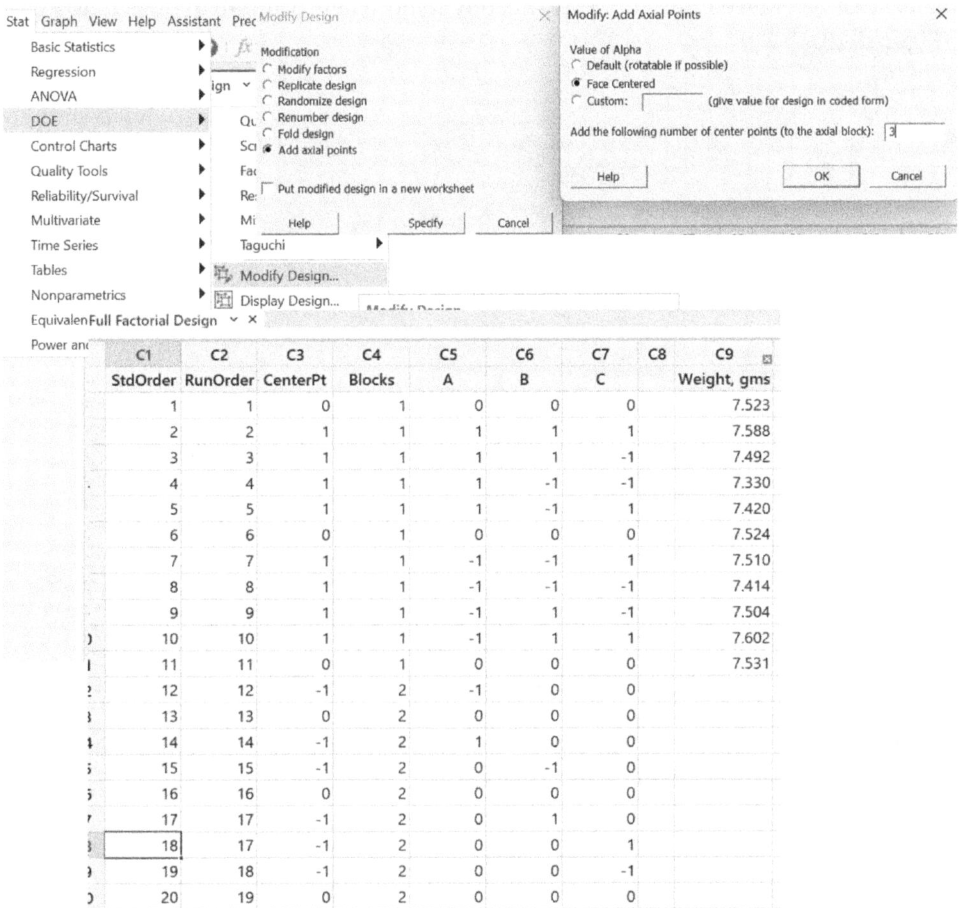

C1 StdOrder	C2 RunOrder	C3 CenterPt	C4 Blocks	C5 A	C6 B	C7 C	C8	C9 Weight, gms
1	1	0	1	0	0	0		7.523
2	2	1	1	1	1	1		7.588
3	3	1	1	1	1	-1		7.492
4	4	1	1	1	-1	-1		7.330
5	5	1	1	1	-1	1		7.420
6	6	0	1	0	0	0		7.524
7	7	1	1	-1	-1	1		7.510
8	8	1	1	-1	-1	-1		7.414
9	9	1	1	-1	1	-1		7.504
10	10	1	1	-1	1	1		7.602
11	11	0	1	0	0	0		7.531
12	12	-1	2	-1	0	0		
13	13	0	2	0	0	0		
14	14	-1	2	1	0	0		
15	15	-1	2	0	-1	0		
16	16	0	2	0	0	0		
17	17	-1	2	0	1	0		
18	17	-1	2	0	0	1		
19	18	-1	2	0	0	-1		
20	19	0	2	0	0	0		

Figure 8.22 Sequence for adding axial points to factorial design

After doing the additional runs, the analysis would be the same as Example 8.5.

When creating DOEs with more than two levels for each factor, a custom response surface design can be used. Example 8.7 will show how to set up and analyze the full-factorial, three-level experiment shown in Example 7.4.

8.2.3 Example 8.7: Creating a Custom Response Surface Design for 2 Factors at Three Levels

The following shows how to create a custom 3^2 design as a response surface. First, the runs must be entered into the worksheet by hand. Then, Minitab defines it as a DOE (Figure 8.23).

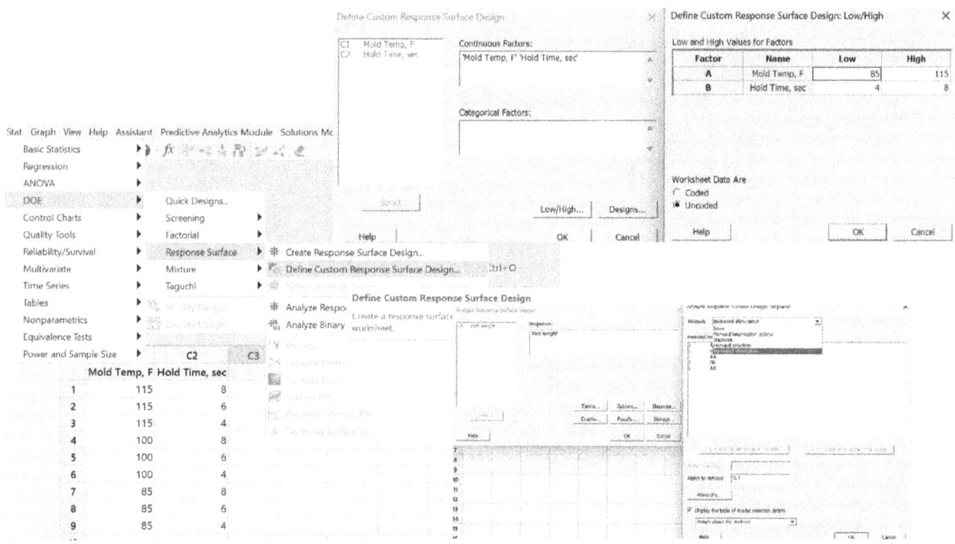

	Mold Temp, F	Hold Time, sec
1	115	8
2	115	6
3	115	4
4	100	8
5	100	6
6	100	4
7	85	8
8	85	6
9	85	4

Figure 8.23 Sequence for creating a custom response surface design in Minitab

The DOE is then modified to add the 4 replicates and the part-weight data from Example 7.4 are entered and then analyzed as shown in Figure 8.24:

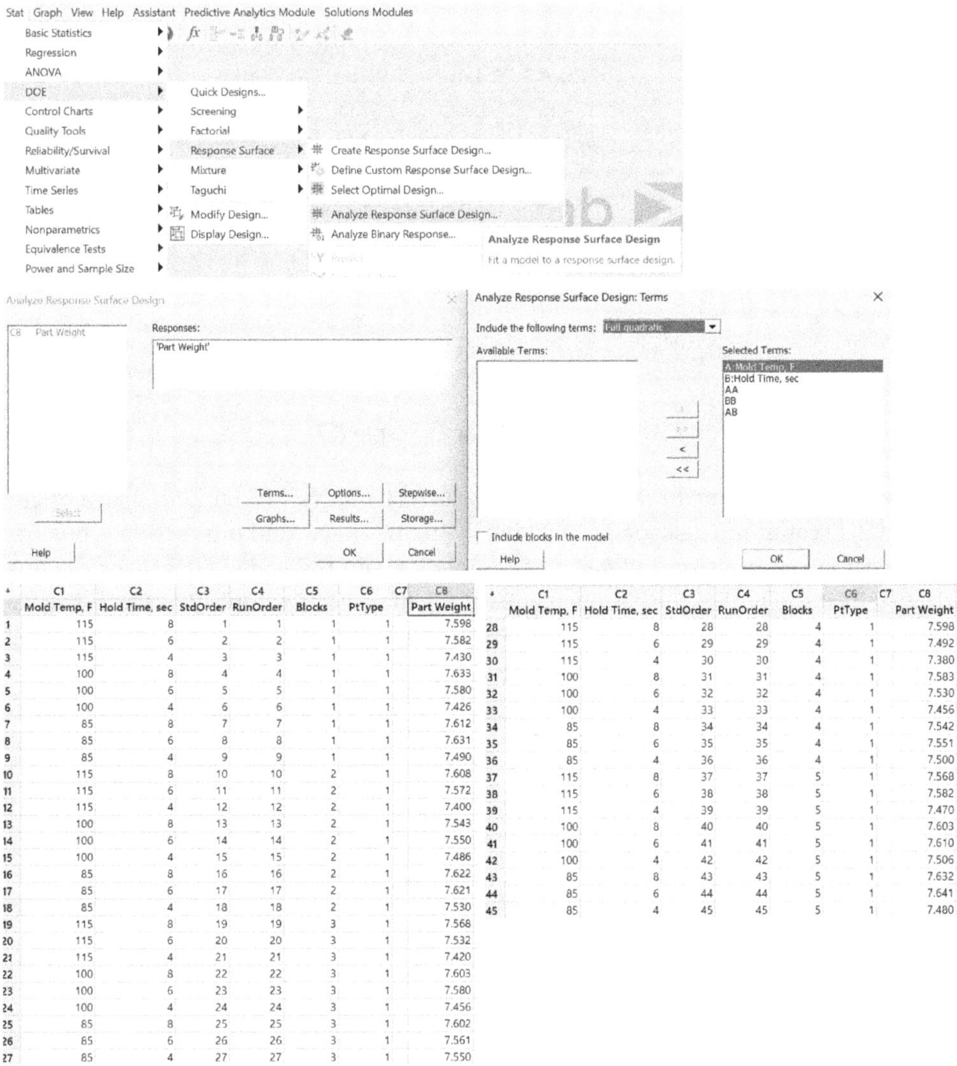

	C1 Mold Temp, F	C2 Hold Time, sec	C3 StdOrder	C4 RunOrder	C5 Blocks	C6 PtType	C7	C8 Part Weight
1	115	8	1	1	1	1		7.598
2	115	6	2	2	1	1		7.582
3	115	4	3	3	1	1		7.430
4	100	8	4	4	1	1		7.633
5	100	6	5	5	1	1		7.580
6	100	4	6	6	1	1		7.426
7	85	8	7	7	1	1		7.612
8	85	6	8	8	1	1		7.631
9	85	4	9	9	1	1		7.490
10	115	8	10	10	2	1		7.608
11	115	6	11	11	2	1		7.572
12	115	4	12	12	2	1		7.400
13	100	8	13	13	2	1		7.543
14	100	6	14	14	2	1		7.550
15	100	4	15	15	2	1		7.486
16	85	8	16	16	2	1		7.622
17	85	6	17	17	2	1		7.621
18	85	4	18	18	2	1		7.530
19	115	8	19	19	3	1		7.568
20	115	6	20	20	3	1		7.532
21	115	4	21	21	3	1		7.420
22	100	8	22	22	3	1		7.603
23	100	6	23	23	3	1		7.580
24	100	4	24	24	3	1		7.456
25	85	8	25	25	3	1		7.602
26	85	6	26	26	3	1		7.561
27	85	4	27	27	3	1		7.550
28	115	8	28	28	4	1		7.598
29	115	6	29	29	4	1		7.492
30	115	4	30	30	4	1		7.380
31	100	8	31	31	4	1		7.583
32	100	6	32	32	4	1		7.530
33	100	4	33	33	4	1		7.456
34	85	8	34	34	4	1		7.542
35	85	6	35	35	4	1		7.551
36	85	4	36	36	4	1		7.500
37	115	8	37	37	5	1		7.568
38	115	6	38	38	5	1		7.582
39	115	4	39	39	5	1		7.470
40	100	8	40	40	5	1		7.603
41	100	6	41	41	5	1		7.610
42	100	4	42	42	5	1		7.506
43	85	8	43	43	5	1		7.632
44	85	6	44	44	5	1		7.641
45	85	4	45	45	5	1		7.480

Figure 8.24 Sequence for analyzing the data from Example 8.7 in Minitab

The following results are obtained (Figure 8.25):

Coded Coefficients

Term	Coef	SE Coef	T-Value	P-Value
Constant	7.57433	0.00814	930.31	0.000
Mold Temp, F	-0.02550	0.00576	-4.43	0.000
Hold Time, sec	0.06450	0.00576	11.20	0.000
Hold Time, sec*Hold Time, sec	-0.04450	0.00997	-4.46	0.000
Mold Temp, F*Hold Time, sec	0.01900	0.00705	2.69	0.010

Regression Equation in Uncoded Units

Part Weight = 7.530 - 0.00550 Mold Temp, F + 0.1024 Hold Time, sec
- 0.01113 Hold Time, sec*Hold Time, sec + 0.000633 Mold Temp, F*Hold Time, sec

Figure 8.25 Regression results in Minitab for Example 8.7

It should again be noted that the response surface analysis in Minitab does not give the "effect" column. The coded "Coef" for the main effects and interaction is one half of the estimated effect calculated in Example 7.4 and the coded "Coef" for the quadratic effect is equal to the estimated effect calculated in Example 7.4. Again, this author prefers the factorial plots shown in Figure 8.26 when discussing the factor effects.

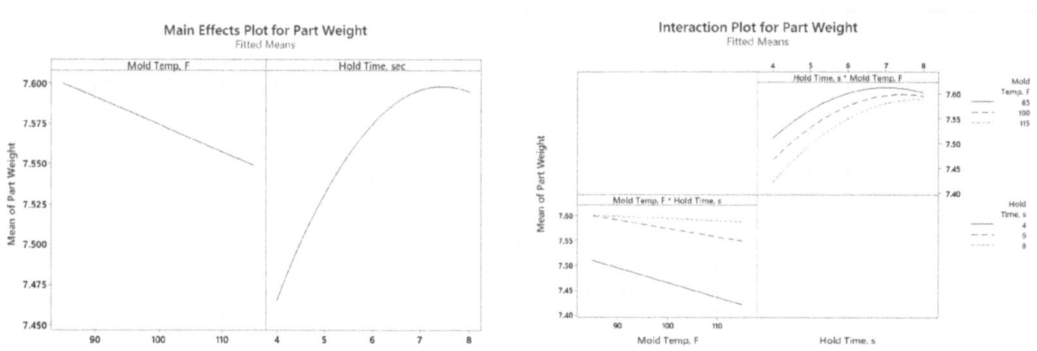

Figure 8.26 Factorial plots from Minitab for Example 8.7

9 Planning and Carrying Out Design of Experiments

Chapters 7 and 8 reviewed some different types of DOEs and how to do the data analysis for each. Prior to even beginning a DOE, a few important questions need to be answered, including:

1. What is the objective of the experiment?
2. What factors should be included in the experiment?
3. How are "non-factors" held constant?
4. What levels should be used for each factor?
5. What responses should be measured and when should they be measured?
6. What process data should be collected while carrying out the DOE?
7. How many parts should be collected?
8. Should replicate runs be made or just repetitions?
9. Should runs be randomized?
10. How much machine time is needed to plan the DOE and to carry out the DOE?
11. Will follow-up studies be needed?

9.1 What is the Objective of the Experiment?

When performing DOEs during the mold development phase, there are several valid objectives that could be stated for every experiment that is done.

- Determine the sensitivity of CTQ characteristics to input factors
- Determine the best or, more likely, better settings for the input factors
- Determine what part or mold design modifications, if any, should be made

If a DOE is deemed to be necessary after the development stage has ended and production has begun, it is usually due to an unforeseen and chronic quality issue that needs to be resolved. DOEs can sometimes be useful for getting at the root cause of issues when typical troubleshooting methods don't, since DOEs can pick up interactions. One caution when such a situation arises is to beware of mission creep. Because of the visibility within an organization (and the presence of individuals not normally involved in setting up injection molding processes) that is created by chronic quality problems, there can be a tendency to try and solve issues that aren't really part of the original problem. When DOE is used as a retroactive troubleshooting tool, it is usually wise to stay very focused on the specific issue that needs to be resolved.

9.2 What Factors Should Be Included in the Experiment?

Some of the typical factors that are included in an injection molding experiment and why they are included are discussed here. It should be noted that some of these factors are not ones that you will want to change to keep the process under control once the mold is released to production. However, better settings than the baseline may be found and learning about the sensitivity of CTQ characteristics may dictate tighter control and/or monitoring of parameters related to those factors. For example, if a CTQ characteristic is very sensitive to changes in mold temperature, putting a thermocouple on or close to the mold surface for monitoring purposes would be prudent. If a CTQ characteristic is sensitive to the %-full at V/P transfer, a method of V/P transfer that is very repeatable should be used (discussed in Chapter 11). Readers not familiar with some of the terminology may want to refer to the Process Set-Up chapter for clarification.

1. Hold pressure(s)
2. Hold time(s)
3. Peak cavity pressure (D2)
4. Injection pressure(s)
5. Injection time(s) or screw travel
6. Barrel and hot-runner temperature(s)
7. Mold temperature(s) or cooling-medium temperature(s)
8. Injection speed(s)

9. Pack speed(s)
10. Cycle time/cooling time
11. % Full at V/P transfer (D2)
12. % Full at V/V transfer (D3 or any injection speed profile)
13. V/P transfer (D1 or D3)
14. Back pressure
15. Screw rotation speed
16. Cushion
17. Barrel residence time
18. Clamp force
19. Ejection speed
20. Material viscosity

This is not a complete list. A few non-conventional factors were included to get the reader to think about the types of things that can be done. DOEs can be used to test out many types of material variations other than viscosity. Tooling variations such as the number of gates and gate locations lead into more-involved DOE set-ups. A sequenced valve gate mold can add several factors, as can the inserts for an insert-molded part. An imbalanced multi-cavity mold adds some complexity and will be discussed for some of the factors.

Hold Pressure(s)

It is very rare that an injection molding DOE would be done that does not include changing the hold pressure. Hold pressure affects the pressure inside the mold (cavity pressure) and the amount of plastic that enters the part cavity. Therefore, it influences most response variables. The simplest process will have only one hold pressure but there are instances where multiple pressures are needed, some of which are discussed in the Process Set-Up chapter. Depending on the reason for the multiple hold pressures, they could be treated as separate factors in the DOE, or they could be one factor that are all changed by the same amount between DOE levels.

Hold Time(s)

The time that the hold pressure is on can also affect the amount of plastic that enters the mold. If working with hot runner valve gates or other hot runners' gates that are operated for a time shorter than gate freeze, hold time is almost always a very important factor to include. For hot runners with gates that are designed to freeze and cold-runner molds, hold time sometimes is not included as a factor. In cases where it is not intended to be used as a factor, it can be set long enough to ensure that it is frozen for every run or the gate freeze time can be determined for each run and the hold time set higher (typically 1 to 3 seconds).

Peak Cavity Pressure

If the mold is instrumented with at least one cavity pressure sensor, peak cavity pressure can be used for a D2 process instead of hold pressure. This will avoid the issue of having four different cavity pressure peaks for two levels each of barrel temperature and hold pressure. It can also be useful with a small window of hold pressures that work for every run. However, peak cavity pressure is not a setting, and it is important that the process stabilizes so that the cavity pressure peak is within at least two standard deviations of the desired value for each run. This can add a considerable amount of time.

Injection Pressure(s)

This setting is just a safety limit in a typical D1, D2 or D3 process and should be set high enough so that it isn't a limiting factor. For an iMFLUX process or other pressure-controlled fill set-ups it can be. For those types of processes, the pressure setting will control the speed profile during fill. The speed will decrease as the plastic is pushed further into the cavity. Sometimes, especially with hot gates, it is necessary to start with a lower pressure to prevent jetting caused by a high initial velocity. The pressure can be increased after the flow front has formed.

Injection Time(s) or Screw Travel

Again, normally a safety limit and not a DOE factor. In cases where injection pressure is a factor, the time that each pressure is maintained can be a factor. As discussed in the Process Set-Up chapter, during fill, it is much better to control the duration of each pressure stage via screw travel rather than by time.

Barrel and Hot-Runner Temperature(s)

Barrel temperature affects melt temperature, which in turn affects material viscosity and material degradation in the barrel. The fact that the material starts at a higher or lower temperature will often affect shrinkage. Although temperature is a factor that takes a long time to stabilize and extends the duration of a DOE, this author typically recommends including most temperature factors in most injection molding DOEs. A high percentage of interactions that appear with injection molding factors involve temperature; consequently, sometimes operating at either a higher or lower temperature has a big effect on the quality of a part. The barrel of an injection unit will have several zones that can be set at different temperatures (more about this in Chapter

11). Typically, with a DOE, all the zones are treated as a single factor and each zone is raised by the same number of degrees to move from a low to a high level. In rare instances, a specific temperature might be treated as a separate factor. Nozzle or hot-drop temperatures may sometimes be separate factors.

Mold Temperature(s) or Cooling Medium Temperature(s)

The temperature of the mold or cooling medium has the greatest impact on the cooling rate of any process input factor. This is particularly important for semi-crystalline materials, where the cooling rate affects crystallinity, but can also be important for amorphous materials, due to different levels of frozen-in orientation at different cooling rates. Some molds are run with just one large cooling circuit, but it is standard practice at many molding facilities to run separate thermolators on each half of the mold. This author has seen molds operating with five separate thermolators hooked up to a mold. These separate thermolators can all be treated as separate factors in a DOE or, like barrel temperatures, they can all be raised by the same number of degrees to go from a low to a high level. Given that temperature gradients are a cause of warpage, different thermolators are likely to be set up as separate factors. Another important property that mold temperature can affect is the surface finish of the part. Typically, a hotter mold results in better replication of the mold surface. Although not mentioned as a separate factor here, coolant-medium flow rate can also affect cooling rate, especially if it is not turbulent flow. In this author's experience, the difference is negligible if each circuit's Reynold's number is above 4,000. However, there are usually exceptions that break any rule.

Injection Speed(s)

Injection speeds can influence molecular orientation and shear heating, both of which can affect shrinkage and other properties. In parts with variations in wall thickness, injection speed can have a big effect on the filling pattern. In a standard D2 process, there is only one injection speed. If multiple speeds must be used during fill, either they can be treated as one factor and all raised the same amount to go from a low to a high level or each speed could be treated as a separate factor.

Pack Speed(s)

If running a D3 process, a separate speed is used to pack out the part. This can affect the flow pattern of the last few percentages of fill and how the part packs.

Cycle Time/Cooling Time

Ideally, the cooling time is set just long enough for the plastic part to hold its shape upon ejection. However, the mold can act as a shrink fixture and sometimes the cooling time needs to be extended to reduce warpage or make dimensions larger. Care must be taken if the mold is used as a shrink fixture, since parts may return to their "un-fixtured" dimensions if they are exposed to high enough temperatures later. Also, although not listed here as a separate factor, shrink fixture use and geometry are sometimes used as DOE factors.

% Full at V/P Transfer (D2)

In a D2 process, transfer to hold typically occurs when the cavity is 95–98% full. Whether it is 95% or 98% can sometimes affect some part characteristics, especially those in areas near the last point to fill. Another method to treat the % full as a factor is to vary the decompression after screw rotation. Without decompression, the % full at transfer is typically lower, due to increased leakage before the nonreturn valve closes.

For an imbalanced multi-cavity mold, maybe one level would be when the first cavity to fill is 95% full and the other would be when the last cavity to fill is 95% full. This is not ideal if the first cavity to fill is starting to pack out before the last cavity is 95% full. Each cavity in a multi-cavity mold is a separate process and an imbalance can lead to significantly different DOE results for each individual cavity.

% Full at V/V Transfer (D3 or any injection speed profile)

In a D3 process, the part is typically 90–98% full when changing from fill velocity to pack velocity. The volume of the part that is filled at the lower velocity can affect part characteristics, even when the part is packed to the same V/P cavity pressure. The sensitivity to changes in % full at any change in velocity can serve as DOE factors. As with the previous factor, decompression can sometimes be adjusted to manipulate the % full. Also, the same issues associated multi-cavity molds as were described for the previous factor apply here.

V/P Transfer (D1 or D3)

As discussed in the Process Set-Up chapter, several different modes can be used to trigger the transfer from velocity control to pressure control. This transfer affects the pack pressure in both D1 and D3 processes, as well as any process in which transfer occurs during or at the end of the pack stage. Two levels of whatever transfer mode is used will result in two different levels of pack pressure.

Back Pressure

Adjusting the back pressure changes the density and consequently the shot weight while simultaneously increasing shear on the material. It will affect % full for the previous two factors and can affect how homogeneous the melt is.

Screw Rotation Speed

Like back pressure, adjusting the RPM will change the shear. However, unlike back pressure, the setting that increases shear decreases the shear time because screw recovery is quicker.

Cushion

A larger cushion will increase the residence time and can affect some thermally sensitive materials. Since "cushion" is not a setting, its two factor levels are usually established by adjusting the shot size and the V/P transfer position that give the same 95% full part. The variations in cushion resulting from other factors are generally not compensated for; the cushion is instead treated as a dynamic output variable.

Barrel Residence Time

This is not a common factor and, ideally, the mold will only run in machines with injection units of identical size and with an acceptable residence time (see the Process Set-Up chapter). This is a difficult factor to adjust without changing a lot of other process conditions. Perhaps the best method of changing this factor is to utilize two different injection units that can be swapped out easily. However, even if the injection units can be changed over in 15 minutes or less, there is always the issue of waiting for thermal stability. The simplest way to increase residence time is to increase mold open time. The problem with that is that it gives the mold more time to cool and reduces mold temperature. Mold open time could be a separate factor, especially if the part is insert-molded and the mold open time is operator controlled.

Clamp Force

This is not a common factor and clamp force should be set (see the Process Set-Up chapter) such that the amount that the mold opens during pack is undetectable in the parts. However, there are times when molds will be run in different size machines and clamp force does have an effect on part characteristics across the parting line.

Ejection Speed

Another uncommon factor. Occasionally, especially if the ejection system design and/or draft of the side walls of the mold are suboptimal, ejection can deform the part.

Material Viscosity

This is a significant variable once a mold starts running in production; therefore, determining the sensitivity of the critical quality characteristics to at least two different levels can provide useful information. Sometimes a facility will buy enough of the given material that they can save samples of higher and lower viscosity lots in order that those lots will be available for DOEs. Sometimes the material supplier will provide samples on each end of the viscosity spec. Another option is to pelletize 3rd or 4th generation regrind of the same material to reduce the viscosity.

9.3 How are "Non-Factors" Held Constant?

When planning a DOE, it is important to think about why factors should be included and how the changes in those factors should be made. Some examples of how the changing of one setting can affect more than one thing were mentioned when reviewing the factors. You should try to change only what you want to change (Yogi Berra didn't say it first). For example, if the effect of injection speed is being looked at, it is usually necessary to have a different V/P transfer for each so that the parts are the same % full when switching to hold pressure. This author has seen several DOEs where this was not done and injection speed was mistakenly identified as an important factor because, at the faster speed, the part was full and packed before transfer to hold pressure took place. If the effect of cooling time is being studied, it is important to remember that the part is cooling during hold time also. It is generally better to make cycle time the factor and change the cooling timer to get the desired cycle time in that case.

In addition to only changing what you want to change, it is important to keep everything else as constant as possible during the DOE. If operating in a shop that is cool in the morning and gets hot in the afternoon, instead of trying to control the shop temperature, the "time of day" could be made a factor – everyone who works in that shop roots for that factor to have a huge effect on the parts. If a material is being dried, it is important that the drying parameters and moisture level are kept as constant as possible. For small parts, this author has used mason jars to seal material directly from the dryer for use during a DOE.

9.4 What Levels Should Be Used for Each Factor?

For two-level experiments, some DOE practitioners recommend using a level below and a level above what the initial process is set up as. For injection molding, it is generally better to use the initial process setting for each factor as one of the two levels. The choice of levels is somewhat subjective. Some practitioners recommend setting the two levels as far apart as the equipment allows. That may work if nothing is permanently broken and your project timeline is infinitely long, but it is not practical for most factors. With wide ranges in the factors, several follow-up experiments are usually required to narrow down behavior around the final process conditions. The factor levels should be within the range of what will make visually acceptable parts and should be far enough apart to make a measurable difference in the response characteristics if they do have an effect. For example, making the low level of hold pressure 3000 psi and the high level 3001 psi would not be very useful. The following is the suggested method of determining the factor levels to use.

1. Set up the baseline process using low-level temperatures
2. Increase temperatures to the high levels and repeat the just-full and just-flash (hold pressure for D2) procedure
3. Choose hold pressures that are well within the window (between dashed lines) for each set of temperatures, as shown in Figure 9.1. In Figure 9.1 the window is 3,000 psi to 7,000 psi.

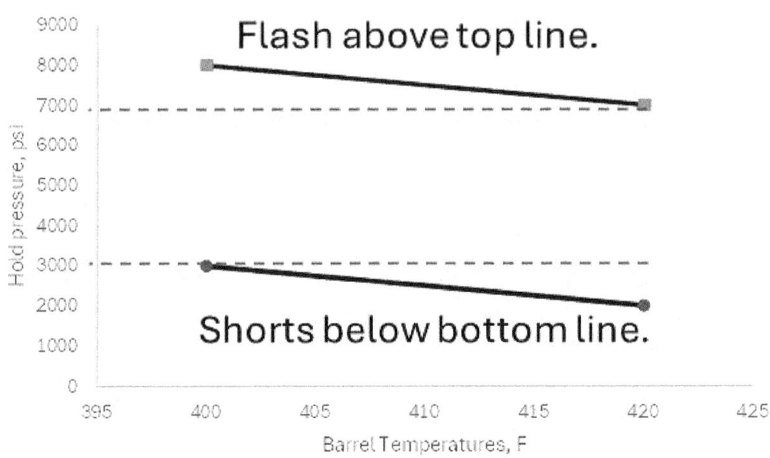

Figure 9.1 DOE window for hold pressure/barrel temperature

It may be necessary to try out the worst-case combinations of every factor for the settings most likely to give a short shot and the settings most likely to give flash, but if the window is not too narrow the preceding procedure is usually sufficient. The following are some guidelines recommended for setting levels for a DOE for some of the factors already discussed for a standard D2 process. There will be special cases where levels should be different than what are recommended here. Also, if the mold is intended to be run on several different machines in production, the speed and pressure settings used should be attainable on every machine.

Table 9.1 List of Potential DOE Factors and Guidelines for Selecting Levels

Hold pressure	At least 750 psi and not more than 2,000 psi difference in the high and low levels. If possible, at least 500 psi from the top and bottom of the window. For the example in Figure 9.1, 4,500 psi for the low and 6,000 psi for the high should work well.
Hold time	If included in a DOE, typically the standard safety factor above gate freeze time (see Process Set-Up chapter) is the high level and about 25% below gate freeze time is the low level. With valve gates, the low level must ensure that the parts are full and packed for a short time and the high level should be below where the gate freezes – usually only a couple of seconds between levels is needed.
Peak cavity pressure	The two peak pressures that correspond to the hold pressure levels recommended for the baseline process would work. It is typically not necessary to develop the window shown in Figure 9.1. 3,000 psi and 4,000 psi at the end of fill work nicely for a lot of parts.
Barrel and hot runner temperature(s)	The baseline process settings usually plus (sometimes minus) 10 to 30 degrees Fahrenheit. 20 degrees is generally recommended, but materials with a small process window may require using only 10 degrees.
Mold temperature(s) or cooling medium temperature(s)	Like barrel temperatures, a change of 20 degrees is usually sufficient. When molds are over 300 degrees Fahrenheit, a 40-degree difference is recommended.
Injection speed(s)	The speed from the baseline process set-up as one level and usually a faster speed as the other. Typically, a faster speed that cuts the fill time in half or a slower speed that doubles the fill time.
Cycle/cooling time	The low level will be the baseline minimum that gives visually acceptable parts and the high level be about 25% higher than the sum of the hold timer and the cooling timer.
% Full at V/P transfer	The baseline (95–98%) is the high and 85–90% is the low. Visual samples should be retained for both levels and for both injection speeds if that is a factor. A fill-only sample should be saved to compare at the start of each run.
Back pressure	The baseline as one level and 50% higher or lower as the other – usually higher.
Screw rotation speed	Typically, only included if the baseline is lower than 90% of the max RPM. Then, the two levels would be the baseline as the low and 90% of max as the high.

Cushion	The baseline as the low and about triple the baseline as the high.
Barrel residence time	Usually included as a factor if the baseline is less than 1½ minutes, which would be the low level. The high level should be about double and in the 2-to-4-minute range.
Material viscosity	If a D2 process, there should typically be a 10 to 20 % difference in the V/P transfer pressure for the baseline process.

9.5 What Responses Should Be Measured and When Should They Be Measured?

It is recommended that all CTQ characteristics and control characteristics be measured. Overall length, overall width and part weight are examples of characteristics that may not be CTQ, but may correlate with CTQ characteristics and be easier to measure during production. Also, sometimes a simple physical test that is not a specification can sometimes correlate with a more difficult to measure characteristic. Ideally, parts will be measured as they would be in production (typically about an hour after molding) and again at least 24 hours later (72 hours is better). Economics and equipment availability will enter the measuring decision. It may not be possible to measure everything an hour after molding and only a few control characteristics may be measurable. The sample size needed and whether the same samples can be used for all measurements will enter the decision also. Any destructive tests usually require larger sample sizes and then cannot be remeasured; therefore, the required sample size must be doubled if they are to be checked at two different times. In addition to the measuring time, large sample sizes can significantly lengthen the molding time.

9.6 What Process Data Should Be Collected While Carrying out the DOE?

It is always wise to document all the settings used for the baseline process and to save the process on the molding machine. Every cycle, other dependent process data from the molding machine that can be either printed as the parts are being molded or recorded by hand include:

1. Cushion
2. V/P transfer pressure

3. Max. pressure
4. Fill time
5. Screw recovery time
6. Cycle time

That list can be increased considerably if using an instrumented mold with things such as:

1. Time stamp
2. Peak cavity pressure
3. Cavity pressure integral
4. Cavity pressure decay rate
5. Minimum mold temperature
6. Mold temperature rise
7. Time to sensor

It is recommended that all the available data be collected during the DOE and that someone is looking at it as it is being generated. This author has personally witnessed several instances of complete DOEs being run before the realization was made that many of the runs had zero cushion – a good reason to always have a minimum cushion alarm on. The data can also be used to pick up mistakes that were made after the fact. A high injection speed having a long fill time or a low hold pressure having a high cavity pressure are two examples. With the time stamp, it is possible to replay process curves on some data acquisition systems. Lastly, correlations of process data with CTQ characteristics can help to set up process monitoring to detect bad parts during production, which is the subject of the next chapter.

9.7 How Many Parts Should Be Collected?

Time, as well as material, is money; therefore, the goal is always to mold and measure as few parts as possible to obtain all the information that is needed. The more parts that are molded and measured, the smaller the differences in the parts that can be detected when analyzing the results. As was found in the chapter on hypothesis testing (Chapter 6), the required sample size is determined by the ratio of the minimum difference you wish to detect to the standard deviation. Table 12.16 from the Appendix is also shown here for discussion.

Power is the probability of detecting a significant effect or difference when one truly exists. The β error is the probability of missing a significant effect. The α error is the probability of "detecting" a significant effect or difference that does not truly exist.

For factorial experiments, the total replicates equal the number of runs multiplied by the replicates per run. For $\alpha = 0.05$, Table 9.2 indicates how many total replicates are needed for different β levels:

Table 9.2 Sample Size Determination for Factorial DOEs (Table 12.16 in the Appendix)

Standard Deviations to Be Able to Detect	$\beta = 0.10$ Power = 0.90	$\beta = 0.20$ Power = 0.80	$\beta = 0.30$ Power = 0.70
0.25	676	508	400
0.50	172	128	104
0.75	80	60	48
1.00	48	36	28
1.50	24	20	16
2.00	16	12	12
2.50	12	12	8
3.00	12	8	4

For example, if historical data or a baseline study shows that a good estimate of standard deviation is 0.004 and it is desired to be able to detect a 0.003 difference, 0.003/0.004 = 0.75. For 90% power, 80 total replicates would be needed. That would be 20 replicates/run of a 4-run experiment, 10 replicates/run of an 8-run experiment, or 5 replicates/run of a 16-run experiment. For a 32-run experiment, either power would have to be sacrificed (2 replicates/run would give a total of 64, just over 80% power) or extra replicates would have to be made (3 replicates/run would give a total of 96, power > 90%).

If the only measurements are dimensions and other non-destructive tests, then saving one shot per replicate is generally sufficient. At least one shot per destructive test, more if to be measured at more than one time after molding, needs to be collected.

If ratios not found in the left-hand column are needed, Figure 9.2 shows how some of the numbers in the table were found in Minitab.

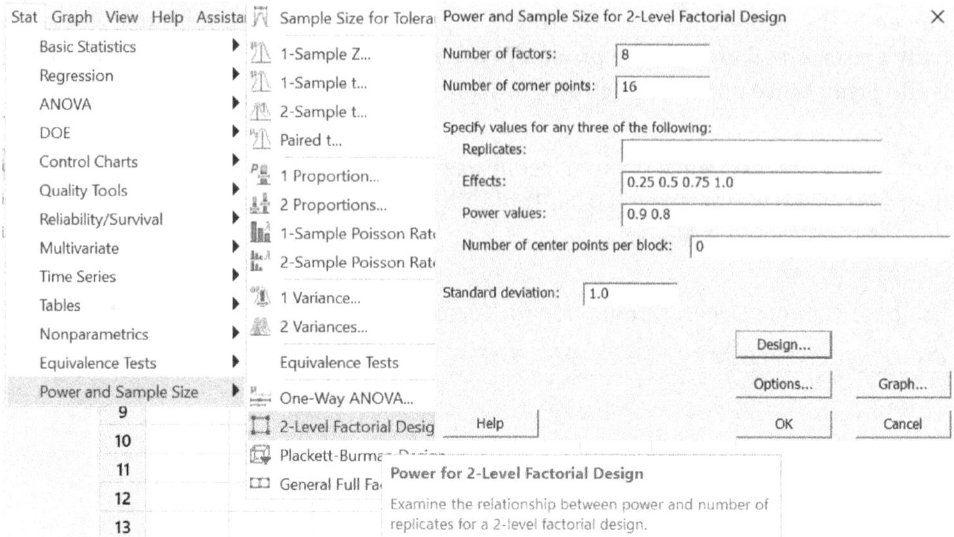

Results

Center Points	Effect	Reps	Total Runs	Target Power	Actual Power
0	0.25	43	688	0.9	0.905588
0	0.25	32	512	0.8	0.805927
0	0.50	11	176	0.9	0.909338
0	0.50	8	128	0.8	0.800703
0	0.75	5	80	0.9	0.910428
0	0.75	4	64	0.8	0.836336
0	1.00	3	48	0.9	0.919009
0	1.00	3	48	0.8	0.919009

Figure 9.2 Using Minitab to find sample size for factorial DOEs

The "Total Runs" column contains the numbers that match Table 12.16. The "Reps" column is "Total Runs" divided by 16 because 16 runs were specified in Minitab.

A question that the reader should have at this point is, "Where do the minimum effect that I want to detect and the standard deviation come from if there is no previous data from this mold?" The answer to the first half of that question is the part specifications. For a given tolerance, a good rule of thumb is to design the DOE to detect an effect equivalent to 20% of the total tolerance width. Thus, if a dimension has a tolerance of ± 0.15 mm or a total tolerance width of 0.30 mm, the sample size should be sufficient to detect a difference of 0.2 × 0.30 mm = 0.06 mm. The answer to the second half of the question is historical data with similar molds and materials. If such data is unavailable, you must generate data to estimate standard deviation – the example at the end of this chapter will use the sample standard deviation of a 30-piece baseline study of parts molded after setting up the initial process.

9.8 Should Replicate Runs Be Made or Just Repetitions?

Repetitions are several consecutive shots from the molding machine. To mold a true replicate of a process set-up, a different process set-up needs to be run in between the replicates. While true replicates provide higher statistical validity, many practitioners in the injection molding industry treat repetitions as replicates to reduce the time and complexity of a DOE. This author generally runs consecutive shots and then uses the average of the repetitions as if having no replicates. The example at the end of this chapter will show this. When true replicates are not performed, the analysis becomes a little more subjective and borderline terms can be left in the regression models. If time permits when not doing true replicates, a more robust approach than collecting consecutive shots is to allow enough cycles between parts to ensure that fresh material is processed for each shot collected. The number of cycles can be estimated with the equations below.

$$N_{\text{cycles}} = 2 \times \frac{\text{Volume}_{\text{Barrel}} + \text{Volume}_{\text{Hot Runner}}}{\text{Volume}_{\text{Shot}}} \tag{9.1}$$

Where:

N_{cycles} = number of cycles to get material out of barrel and hot runner

$\text{Volume}_{\text{Barrel}}$ = maximum capacity of barrel, cm^3

$\text{Volume}_{\text{Hot Runner}}$ = volume of hot runner, cm^3

$\text{Volume}_{\text{Shot}}$ = volume of shot, cm^3

$$\text{Volume}_{\text{Shot}} = \text{Weight of shot(g)} \times \text{Melt density}(\text{g/cm}^3) \tag{9.2}$$

The shot is everything that comes out of the mold when it opens. Melt densities for several materials are presented in Table 12.17 in the Appendix. Numbers for specific grades can usually be obtained from the material supplier or other database. Also, the N_{cycles} will not displace all the old material out of the barrel or hot runner. Anyone who has ever done a color change knows that it can take many more cycles to flush all the first colors out. But, if nothing in the material formulation is changing, the N_{cycles} should be sufficient.

9.9 Should Runs be Randomized?

The short answer is no, at least not when temperature is a factor. The problem with randomizing is the time that it adds to molding the parts when waiting for temperatures to stabilize. The general rules for randomizing runs are:

1. If the process is so unstable that, without randomization, it is believed that results would be useless and randomization would be very difficult, then don't do the experiment. Work on stabilizing the process instead.
2. If you are making replicates and randomization only slightly complicates things, always randomize. It is slightly less desirable to block the difficult-to-change factors (temperature or material change, for example).
3. If you are not performing replicates, block the difficult-to-change factors and randomize only what's easy to adjust.
4. If you are performing replicates, if randomization is very difficult or if trends or other special causes of variation are a concern, make several added "check" runs of the same set-up to monitor for trends.

This author learned about scenario 1 the hard way. There was a reason that the production manager gave a twenty-something-year-old engineer an old machine that was not scheduled to run production very often to run his first DOE. After struggling through the DOE and getting worthless results, several mechanical issues discovered during the initial attempt were fixed for the second DOE, which proved successful – and production got a much better machine back when the DOE was over.

The blocking that scenarios 2 and 3 mention has to do with grouping runs together that have common settings for hard-to-change factors. For example, if conducting a 16-run experiment with two temperature factors at two levels each, there are four possible combinations that serve as the blocks: 1) both temperatures are at the low level; 2) both temperatures are at the high level; 3) One temperature is high and the other is low; and 4) The inverse of scenario 3. Each block would consist of the four runs performed at each of those four temperature combinations.

Scenario 4 would apply when center runs are used for the added runs, as was shown in Example 7.7. If the measurements obtained from these repeated runs were found to be statistically different, then the rest of the data would be questionable.

9.10 How Much Machine Time is Needed to Plan the DOE and to Carry out the DOE?

Assuming that any mechanical issues have been worked out of the mold before starting, Table 9.3 has recommended guidelines for scheduling time. For these guidelines, the following equation will be needed:

$$T_{\text{residence}} = N_{\text{cycles}} \times \text{Cycle time(s)} \tag{9.3}$$

For a more accurate residence time, the time needed for darker colored pellets (or colorant) to make their way from the top of the turning screw to the parts will give the actual residence time. Equation 9.3 tends to be a conservative estimate.

Table 9.3 Guidelines for Process Stabilization Times

Step or Process Change	No. of Cycles or Time
Barrel and mold heat-up time	45 minutes
Baseline process set-up	120 minutes
Extra trials to determine factor levels	60 minutes (includes temperature change)
Start-up	20 minutes
Temperature increase	15 minutes
Temperature decrease	30 minutes (shut down for the first 10 minutes)
Cycle time change	$T_{residence}$
Back pressure or RPM change	$T_{residence}$
Short downtime (less than 1 minute)	$T_{residence}$
Material change	20 shots

To ensure a consistent barrel residence time and a consistent mold temperature during a DOE, it is important to keep the machine running automatically or to maintain a very consistent mold open time if semi-automatic operation is required. For semi-automatic operation, the sequence of many machines can be set up to keep a constant mold open time provided that the operator door is shut before a mold-open timer times out.

9.11 Will Follow-Up Studies Be Needed?

An initial injection molding DOE typically includes at least 6 factors for a basic mold and process and sometimes up to 15 for more complex molds and processes. As mentioned earlier, this author recommends starting with a Resolution IV fractional factorial design. For factors identified with very large effects, it is sometimes prudent to run follow-ups to check for non-linear effects.

What you learn from the DOE may also influence follow-up studies. For example, in the case discussed in Chapter 7, warpage was drastically improved at the high levels of barrel temperature and hold pressure; consequently, a follow-up to look at higher levels of both might be in order – if there is room in the hold pressure/melt temperature window. In the following example, it is discovered that the flow pattern near the last place to fill is important. Therefore, adding % full at V/P transfer as a factor in a follow-up would be a good idea.

The remainder of this chapter will be an example of carrying out a DOE.

9.12 Example 9.1: Carrying out a 16-Run 2-Level DOE with 8 Factors

Figure 9.3 shows the part that will be used for this study and is called the "small frame". The part is made from ABS and the main concern is the weld line shown is weak and will crack during assembly. The weld line strength will be tested by cutting the end off the part, gripping the thick sides and pulling it apart at 0.2 in/s on a tensile testing machine. The primary objective of this DOE is to maximize the weld line strength. The overall length will also be measured to determine how sensitive it is to the input factors and the part weight will be measured to see how well it correlates to the weld line strength and length.

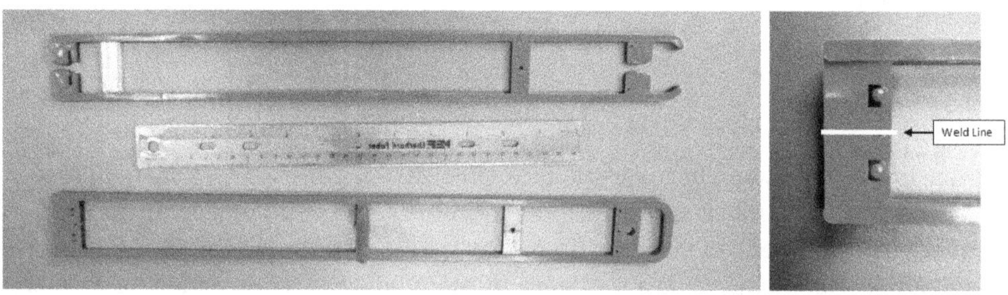

Figure 9.3 "Small Frame" with runner attached, 95%-full part, and end with tested weld line

The eight factors chosen are those which the team carrying out the DOE thought would most likely affect weld line strength or dimensions. Some considerations regarding weld lines are:

1. Hold pressure – A higher hold pressure may cause a better bond at the weld line, because it pushes the flow fronts together harder.
2. Barrel temperatures – A hotter melt may also give a stronger weld line, because the flow fronts should be hotter when they meet and that might promote bonding.
3. Thermolator temperature – A hotter mold will have a thinner frozen layer, which could allow a longer gate-freeze time and a larger effective flow thickness, allowing pressure to be transmitted better to the weld line and for longer.
4. Material dryness – The ABS was not typically dried in this facility, as it was environmentally controlled, and they rarely saw splay with this ABS. One level will be undried, and the other level will be material dried for 4 hours at 175°F. The moisture level was not measured, but should have been. For the sake of this exercise, let it be said that the undried was 0.08% (99.92% dry) and the dried was 0.03% (99.97% dry).

5. Injection velocity – A higher velocity will provide a hotter melt when the two flow fronts meet and a thinner frozen layer. However, if there is significant shear heating through the part, the gate may freeze off before the hotter areas can be packed well enough.
6. Hold time – A hold time shorter than gate freeze could affect how long the pressure is maintained at the weld line. However, since the weld line is at the last place to fill, the effect should be small, even if it exists.
7. Cycle time – The thought here is that a longer cycle time may be worse because it gives the material more time to shrink onto the core which makes the small hole right in the weld line and that could stress the material locally.
8. Back pressure – A higher back pressure will be tried to see if a more homogeneous melt will improve the weld line.

The preceding considerations are a typical thought process when setting up a DOE. It will be discovered after this DOE is analyzed that, in fact, none of these considerations yields the strongest weld lines.

The following exercises were performed to determine the baseline injection velocity and hold time (see the Process Set-Up chapter).

Determination of Initial Injection Velocity

Note: Yellow highlighted areas are user-defined.

Transfer Pressure		Fill Time		Injection Velocity		Relative Shear Rate	Relative Viscosity
13600	psi	0.21	s	11.80	in/s	4.76	2856
12320	psi	0.30	s	7.87	in/s	3.33	3696
10650	psi	0.44	s	5.25	in/s	2.27	4686
9500	psi	0.65	s	3.50	in/s	1.54	6175
8450	psi	0.98	s	2.34	in/s	1.02	8281
7850	psi	1.41	s	1.56	in/s	0.71	11069
7120	psi	2.25	s	1.04	in/s	0.44	16020
6610	psi	3.22	s	0.69	in/s	0.31	21284
6350	psi	4.50	s	0.46	in/s	0.22	28575
6550	psi	7.54	s	0.31	in/s	0.13	49387
7220	psi	11.24	s	0.21	in/s	0.09	81153
	psi		s	0.14	in/s	#N/A	FALSE
	psi		s	0.09	in/s	#N/A	FALSE

HOLD TIME (GATE FREEZE) DETERMINATION

Hold Time	Part Weight
20	28.32
18	28.33
16	28.32
14	28.33
12	28.3
10	28.21
8	28.14
6	28.06
4	27.93
2	27.68

SELECTED VELOCITY = 5.1 in/s

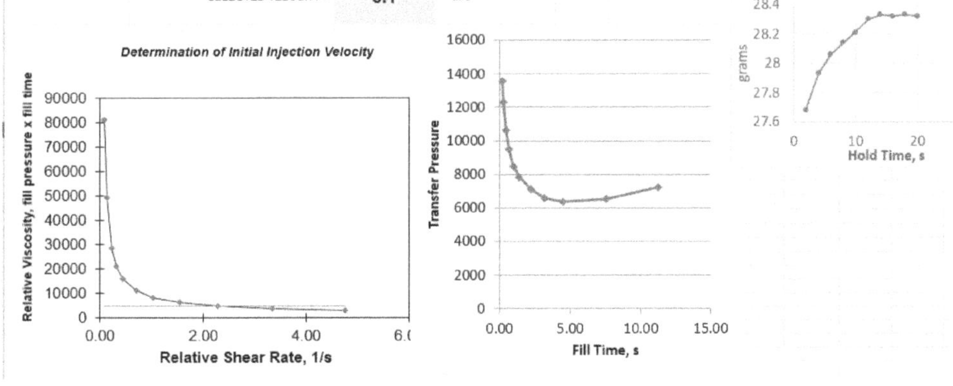

Figure 9.4 Process for determining injection velocity and hold time baseline for Example 9.1

The injection velocity study was run at the low levels of the temperature, as was the gate-freeze study shown. The gate-freeze study was conducted at the low temperatures and under a hold pressure of 3,250 psi. It was repeated at the high temperatures under a hold pressure of 2,750 psi and only increased the gate freeze by about 1 second. The hold pressure/barrel temperature was determined to be between 1500 psi and 4500 psi, as shown in Figure 9.5.

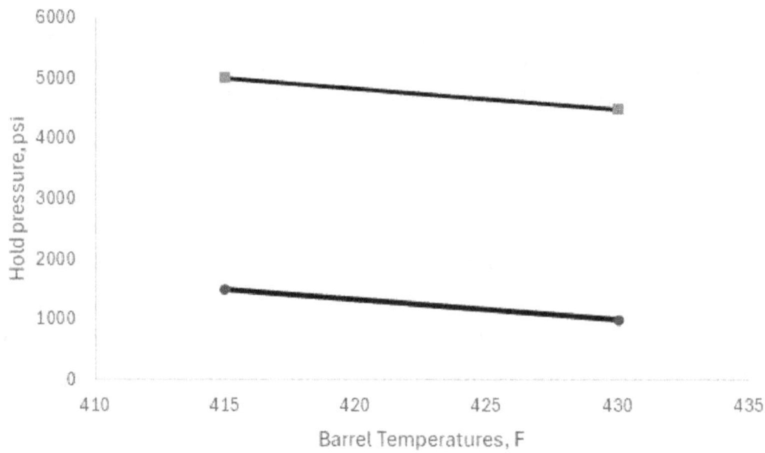

Figure 9.5 Hold pressure/barrel temperature window for Example 9.1

The two DOE levels decided on for each DOE factor are as follows:

Table 9.4 DOE Factor Levels for Example 9.1

Factor	Low Level	High Level
Hold pressure	2,000 psi	4,000 psi
Barrel temperature	415°F	430°F
Thermolator temperature	95°F	115°F
Material dryness	99.92%	99.97%
Injection velocity	5.1 in/s	10.2 in/s
Hold time	14 s	19 s
Cycle time	29 s	39 s
Back pressure	800 psi	1200 psi

The DOE matrix in Minitab is shown in Table 9.5.

Table 9.5 DOE Settings for Each Run for Example 9.1

	C1	C2	C3	C4	C5	C6	C7	C8	C9	C10	C11	C12	C13	C14	C15
↓	Std	Run	Ce	Bl	Thermo-lator	Barrel	Material	Hold P	Inject V	Hold T	Cycle T	Back P		Cool Time	V/p T
1	1	1	1	1	95	415	99.92	2000	5.1	14	29	800		8.8500	0.95
2	2	2	1	1	95	415	99.92	2000	10.2	19	39	1200		14.0800	1.23
3	3	3	1	1	95	415	99.97	4000	10.2	19	29	800		4.0800	1.23
4	4	4	1	1	95	415	99.97	4000	5.1	14	39	1200		18.8500	0.95
5	5	5	1	1	115	415	99.97	2000	5.1	19	39	800		13.8500	0.95
6	6	6	1	1	115	415	99.97	2000	10.2	14	29	1200		9.0800	1.23
7	7	7	1	1	115	415	99.92	4000	10.2	14	39	800		19.0800	1.23
8	8	8	1	1	115	415	99.92	4000	5.1	19	29	1200		3.8500	0.95
9	9	9	1	1	115	430	99.92	2000	5.1	14	39	1200		18.8500	0.95
10	10	10	1	1	115	430	99.92	2000	10.2	19	29	800		4.0800	1.23
11	11	11	1	1	115	430	99.97	4000	10.2	19	39	1200		14.0800	1.23
12	12	12	1	1	115	430	99.97	4000	5.1	14	29	800		8.8500	0.95
13	13	13	1	1	95	430	99.97	2000	5.1	19	29	1200		3.8500	0.95
14	14	14	1	1	95	430	99.97	2000	10.2	14	39	800		19.0800	1.23
15	15	15	1	1	95	430	99.92	4000	10.2	14	29	1200		9.0800	1.23
16	16	16	1	1	95	430	99.92	4000	5.1	19	39	800		13.8500	0.95

In every case other than hold pressure and material dryness, the low level was the baseline process. Neither of the hold pressures was from the baseline; these were determined from the hold pressure/barrel temperature window. Dry material was used when setting up the baseline process. For hold time, the team used gate freeze at the high temperatures as the low level and the longest time that wouldn't extend the low level cycle time as the high level. This necessitated running the screw recovery speed at 90% of maximum for the entire experiment or adjusting the screw speed every run so that the screw would finish recovery about two seconds before the mold opened. It was decided to use 90% of maximum screw speed for all runs.

It should also be noted in the above DOE matrix that two additional columns were added for reference during execution of the DOE. Column 14 specifies what the cooling timer setting will need to be to achieve the desired cycle time, and Column 15 is the V/P transfer position necessary to get a 95%-full part before transferring to hold pressure.

Additionally, 30-pieces of the baseline process were collected and measured 24 hours after molding to establish an estimated standard deviation for each of the three measurements.

Table 9.6 Measurements for Example 9.1

Sample #	Length, in	Weight, g	Weld Pull, lb
1	17.088	28.114	175.9
2	17.088	28.104	198.1
3	17.0875	28.05	211.5
4	17.0875	28.043	207.2
5	17.088	28.037	204.8
6	17.0875	28.095	186
7	17.087	28.064	179.8
8	17.087	28.085	204.8
9	17.087	28.067	167
10	17.0875	28.034	161.9
11	17.087	28.103	209.8
12	17.087	28.139	196.2
13	17.0875	28.083	193.6
14	17.087	28.077	206.1

Sample #	Length, in	Weight, g	Weld Pull, lb
	17.088	28.026	187.7
16	17.0875	28.08	188.2
17	17.0865	28.061	205.1
18	17.087	28.099	178.5
19	17.0875	28.104	195.5
20	17.087	28.028	205.6
21	17.087	28.085	184.8
	17.0875	28.124	182.5
23	17.087	28.06	183.1
24	17.087	28.042	168.1
25	17.087	28.003	175.7
26	17.087	28.077	167.9
27	17.0875	28.103	162.2
28	17.087	28.063	193.2
29	17.087	28.079	191.8
30	17.088	28.013	162.8
Sample standard deviation	**0.000407**	**0.03344**	**15.46**
Max	17.088	28.139	211.5
Min	17.0865	28.003	161.9

In addition to the sample standard deviation, the maximum and minimum of each column is shown. This is a good way to catch data-entry errors. Another thing to note is the small variation for the part length. The digital gauge used to determine the length measured to the nearest 0.0005 inches. Half of the readings were the same number, 17.0870 inches. This gauge would most likely not be suitable for control charts, and resolution to 0.0001 inches would be better. However, if the length spec is 17.000 ± 0.015 inches, 20% of the tolerance width is 0.006 inches, which is 15 (0.006/0.000407) standard deviations and just two replicates could easily pick up that difference in a DOE. Two replicates could pick up a 0.0005-inch difference with a 16-run experiment. Figure 9.6 shows how to determine the number of replicates needed to pick up 5-, 10-, and 15-lb difference in weld pull strength using Minitab with a 90% power.

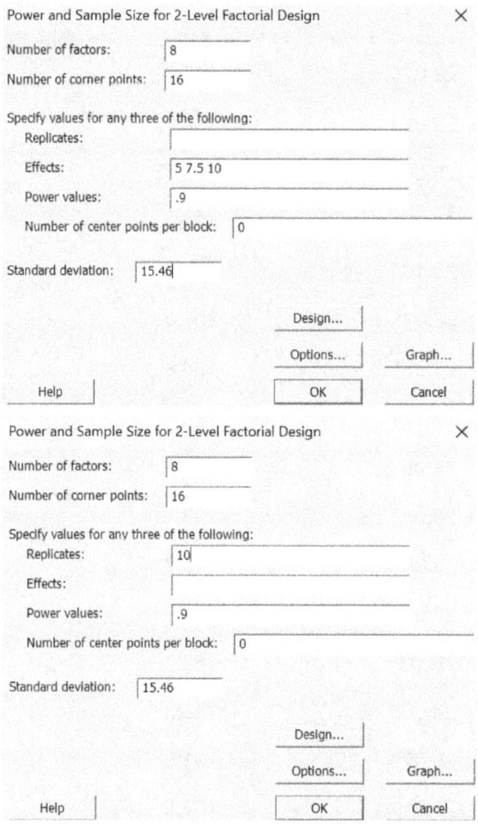

Power and Sample Size

2-Level Factorial Design
α = 0.05 Assumed standard deviation = 15.46

Method

Factors: 8 Base Design: 8, 16
Blocks: none

Results

Center Points	Effect	Reps	Total Runs	Target Power	Actual Power
0	5.0	26	416	0.9	0.908292
0	7.5	12	192	0.9	0.916617
0	10.0	7	112	0.9	0.923398

Results

Center Points	Reps	Total Runs	Power	Effect
0	10	160	0.9	7.97708

Figure 9.6 Determining the sample size for the weld pull strength of the small frame

The decision was made to go with 10 parts per run (160 total), which should be able to pick up about an 8-lb difference in the weld pull strength. The next decision involved how to collect the parts and whether to use repetitions or replicates. The following time study looked only at the running of the DOE, as the preliminary steps of setting up the process and determining factor levels had been completed (it took about 3 hours of machine time). Three scenarios will be looked at:

1. True replicates with eight blocks of common temperature and material (Runs 1–2, 3–4, 5–6, 7–8, 9–10, 11–12, 13–14, and 15–16). For example, take one part each from Runs 1 and 2 before taking the second parts from 1 and 2 and so on.

2. Taking all 10 replicates from each run before going to the next run, but letting enough cycles run between parts, N_{cycles}, to allow fresh material for each part collected.

3. Collecting 10 consecutive parts.

Equation 9.1 – Equation 9.3 are used to determine $T_{residence}$.

$$N_{cycles} = 2 \times \frac{\text{Volume}_{\text{Barrel}} + \text{Volume}_{\text{Hot Runner}}}{\text{Volume}_{\text{Shot}}} = 2 \times \frac{83\ \text{cm}^3 + 0}{33\ \text{cm}^3} = 5\ \text{cycles}$$

$$T_{residence} = N_{cycles} \times \text{Cycle Time(s)} = 5\ \text{cycles} \times 34\ \text{s/cycles} = 170\ \text{seconds}$$

Where

$\text{Volume}_{\text{Barrel}}$ = Maximum capacity of barrel = 83 cm^3 (from machine spec sheet)

$\text{Volume}_{\text{Hot Runner}}$ = Volume of hot runner = 0

$\text{Volume}_{\text{Shot}}$ = Weight of shot (g) × Melt density (g/cm^3)

= 34 g (part & runner) × 0.97 g/cm^3 (from Table 12.17 in the Appendix)

= 33 cm^3

Table 9.7 Determining Experimental Times for 3 Scenarios for Example 9.1

	Scenario 1	Scenario 2	Scenario 3
Barrel and mold heat-up	45 min.	45 min.	45 min.
Start-up stabilization	20 min.	20 min.	20 min.
Runs 1 & 2	170 s × 20 = 3400 s = 56.7 min.	170 s × 20 = 3400 s = 56.7 min.	29 s × 10 + 170 s + 39 s × 10 = 850 s = 14.2 min.
Material change	29 s × 20 = 580 s = 9.7 min.	29 s × 20 = 580 s = 9.7 min.	29 s × 20 = 580 s = 9.7 min.
Runs 3 & 4	170 s × 20 = 3400 s = 56.7 min.	170 s × 20 = 3400 s = 56.7 min.	29 s × 10 + 170 s + 39 s × 10 = 850 s = 14.2 min.
Temperature increase	15 min.	15 min.	15 min.
Runs 5 & 6	170 s × 20 = 3400 s = 56.7 min.	170 s × 20 = 3400 s = 56.7 min.	29 s × 10 + 170 s + 39 s × 10 = 850 s = 14.2 min.
Material Change	29 s × 20 = 580 s = 9.7 min.	29 s × 20 = 580 s = 9.7 min.	29 s × 20 = 580 s = 9.7 min.

Table 9.7 Determining Experimental Times for 3 Scenarios for Example 9.1 (*continued*)

	Scenario 1	Scenario 2	Scenario 3
Runs 7 & 8	170 s × 20 = 3400 s = 56.7 min.	170 s × 20 = 3400 s = 56.7 min.	29 s × 10 + 170 s + 39 s × 10 = 850 s = 14.2 min.
Temperature increase	15 min.	15 min.	15 min.
Runs 9 & 10	170 s × 20 = 3400 s = 56.7 min.	170 s × 20 = 3400 s = 56.7 min.	29 s × 10 + 170 s + 39 s × 10 = 850 s = 14.2 min.
Material Change	29 s × 20 = 580 s = 9.7 min.	29 s × 20 = 580 s = 9.7 min.	29 s × 20 = 580 s = 9.7 min.
Runs 11 & 12	170 s × 20 = 3400 s = 56.7 min.	170 s × 20 = 3400 s = 56.7 min.	29 s × 10 + 170 s + 39 s × 10 = 850 s = 14.2 min.
Temperature decrease	30 min.	30 min.	30 min.
Runs 13 & 14	170 s × 20 = 3400 s = 56.7 min.	170 s × 20 = 3400 s = 56.7 min.	29 s × 10 + 170 s + 39 s × 10 = 850 s = 14.2 min.
Material Change	29 s × 20 = 580 s = 9.7 min.	29 s × 20 = 580 s = 9.7 min.	29 s × 20 = 580 s = 9.7 min.
Runs 15 & 16	170 s × 20 = 3400 s = 56.7 min.	170 s × 20 = 3400 s = 56.7 min.	29 s × 10 + 170 s + 39 s × 10 = 850 s = 14.2 min.
Total	617.4 min. = 10.3 h	617.4 min. = 10.3 h	277.4 min. = 4.6 h

Because of the difference in DOE completion time – 4.6 hours versus 10.3 hours – Scenario 3 was chosen. Also, since the only practical randomization allowed for two runs at a time, the decision was made not to randomize. The results will be analyzed using the average as a single value per run. As can be seen, Scenarios 1 and 2 would each take 10.3 hours. However, Scenario 1 would have a higher risk of experimental error because factors are being changed between every collected part. It should also be noted that if only three replicates were required instead of ten, the difference in time would not be as large. The first 2 scenarios could be completed in 5 machine hours and the 3rd scenario in 3.3 machine hours.

In addition to collecting the parts, the following dependent process data will be collected for each part made: time molded, fill time, cushion, V/P transfer pressure, screw recovery time, cycle time, and peak cavity pressure. The cavity pressure was measured about 0.5 inches from the gate. This data will be reviewed in the Process Monitoring chapter (Chapter 10).

The DOE was carried out, the parts were measured 24 hours later, and the results shown in Figure 9.7 were obtained. Only five parts per run were weighed and measured for length. After a review of that data to make sure nothing needed to be remeasured, the weld pull test was performed on all ten parts.

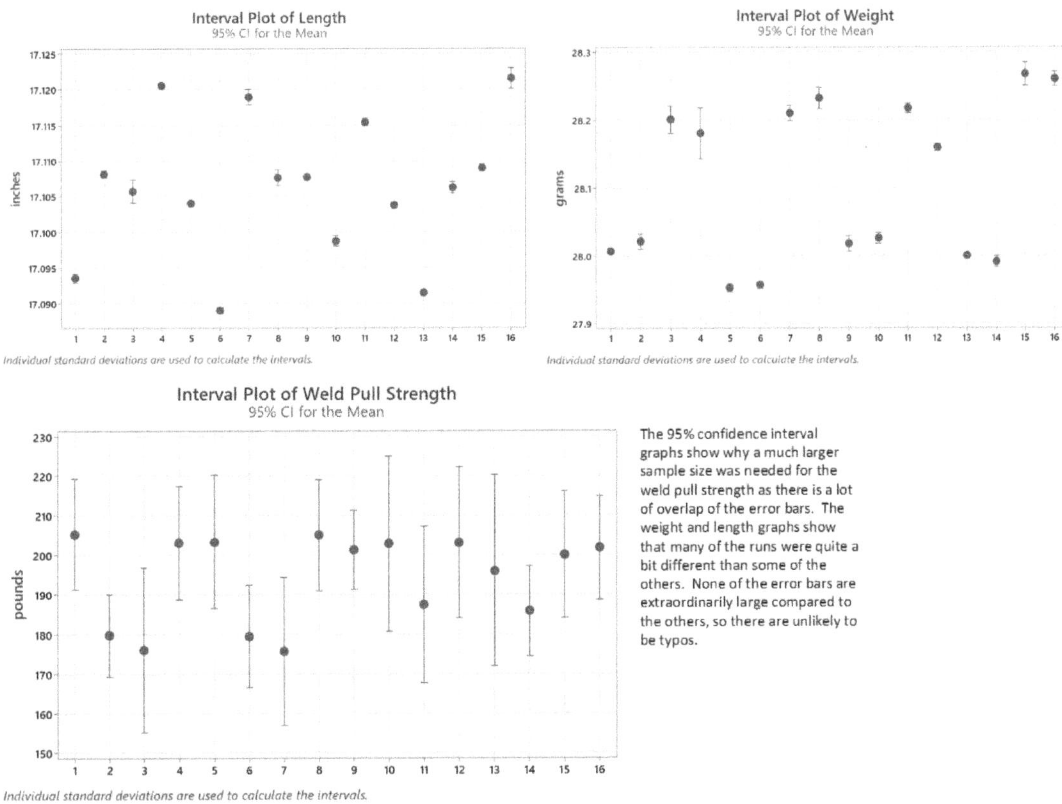

Figure 9.7 05% confidence intervals of the means for all 3 responses for Example 9.1

The 95% confidence interval graphs show why a much larger sample size was needed for the weld pull strength, as there is a lot of overlap of the error bars. The weight and length graphs show that many of the runs were quite a bit different than some of the others. None of the error bars are extraordinarily large compared with the others, so there are unlikely to be data-entry errors.

When developing a mold, the 95% confidence intervals of the runs are useful for evaluating how sensitive dimensions are to process variations compared with the spec width. The only dimension analyzed here was length. The upper confidence limit of Run 16 was 17.123 and the lower confidence limit of Run 6 was 17.089, yielding a range of 0.034 inches. The smaller that range is compared with the spec limits, generally the fewer problems that will be experienced with keeping the dimension within tolerance once it is centered. Ideally, the range should not be more than twice the spec limit width. So, with a range of 0.034 inches, problems are more likely if the spec limits are tighter than ± 0.0085 inches (0.017 inches wide). It should be noted that, if only the low (29 second) cycle runs are considered, Run 15 has the highest confidence interval at about 17.109 and the range is reduced to about 0.020 inches.

The average for each of the three responses is entered into the Minitab workshop for all 16 runs. The first analysis does not include a stepwise regression and yields the results shown in Figure 9.8.

Factorial Regression: Weld Avg, pounc

Term	Effect
Constant	
Thermolator Temp, F	1.3162
Barrel Temp, F	6.381
Material Dryness, %	-4.661
Hold Press, psi	-0.2213
Inject Velocity, in/sec	-16.471
Hold Time, sec	-0.19625
Cycle Time, sec	-3.689
Back Press, psi	-0.2188
Thermolator Temp, F*Barrel Temp, F	1.4438
Thermolator Temp, F*Material Dryness, %	1.8013
Thermolator Temp, F*Hold Press, psi	-3.709
Thermolator Temp, F*Inject Velocity, in/sec	-0.3537
Thermolator Temp, F*Hold Time, sec	9.971
Thermolator Temp, F*Cycle Time, sec	-1.9413
Thermolator Temp, F*Back Press, psi	-2.711

Factorial Regression: Length Avg, in \

Term	Effect
Constant	
Thermolator Temp, F	-0.001363
Barrel Temp, F	0.000788
Material Dryness, %	-0.003637
Hold Press, psi	0.012963
Inject Velocity, in/sec	0.000138
Hold Time, sec	0.000488
Cycle Time, sec	0.012963
Back Press, psi	-0.000438
Thermolator Temp, F*Barrel Temp, F	0.000713
Thermolator Temp, F*Material Dryness, %	-0.001563
Thermolator Temp, F*Hold Press, psi	-0.001413
Thermolator Temp, F*Inject Velocity, in/sec	-0.000388
Thermolator Temp, F*Hold Time, sec	0.001112
Thermolator Temp, F*Cycle Time, sec	-0.001212
Thermolator Temp, F*Back Press, psi	-0.000962

Factorial Regression: Weight Avg, gms

Term	Effect
Constant	
Thermolator Temp, F	-0.019400
Barrel Temp, F	0.02210
Material Dryness, %	-0.04785
Hold Press, psi	0.2195
Inject Velocity, in/sec	0.010050
Hold Time, sec	0.014850
Cycle Time, sec	0.000100
Back Press, psi	0.010500
Thermolator Temp, F*Barrel Temp, F	-0.004900
Thermolator Temp, F*Material Dryness, %	-0.002050
Thermolator Temp, F*Hold Press, psi	-0.002950
Thermolator Temp, F*Inject Velocity, in/sec	0.001850
Thermolator Temp, F*Hold Time, sec	0.006050
Thermolator Temp, F*Cycle Time, sec	0.005700
Thermolator Temp, F*Back Press, psi	0.008300

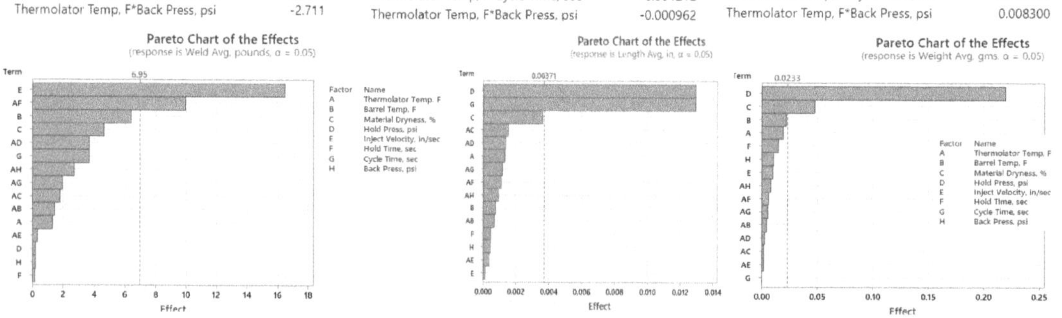

Figure 9.8 Estimated effects from Minitab for all 3 responses for Example 9.1

The factors with estimated effects above the dotted line (weld effect >6.95 lb, length effect > 0.00371 inches, and weight effect >0.0233 g) are statistically significant. Some of the factors that are just below the dotted line will be found to be significant once the insignificant factors are removed from the model during regression analysis. It is also evident that there was a significant two-factor interaction affecting weld pull strength. Minitab always shows the interaction as the first entry in the alias table, as shown in Figure 9.6. After examining the fracture surfaces and because injection velocity and barrel temperatures are the two dominant main factor effects, the team had a high degree of confidence that the significant interaction was BE (barrel temperature × injection velocity). At the low barrel temperatures and high injection velocity, the flow pattern shifted slightly, causing the weld/meld lines originating from the square openings (see Figure 9.9) to merge with the center weld line differently. The higher weld pull strength could be due to the fill pattern and polymer orientation at lower velocities. Another contributing factor could be better venting at the low velocity.

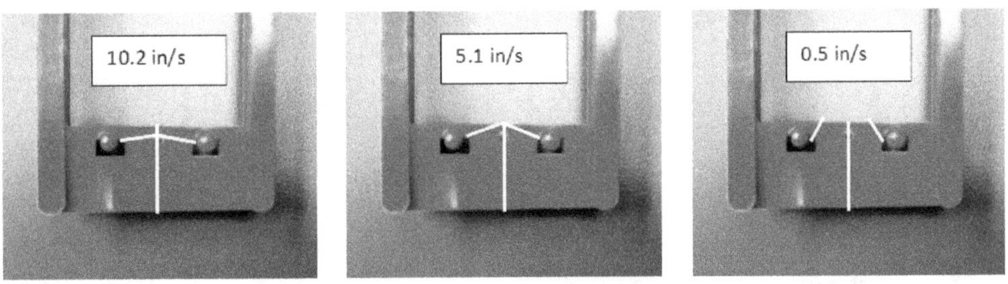

Figure 9.9 Weld line location for 3 injection speeds for the part used in Example 9.1

The lines in Figure 9.9 show approximately how the weld/meld lines moved. As the flow front slowed, there was more hesitation across the top of the rectangular openings. Although not part of this experiment, the 0.5 in/s injection velocity gave significantly stronger pull strength but added a few seconds to the cycle. This part would be a good candidate for a couple of process variations that are discussed in the Process Set-Up chapter: the iMFLUX process or a D3 process. iMFLUX is pressure controlled and typically the flow front is moving very fast when filling the first portion of the part but would slow most likely to at least the 0.5 in/s by the time the flow front reached the rectangular openings. With the D3 process, it would be possible to transfer to a slower velocity just before the flow front reached the rectangular openings and the slow velocity could be used to finish the fill, and to pack out the parts. This DOE is a good example of one that would justify follow-up studies. A few possibilities are one-factor-at-a-time studies looking at several different injection velocities, including % full at V/P transfer as a factor, and alternate process set-up strategies (iMFLUX or D3), using a smaller diameter sprue with one of Beaumont's flow rotation devices to change the fill pattern.

When a reduced regression model (using backward elimination of insignificant factors) is requested during analysis of the DOE for weld pull strength, the results shown in Figure 9.10 are obtained.

Regression Equation in Uncoded Units

Weld Avg, pounds = 378 - 3.225 Thermolator Temp, F + 0.425 Barrel Temp, F
- 3.230 Inject Velocity, in/s - 20.98 Hold Time, s
+ 0.1994 Thermolator Temp, F*Hold Time, s

Figure 9.10 Regression results from Minitab for Example 9.1

This model will obviously not yield correct results for other process settings if the team's hypothesis about the barrel temperatures/injection velocity interaction is accurate. A follow-up study could confirm their conclusion.

Aliases

I
A
B
C
D
E
F
G
H
AB + CD + EF + GH
AC + BD + EH + FG
AD + BC + EG + FH
AE + BF + CH + DG
AF + BE + CG + DH
AG + BH + CF + DE
AH + BG + CE + DF

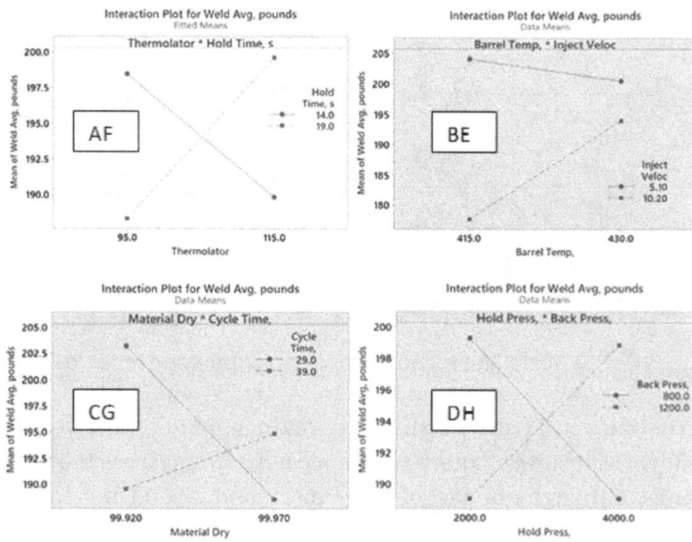

Figure 9.11 Evaluating a significant 2-factor interaction. The most likely interaction is BE, not AF

To generate a better regression model in Minitab when the DOE analysis uses the wrong 2-factor interaction, you must perform the analysis outside of the standard DOE menu. There are several ways to access linear regression in Minitab. Figure 9.12 shows how to do it under the Predictive Analytics tab.

As can be seen in Figure 9.12 all eight of the main factors plus the barrel temperatures/hold pressure interaction are possible terms for the regression equations. The equations and statistics shown in Figure 9.13 are obtained.

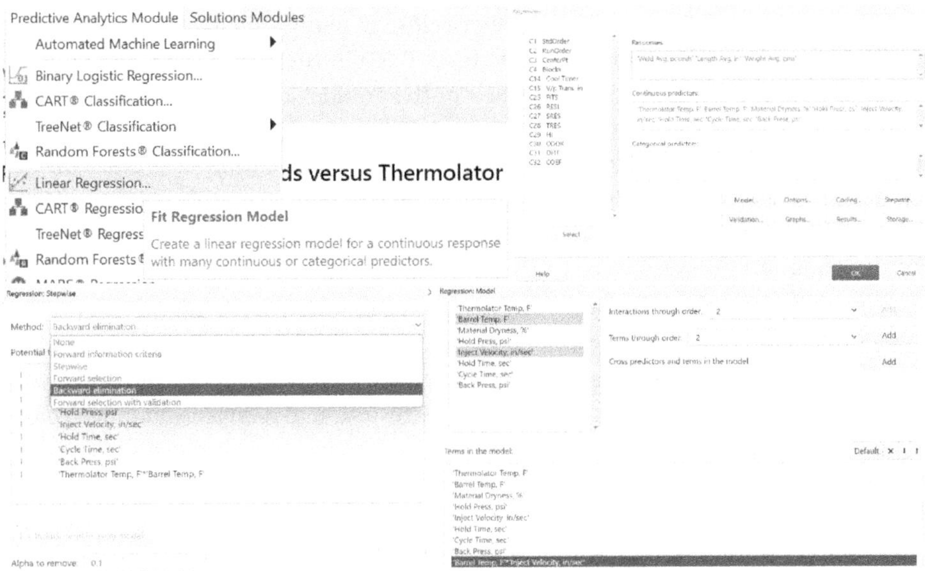

Figure 9.12 Set-up of a linear regression in Predictive Analytics

Weld Avg. pounds = 10199 - 1.569 Barrel Temp, F - 93.2 Material Dryness, % - 113.4 Inject Velocity, in/sec + 0.2607 Barrel Temp, F*Inject Velocity, in/sec

Coefficients

Term	Coef	SE Coef	T-Value	P-Value
Constant	10199	4083	2.50	0.030
Barrel Temp, F	-1.569	0.430	-3.65	0.004
Material Dryness, %	-93.2	40.8	-2.28	0.043
Inject Velocity, in/sec	-113.4	22.5	-5.03	0.000
Barrel Temp, F*Inject Velocity, in/sec	0.2607	0.0533	4.89	0.000

Model Summary

S	R-sq	R-sq(adj)	R-sq(pred)
4.08113	90.44%	86.96%	79.77%

Weight Avg, gms = 122.9 - 0.000970 Thermolator Temp, F + 0.001473 Barrel Temp - 0.957 Material Dryness, % + 0.00010973 Hold Press, psi + 0.00297 Hold Time, sec

Coefficients

Term	Coef	SE Coef	T-Value	P-Value	VIF
Constant	122.9	12.5	9.85	0.000	
Thermolator Temp, F	-0.000970	0.000312	-3.11	0.011	1.00
Barrel Temp, F	0.001473	0.000416	3.54	0.005	1.00
Material Dryness, %	-0.957	0.125	-7.67	0.000	1.00
Hold Press, psi	0.000110	0.000003	35.16	0.000	1.00
Hold Time, sec	0.00297	0.00125	2.38	0.039	1.00

Model Summary

S	R-sq	R-sq(adj)	R-sq(pred)
0.0124833	99.25%	98.87%	98.08%

Length Avg, in = 24.31 - 0.0728 Material Dryness, % + 0.000006481 Hold Press, psi + 0.0012963 Cycle Time, sec

Coefficients

Term	Coef	SE Coef	T-Value	P-Value	VIF
Constant	24.31	1.97	12.34	0.000	
Material Dryness, %	-0.0728	0.0197	-3.69	0.003	1.00
Hold Press, psi	0.000006	0.000000	13.15	0.000	1.00
Cycle Time, sec	0.001296	0.000099	13.15	0.000	1.00

Model Summary

S	R-sq	R-sq(adj)	R-sq(pred)
0.0019709	96.77%	95.96%	94.26%

Figure 9.13 Regression results using Predictive Analytics in Minitab for Example 9.1

As can be seen, some of the results that were close to being statistically significant in the DOE are statistically significant (p value < 0.05) in these analyses. The elimination of insignificant factors allowed a few of the smaller effect factors to make the cut. It should also be noted here that the predictive analytics results are identical with what would have been obtained if the backward elimination regression had been done for length and weight in the DOE analysis. However, substituting what is believed to be the correct 2-factor interaction for weld pull strength resulted in a different equation.

One factor to note is the moisture level's influence on all three responses. This effect is most likely due to water acting as a plasticizer and reducing the viscosity of the ABS. The low moisture level yielded slightly heavier, longer, and stronger weld pulls. With lower viscosity, the parts tend to pack out more, resulting in heavier and longer parts. The better weld pull strength correlates with the higher melt temperature, which would have a lower viscosity, leading higher weld pull strength – at least at the fast injection velocity. It is a little harder to justify this based on the hesitation theory, since lower viscosities usually hesitate less. Regardless, the fact that moisture content appears as a significant factor in all three responses is a good reason to always dry the material consistently.

A sometimes-useful Minitab tool is Response Optimizer. It is available in both the DOE tabs and in Predictive Analytics (if the module is installed). It will attempt to find factor settings that will either hit a target, minimize a response (such as warp), or maximize a response (weld pull strength). It will attempt to do this to meet the requirements of all the responses. You can also specify that you want certain factors (cycle time and dryness in this example) to be at a fixed value and you can specify a range of allowable values for a factor. The following shows how to do this for the models that were found in Predictive Analytics. The Response Optimizer can be opened by clicking on the gold star in the toolbar directly above the main tabs, as shown in Figure 9.14.

For this optimization, weld was set to "maximize" and a couple of values near the middle of the DOE values were chosen as targets for the other two responses. Note in Figure 9.8 that values for "weight" and "importance" are defined for each response. Weights control how closely a response must match its target, while importance determines how much a response contributes to the overall optimization. For this experiment, maximizing the weld pull strength was the most important, so it was given the maximum value of 10 for each. The weight response almost always gets 0.1, because it is primarily tracked for process monitoring (see Chapter 10) since part weight is easy to measure and can be checked shortly after molding. Typically, for a DOE done during mold development, the dimensions are assigned a weight of 0.1 and the mold dimensions are modified to hit the target.

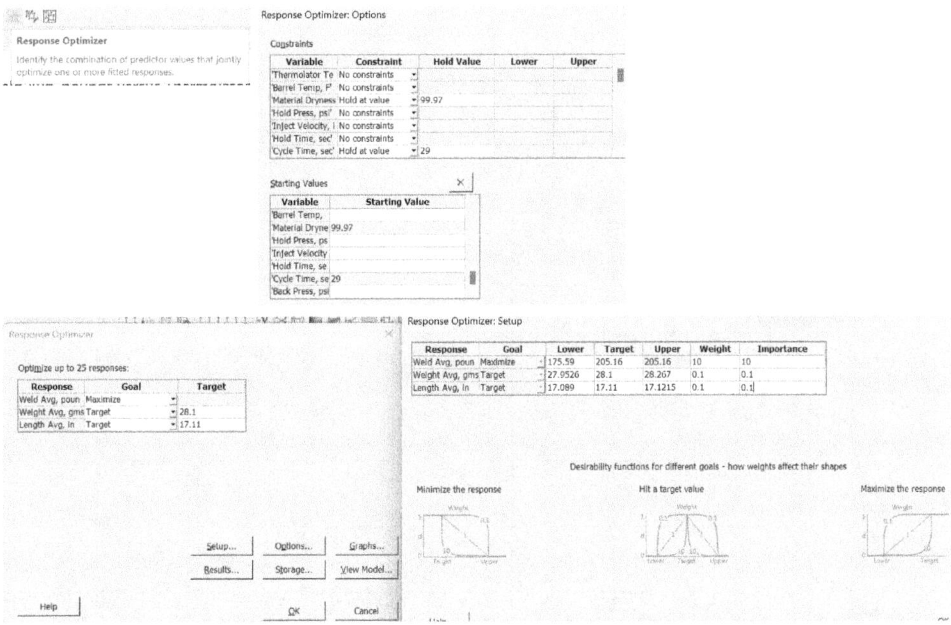

Figure 9.14 Opening the Response Optimizer for Predictive Analytics

The following shows the results of the Response Optimizer analysis

Solution

Solution	Thermolator Temp, F	Barrel Temp, F	Material Dryness, %	Hold Press, psi	Inject Velocity, in/sec	Hold Time, sec	Cycle Time, sec	Back Press, psi
1	115	430	99.97	3256.38	5.1	19	29	800

Solution	Weld Avg, pounds Fit	Weight Avg, gms Fit	Length Avg, in Fit	Composite Desirability
1	214.086	28.1000	17.1010	0.999452

Figure 9.15 Results of the Minitab Response Optimizer for Example 9.1

A composite desirability value would be 1.0 if all the targets were hit. In this case, the length was 0.009 inches off, so it is less than 1.0.

There is an interactive optimization plot included in the results of Response Optimizer. It allows the user to use slide bars to adjust settings in real-time, providing instant updates on the predicted responses and desirability scores.

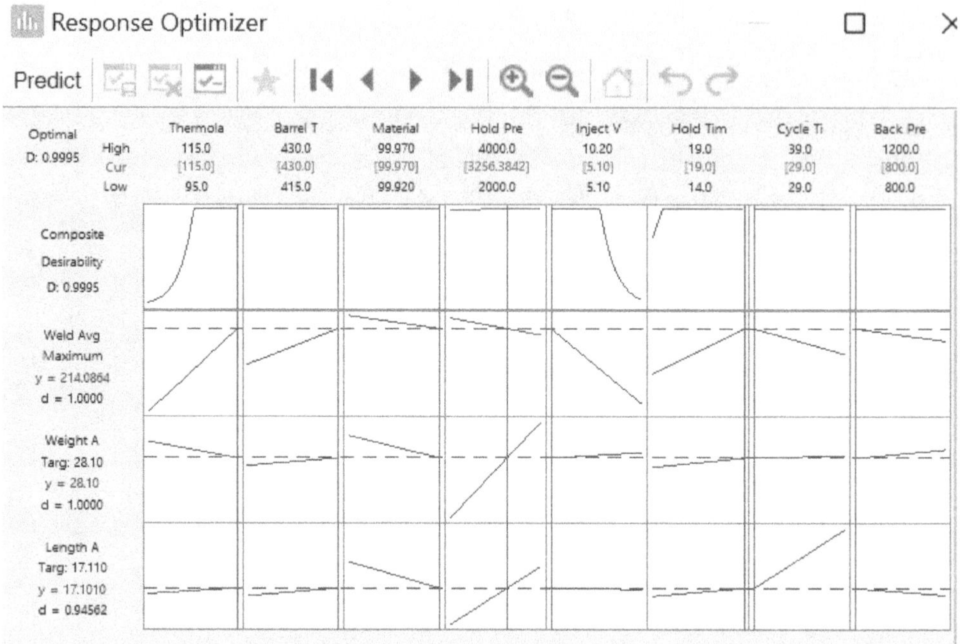

Figure 9.16 Minitab Response Optimizer slide bar

The true test of the models is to mold parts and see how well the predictions compare. Parts were molded with the above settings and measured 24 hours after molding. Ten parts were subjected to the weld pull test while five parts each were used for weight and length measurements. The results at the 95% confidence interval are as follows.

Weld pull strength = 203 ± 13 lb (190 to 216 lb)

Length = 17.1045 ± 0.0006 in (17.1039 to 17.1105 in)

Weight = 28.137 ± 0.011 g (28.126 to 28.148 g)

The weld pull strength matched the prediction, as it fell within the 95% confidence interval. The length and weight predictions were both slightly low. Again, as George Box said, “all models are wrong, but some are useful.” These models are close, so they remain useful. One possible explanation for the discrepancies in length and weight is that the hold pressure response has some curvature. A slightly lower hold pressure would be needed to match the prediction. As can be seen in Figure 9.17, a 2,000 psi increase in hold pressure resulted in an estimated effect was 0.012963 inches on length and 0.2195 grams on weight. Therefore, if hold pressure were changed by 10% of 2,000 psi (200 psi) and the response were linear (likely close for small changes), the length would be expected to change by 0.0013 inches and the weight by 0.022 grams. The average weight of our sample was off the prediction by 0.037 grams (28.137−28.1) and the average length of our sample was off the prediction by 0.0035 inches

(17.1045–17.101). If the hold pressure were decreased by 200 psi to 3056 psi, the length would likely drop to about 17.1032 inches and the weight to about 28.115 grams. Another good follow-up study would be an OFAT of hold pressure on the effect on length with the cycle time at 29 seconds, dry material, and keeping the other settings that maximized weld pull strength. If 2,000, 2,500, 3,000, 3,500 and 4,000 psi were run, a very nice non-linear model could be made. An other point to note here is that if the length were a CTQ dimension, it would likely be necessary to revise the tooling to keep the hold pressure close to the center of the window.

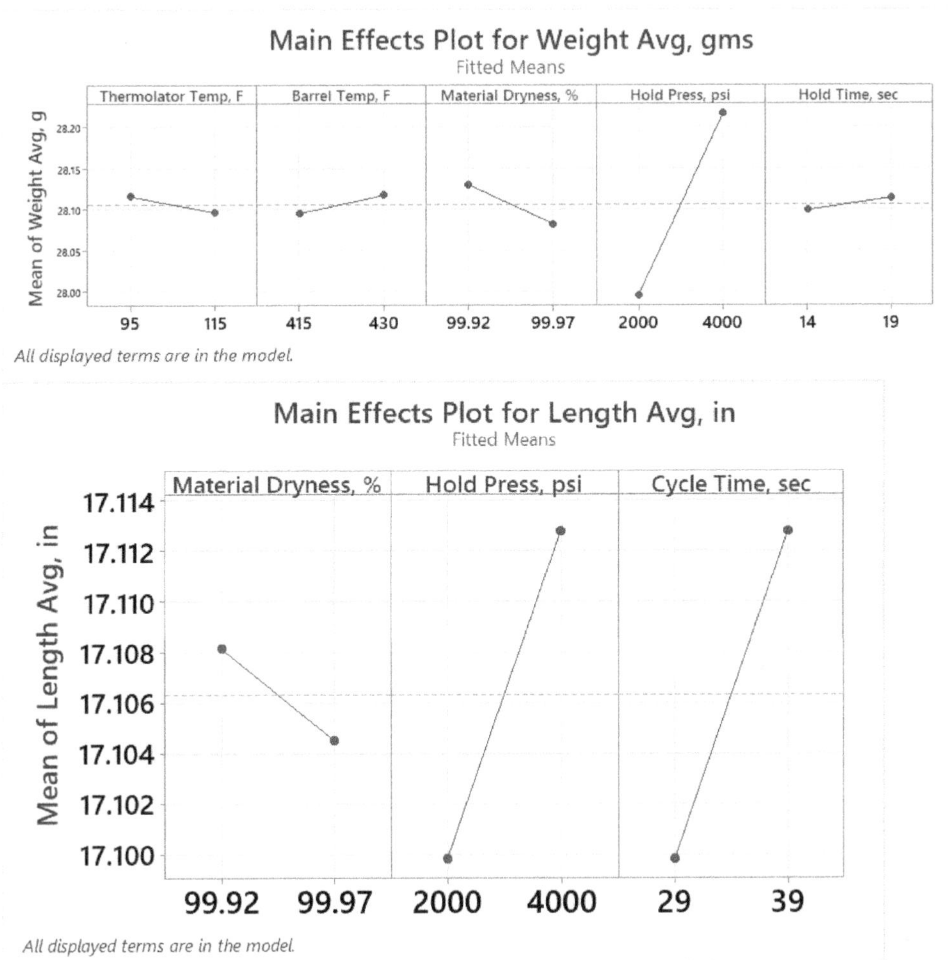

Figure 9.17 Minitab factorial plots for length and weight for Example 9.1

Since weight and length are highly sensitive to hold pressure at a constant cycle time, it would be good to monitor part weight during production as a way of controlling length. As mentioned earlier, weight can be measured very shortly after molding and is a very easy (inexpensive) measurement. This, and the use of dependent process variables to monitor production and to try to determine whether parts are good before they are ejected from the mold, is the subject of the following chapter.

10 Process Monitoring

Chapter 3 reviewed the common methods used to make control charts to monitor the CTQ characteristics or control characteristics of molded parts and assemblies. This chapter will discuss monitoring the CTQ characteristics indirectly via dependent process variables or part characteristics that are not CTQ characteristics. There are several reasons that one would want to do this.

1. The CTQ characteristic typically is not checked until, at a minimum, the part has been cooling for twenty minutes to an hour. In the meantime, more parts that could be suspect are being molded
2. The CTQ characteristic may be difficult or time-consuming to measure
3. Parts may have to be destroyed to measure the CTQ characteristic
4. In most cases, every part or at least every shot can be monitored through process monitoring

Any method of process monitoring will have the risks that were shown in Figure 3.9... some good parts will be called bad (α error) and some bad parts will be called good (β error). Those deemed to be bad can be segregated by a part-removal robot, part-deflector chute, reversing conveyor, or other method. Process monitoring is a common feature on most molding machines and there are also auxiliary systems that can be used, especially in the case of non-machine variables like cavity pressure and mold temperature. Some of the common process monitoring variables that will be discussed in this chapter are:

Machine variables:

- Cushion
- V/P transfer pressure
- Max. pressure

- Fill time
- Screw-recovery time
- Cycle time

Cavity-pressure transducer variables:

- Peak cavity pressure
- Cavity-pressure integral
- Cavity-pressure-decay rate

Cavity surface thermocouple variables:

- Minimum mold temperature
- Mold-temperature rise
- Temperature-rise integral
- Time to sensor

Part variables:

- Part weight
- "Hot" dimensions

10.1 Cushion

Cushion is reported differently on different machines. Some machines capture the screw position at the end of hold time, others capture the minimum screw-forward position, and still others do both. Usually, with a correctly set-up process, the two values are the same. Occasionally, they will be different. One example would be a D3 process where the minimum screw-forward position is the V/P transfer position and, for some reason, the hold pressure is not set high enough to recover the complete screw bounce (see Chapter 11).

From what this author has seen, cushion is the most widely monitored process variable in the injection molding industry, even if it does not correlate with any CTQ characteristic during a typical shot. If it goes to zero or is three times what its usual value for a single shot, those parts are most likely bad and should not be put with production. A one-off anomaly may simply be some contamination causing a check-ring issue. If it repeats over several shots, the machine should shut down so that it can be checked out. At the very least, machines can be set up to detect a bad cushion, defined as being below 0.05 inches and above a value just below the V/P transfer position. Using techniques discussed later in this chapter, these limits could be tightened if cushion show good correlation with a CTQ characteristic.

10.2 V/P Transfer Pressure

This pressure is usually calculated as nozzle pressure from the pressure or force behind the injection screw. Some set-ups utilize a melt pressure transducer in the nozzle to measure the actual melt pressure during injection. While a nozzle pressure transducer is more accurate when working correctly, it tends to be a high-maintenance component. For a D2 process running on a closed-loop velocity-controlled molding machine, the injection velocity should remain constant even if the viscosity of the material changes. Therefore, as can be seen with Equation 10.1 and Equation 10.2 (Hagen-Poiseuille pressure drop equations), the pressure drop (ΔP) will go up or down with the material viscosity. ΔP is the V/P transfer pressure in a D2 process, as the pressure at the flow front is zero.

For a rectangular flow channel,

$$\Delta P = \frac{12\eta LQ}{Wh^3} \tag{10.1}$$

where η is viscosity, L is the flow length, Q is the volumetric flow rate, which equals the injection velocity multiplied by the cross-sectional area of the injection screw or plunger, W is the channel width, and h is the channel thickness ($h < W$).

For a cylindrical flow channel,

$$\Delta P = \frac{128\eta LQ}{\pi d^4} \tag{10.2}$$

Where:

d = the diameter of the flow channel

For a D2 process, setting monitoring limits on the V/P transfer pressure allows for the detection of viscosity changes, which are a common reason to investigate or change the process. Viscosity changes may result from a new batch of material, but they can also mean that there is a problem with temperature control or the moisture content of the material. V/P transfer pressure will also drop if the check ring starts to leak.

10.3 Maximum Pressure

For a basic D2 process with a single injection velocity, the maximum pressure should be the same as the V/P transfer pressure. In addition to the reasons discussed for V/P transfer, a cold slug or other blockage could cause a spike in pressure. For D3 or other processes that require a profiled injection velocity (usually to resolve surface blemishes), the peak pressure may not occur at V/P transfer and maximum pressure may be a more effective variable to monitor.

10.4 Fill Time

The fill time here refers to the time required to reach the V/P transfer position, which is not “full” for a D2 process. For any velocity-controlled filling process, the fill time should not change unless the pressure limit setting is reached. Especially if a velocity profile is used, there is a chance that the fill time will change if tuning parameters have been altered and the machine does not change from one velocity to the next in the same manner. This is more of a process set-up and documentation issue; therefore, fill time is not normally monitored with velocity-controlled processes.

With pressure-controlled filling processes (like iMFLUX), a change in fill time is an indication that the viscosity has changed. Examining Equation 10.1 and Equation 10.2, it can easily be seen that for a constant pressure (ΔP), as viscosity (η) goes up, the flow rate (Q) must decrease by the same percentage. A lower flow rate leads to a longer fill time. In the same way, a lower viscosity will lead to a faster fill time. Because of this relationship, fill time is a good variable to monitor in pressure-controlled fill processes.

10.5 Screw-Recovery Time

The characteristic equation for an extruder is shown in Equation 10.3:

$$Q = k1 \times N - (k2 \times P/\eta) \tag{10.3}$$

Where:

Q = volumetric flow rate of the plastic through the extruder die

N = screw rotational speed (RPMs)

P = pressure in the die head

η = viscosity

$k1$ and $k2$ = constants which depend on the screw geometry

For injection molding, the volumetric flow rate of the plastic to the front of the screw is equal to the distance that the screw travels while turning (shot size minus cushion) divided by the screw-return time (SRT). Thus, Equation 10.4 can be derived from Equation 10.3.

$$SRT = \frac{\text{Shot Size} - \text{Cushion}}{k1 \times N - (k2 \times P/\eta)} \tag{10.4}$$

For injection molding, *P* is the back-pressure setting. As can be seen in this equation, if everything else is constant, an increase in viscosity would lead to a decrease in SRT. It should also be noted that an increase in viscosity would also likely increase cushion, which would also decrease SRT. In the same way, a lower viscosity would lead to a higher SRT. Although not specifically shown in Equation 10.3 and Equation 10.4, how well the plastic pellets or regrind particles feed onto the back of the screw and how they melt once on the screw can also affect SRT. Larger "chunks" that typically occur in regrind sometimes do not feed as well and do not melt as quickly on the screw and lead to an increased SRT.

10.6 Cycle Time

Cycle time should not normally change unless a setting is altered. If SRT increases enough to exceed the cooling time setting or if viscosity increases in a pressure-controlled process, the cycle time may also increase. Another indicator is an increase in the amount of time to travel the mold protection distance. This change can indicate that the mold temperature has increased or that something is on the parting line.

10.7 Peak Cavity Pressure

If sensors are installed in the mold, whether directly on the cavity surface or indirectly behind ejector pins or other pins, cavity pressure curves can give a great deal of information about the process.

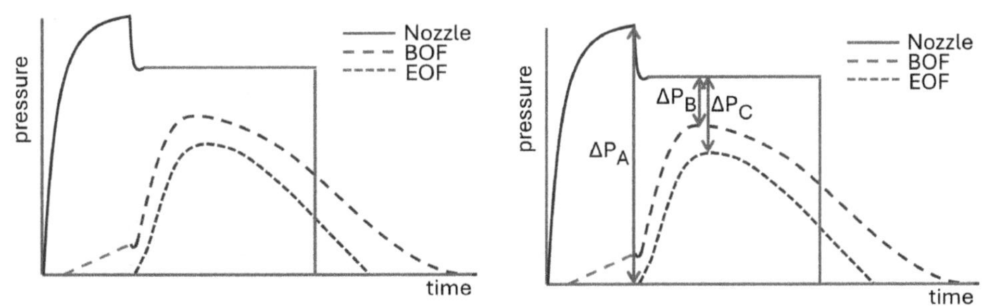

Figure 10.1 Representative pressure curves for a D2 process: nozzle, beginning-of-fill (BOF), & end-of-fill (EOF). The chart on the right shows pressure drops from the nozzle to the cavity

Three pressure drops are labeled in Figure 10.1. Pressure drop A is the one previously discussed as the V/P transfer pressure for a D2 process. Pressure drops B and C are the pressure drops from the constant hold pressure to a beginning-of-fill and an end-of-fill pressure sensor, respectively. All three pressure drops will increase if the viscosity of the material increases. As discussed earlier, that will lead to an increase in V/P transfer pressure. Since, for a D2 process, the hold pressure is held constant on a closed-loop controlled machine, the peak cavity pressure will decrease when the pressure drop increases. The extent of the decrease under a constant nozzle pressure is complicated by the fact that several variables will change in Equation 10.1 when the pressure drop increases. A higher pressure drop causes the flow rate (Q) to decrease, which will cause the frozen-layer thickness to increase, which effectively reduces flow thickness (h), which causes the pressure drop to increase. The decreasing flow rate also reduces the shear rate, which increases the viscosity, which causes the pressure drop to increase. However, a decrease in flow rate on its own decreases the pressure drop. The extent of these pressure changes can be predicted by injection molding simulation programs – though, as George Box said, "all models are wrong, but some are useful."

It may not be easy to predict, but monitoring of the peak cavity pressure can be a good indicator of when parts need to be segregated and/or when the process requires adjustment. Like cushion, even if CTQ characteristics do not correlate perfectly with peak cavity pressure, limits can be established to prevent short shots and, potentially, gross flash. Installing an end-of-fill sensor in every cavity or at least in the last cavity to fill in a mold and setting a limit of 100 psi (normal pressures are several thousand psi) or some very low number and using an automatic sorting method should prevent short shots from reaching the customer. A high limit, set well above normal, can catch flashed parts.

10.8 Cavity Pressure Integral

The area under a cavity pressure curve is somewhat of a catch-all for monitoring the process.

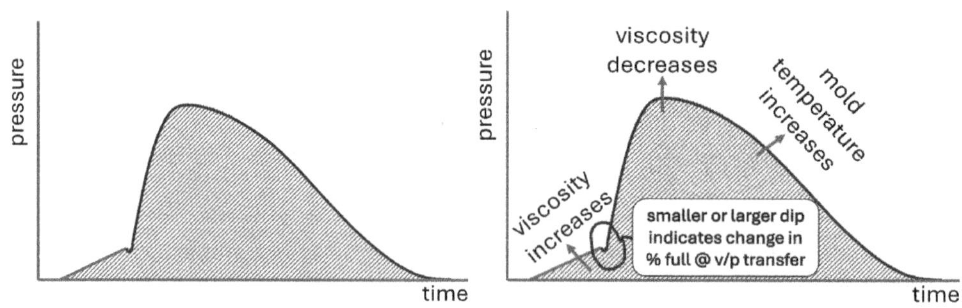

Figure 10.2 Cavity pressure integral of a typical BOF curve for a D2 process. The chart on the right shows some of the variations that would change the integral

Some of the variations that would change the integral or area under the curve are shown in Figure 10.2. The referenced "dip" occurs when switching over from velocity control to pressure control. A larger dip would reduce the cavity pressure integral and possibly indicate that material is leaking past the check ring, meaning that the part is not as full at the point of transfer. If the part is less full at transfer, the peak cavity pressure may also end up being lower.

The cavity pressure integral can be calculated for the entire cycle, as shown. It could also just be isolated to the first stage of injection (before V/P transfer). The "fill" integral will generally pick up both viscosity changes and changes in the %-full at transfer.

10.9 Cavity Pressure Decay Rate

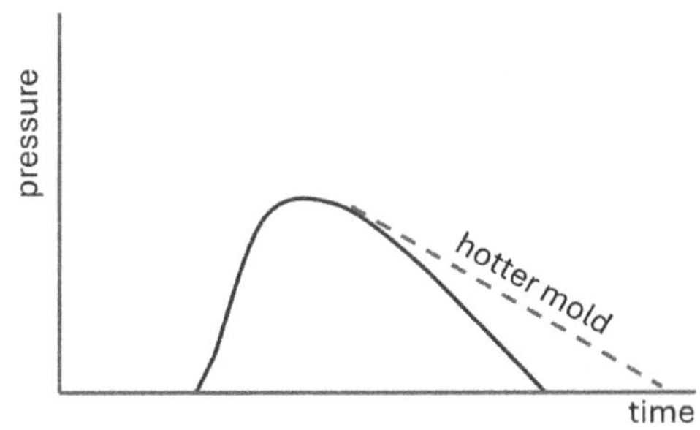

Figure 10.3
How EOF cavity pressure changes with a hotter mold

How fast the cavity pressure decays (psi/s) after the part is packed out is an indication of how quickly the plastic is cooling in the cavity. RJG calls this parameter the cooling rate. The dashed line in Figure 10.3 shows how the pressure would decrease more slowly with a hotter mold. If a part has CTQ characteristics that are sensitive to mold temperature changes, the cooling rate can be an important variable to monitor. However, in such cases, it would be better to monitor the variable discussed in the following section.

10.10 Minimum Mold Temperature

A typical surface thermocouple temperature curve is shown in Figure 10.4.

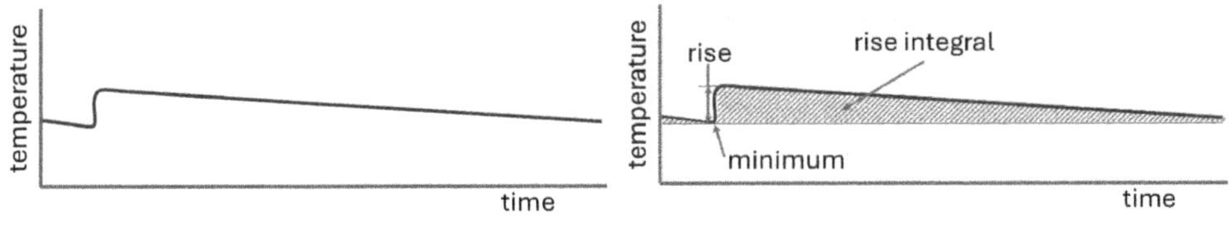

Figure 10.4 Typical mold surface thermocouple temperature curve. The chart on the right shows three commonly monitored variables

The minimum mold temperature typically occurs in the cycle just before the plastic melt hits it. Monitoring it is essential when there are CTQ characteristics that are sensitive to temperature fluctuations. It is also a very good metric for determining when to start collecting parts after a mold has been idle for a few minutes or during first start-up. It is common practice for process technicians to have procedures in place to delay parts collection for production until several cycles have been run. The two main reasons are 1) to allow enough shots to ensure a steady-state melt residence time (see Equation 9.1) and 2) to allow the mold temperature to reach its steady state. Usually, reason number 2 requires more cycles and an example of what is happening to the temperature is shown in Figure 10.5.

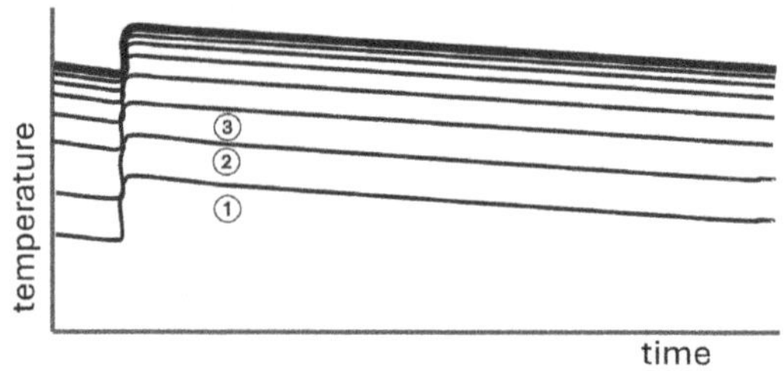

Figure 10.5
Start-up surface temperature curves (15 shots). The first three shots are labeled. The temperature appears to reach a steady state at about shot 8 or 9

Different lengths of downtime will require different numbers of shots to reach steady-state temperature. It is also not uncommon in a busy production environment for a manual count of parts to be shortened. An automated process-monitoring system should ensure that the mold is always up to temperature before parts are saved for production. The minimum mold temperature is also an early detector of when a thermolator kicks off or of other cooling issues that cause the mold to get too hot.

10.11 Mold-Temperature Rise

Some of the things that could cause a change in the melt temperature rise shown in Figure 10.5 include a thermocouple problem, a cooling circuit issue, or a change in the melt temperature. For the last issue, the next variable discussed is generally better. Historically, measuring the actual melt temperature has proved difficult. There are sensors that can measure the actual melt temperature, but repeatability and production-robustness are usually an issue.

10.12 Temperature-Rise Integral

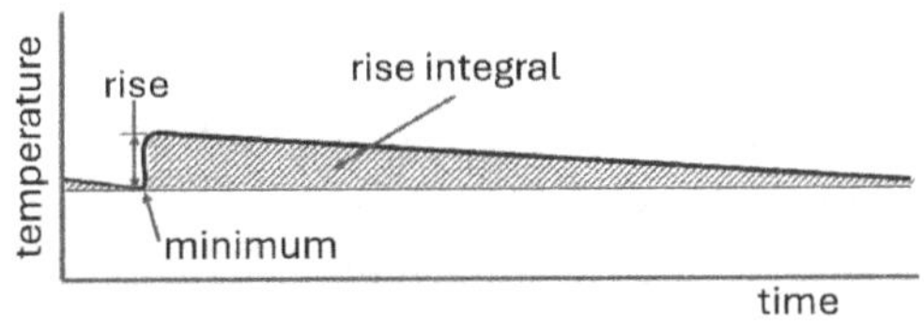

Figure 10.6
Mold-temperature rise integral

A research paper by Johnston et al., presented at the 2006 SPE ANTEC, detailed a method for predicting melt temperature based on information garnered from mold-surface temperature data – specifically, the integral shown in Figure 10.6. Equation 10.5 was derived from that paper to predict the melt temperature, T_2.

$$T_2 = T_1 + \frac{2}{h\rho C_p} \int_0^t \frac{k(T_S(t) - T_1)}{\sqrt{\pi \alpha t}} dt \quad (10.5)$$

Where:

T_1 = initial mold temperature

t = time

T_s = sensed mold surface temperature

h = part thickness

ρ = polymer melt density

C_p = polymer melt specific heat

k = thermal conductivity of the mold material

α = thermal diffusivity of the mold material

The equation demonstrated a good correlation with the melt temperature at a given mold temperature. The integral shown in Figure 10.6 also only accounts for $(T_S(t) - T_1)dt$ and does not account for the time shown in the denominator. Therefore, the correlation between the mold temperature-rise integral and the melt temperature is only valid if the cycle time remains constant.

10.13 Time-to-Sensor

This value can be derived from either a pressure sensor or a surface-temperature sensor. Comparison of Figure 10.1 and Figure 10.4 illustrates why temperature is preferred. Unless the pressure sensor is located right at the end-of-fill or in an area where the melt is converging quickly, the pressure rise during fill is much more gradual than the temperature rise. Another option that has been shown to perform as effectively as temperature is the use of an in-mold contact switch (MeltSwitch).

If the time-to-sensor fluctuates too much, this can be an indication that there is a problem with the check-ring shut-off. That can cause variations in %-full at V/P transfer or, in the case of a profiled injection velocity, variations in how full the part is when the velocity changes. The example in the preceding chapter is one where maintaining a consistent transfer to a slow fill at the end of the cavity appears to be important. For multi-cavity molds, placing a sensor or switch in each cavity allows the balance to be monitored. If a gate becomes partially plugged and one cavity doesn't fill, it is generally good practice to reject the entire shot. If the problem persists, the machine could be shut down until someone checks it out.

10.14 Part Weight

Weight is included in this section because, it typically isn't a specified part characteristic, it sometimes correlates well with one or more CTQ dimensions of the part. Two advantages of weight measurements are: 1) that it is a relatively inexpensive measurement from both an equipment and a time standpoint and 2) parts can be measured before they have cooled enough for the dimensions to stabilize. Some commercial applications weigh every part immediately after molding and predetermined monitoring limits dictate whether the part is rejected or accepted.

10.15 "Hot" Dimensions

As is the case with weight, parts are measured immediately or shortly after molding. It is not uncommon to check every part with a camera to ensure that all features are present and that the part is full and does not have gross flash. When dimensions are checked, the required measurement resolution and precision can cause the cost to rise significantly. The advantage over weight is that "hot" dimensions almost always correlate much more accurately with cold dimensions than weight does.

The variable discussed represent just a partial list of what can be monitored during every shot. A paper by Johnson & Johnson (presented at the 2016 SPE ANTEC) detailed the information that can be obtained from an infrared temperature sensor embedded in the mold. Some other variables not discussed include water or other coolant flow rates through each circuit in the mold, coolant temperature into and out of the mold, material dryness and dryer dewpoint. Some practitioners like to put sensors in every mold and some like to only use sensors on the machine and auxiliary equipment. In addition to the initial cost of sensors, the added ongoing cost of maintaining them in good working order needs to be factored into the decision-making process.

10.16 Setting Monitoring Limits

Setting up of monitoring limits for gross defects has already been discussed for some of the monitoring variables. If the goal is to catch smaller deviations in part CTQ characteristics, there are several ways to set up initial limits. Monitoring limits should never be treated as "set and forget." There should be some method to occasionally check the percentage of rejected parts that actually fall within the operating specification limits for each CTQ characteristic. The strategy for adjustment depends on the consequences of undesirable parts reaching the customer but, in general, if all the rejected parts are "bad", the monitoring limits need to be tightened and, if all the rejected parts are "good", the limits need to be widened.

Three methods of setting up monitoring limits are:

1. Guess
2. Use data collected during a process capability study
3. Use data collected during a DOE

The guess is used a lot. It is usually an educated guess based on similar parts and what works for those parts. If the process monitoring data is collected during a capability study, limits can be set based on the estimated standard deviation observed in the monitoring variable during the study. If the data is collected during a DOE, the average variation observed within each run can be used to estimate the standard deviation. Using data from the DOE example from the preceding chapter, starting widths will now be determined for monitoring the process. Before setting these limits, it is a good idea to identify which variables make sense to monitor based on how well they correlate with the CRT characteristics.

The correlation coefficient (r), is a statistic which measures the relationship between a variable (x) and a response variable (y) and is calculated as shown in Equation 10.6.

$$r = \frac{\sum_{i=1}^{n} y_i \left(x_i - \overline{x}\right)}{\sqrt{\sum_{i=1}^{n} \left(x_i - \overline{x}\right)^2 \sum_{i=1}^{n} \left(y_i - \overline{y}\right)^2}} \qquad (10.6)$$

Correlation coefficients of 1.0 and −1.0 are perfect correlations. This means that if you are monitoring variable x, you would know what variable y is going to be before ever checking it. However, this never happens in practice. r^2 is the coefficient of determination that was discussed in Chapter 7. A correlation coefficient of 1.0 or -1.0^2 is equal to 1.0, so it can be said that 100% of the variation in response y could be explained by the variation in x. Perfect correlations never occur in injection molding and are rare om any process. If they did, process monitoring limits could set up at the same width as the control limits discussed in Chapter 3 (± 3 standard deviations). Equation 10.7 is what this author recommends using for starting monitoring limits.

$$\text{Half of monitoring limits width} = \min\left(\frac{3}{r^2}\tilde{\sigma}_X, 6\tilde{\sigma}_X\right) \tag{10.7}$$

The limits would be set at a baseline average ± the result of Equation 10.7. Whichever CTQ characteristic that the monitoring variable correlates best with determines the width to use.

10.16.1 Example 10.1: Setting Up Process Monitoring Limits Using DOE Data

The average for each of the responses and the process monitoring variables are shown in Figure 10.7, as well as how to do the correlation analysis.

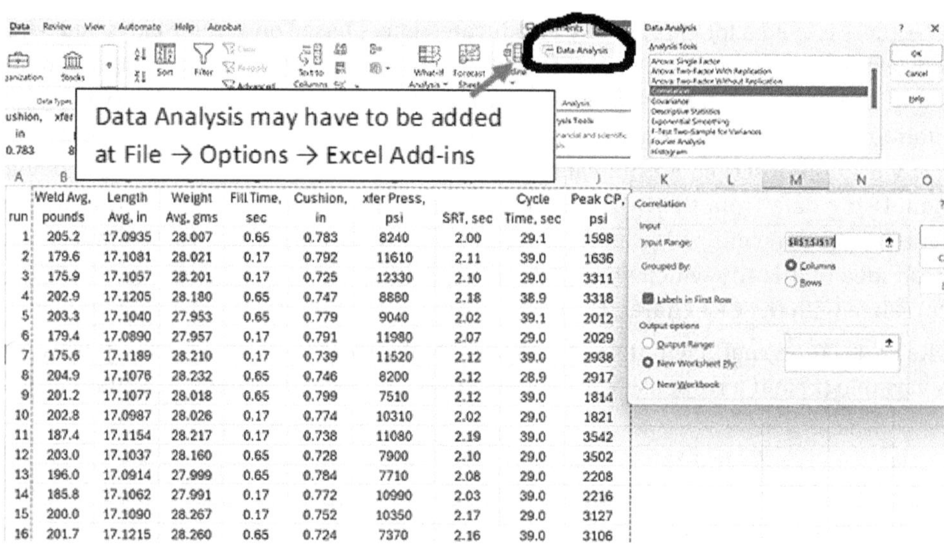

run	Weld Avg, pounds	Length Avg, in	Weight Avg, gms	Fill Time, sec	Cushion, in	xfer Press, psi	SRT, sec	Cycle Time, sec	Peak CP, psi
1	205.2	17.0935	28.007	0.65	0.783	8240	2.00	29.1	1598
2	179.6	17.1081	28.021	0.17	0.792	11610	2.11	39.0	1636
3	175.9	17.1057	28.201	0.17	0.725	12330	2.10	29.0	3311
4	202.9	17.1205	28.180	0.65	0.747	8880	2.18	38.9	3318
5	203.3	17.1040	27.953	0.65	0.779	9040	2.02	39.1	2012
6	179.4	17.0890	27.957	0.17	0.791	11980	2.07	29.0	2029
7	175.6	17.1189	28.210	0.17	0.739	11520	2.12	39.0	2938
8	204.9	17.1076	28.232	0.65	0.746	8200	2.12	28.9	2917
9	201.2	17.1077	28.018	0.65	0.799	7510	2.12	39.0	1814
10	202.8	17.0987	28.026	0.17	0.774	10310	2.02	29.0	1821
11	187.4	17.1154	28.217	0.17	0.738	11080	2.19	39.0	3542
12	203.0	17.1037	28.160	0.65	0.728	7900	2.10	29.0	3502
13	196.0	17.0914	27.999	0.65	0.784	7710	2.08	29.0	2208
14	185.8	17.1062	27.991	0.17	0.772	10990	2.03	39.0	2216
15	200.0	17.1090	28.267	0.17	0.752	10350	2.17	29.0	3127
16	201.7	17.1215	28.260	0.65	0.724	7370	2.16	39.0	3106

Figure 10.7 Correlation analysis in Excel

The results of the analysis are shown in Table 10.1. The two CTQ characteristics that are of interest in this example are weld pull strength and length. The highlighted correlation coefficients will be used to set monitoring limits other than for fill time and cycle time. Fill time and cycle time should not change and their limits will be set in the standard way discussed in Chapter 11.

Table 10.1 Correlation Coefficients from Excel for Example 10.1

	Weld Avg, pounds	Length Avg, in	Weight Avg, g	Fill Time, s	Cush-ion, in	Transfer Press, psi	SRT, s	Cycle Time, s	Peak CP, psi
Weld avg, lb	1.00								
Length avg, in	-0.03	1.00							
Weight avg, g	0.03	0.71	1.00						
Fill time, s	0.75	-0.01	-0.04	1.00					
Cushion, in	0.04	-0.61	-0.87	0.02	1.00				
Transfer press, psi	-0.85	-0.01	-0.02	-0.93	-0.02	1.00			
SRT, s	-0.07	0.72	0.78	-0.03	-0.54	-0.01	1.00		
Cycle time, s	-0.17	0.68	0.00	0.00	0.02	0.04	0.29	1.00	
Peak CP, psi	-0.04	0.60	0.87	-0.01	-0.92	0.03	0.72	0.00	1.00

The correlation analysis could have also been done in Minitab, as shown in Figure 10.8. The results are not shown, but are the same as those in Excel.

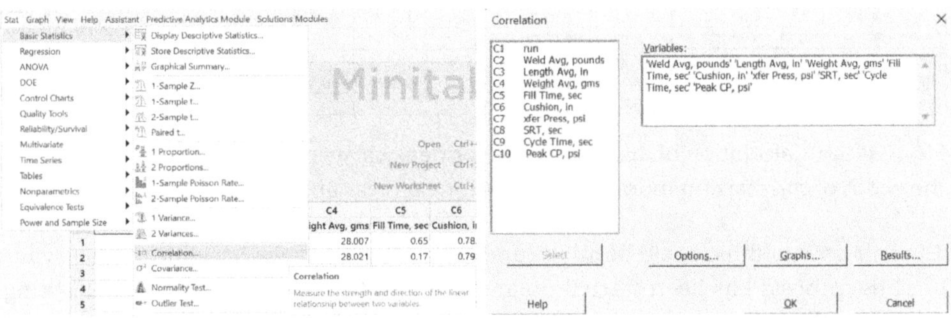

Figure 10.8 Correlation analysis in Minitab

The correlation numbers all align with the DOE results. The injection velocity was the main factor affecting the weld pull strength and that would change both fill time and

V/P transfer pressure. Analysis of the fill pattern suggests that the speed of the plastic melt front as it moves around the rectangular openings is critical. Therefore, adding a surface thermocouple or contact switch to make sure that the slower speed starts at the correct position would likely be more effective than monitoring in this case. Hold pressure and cycle time were the biggest effects for the length dimension. Cycle time has already been discussed and will be held constant. Weight was also most affected by hold pressure, explaining the good correlation between length and weight. Hold pressure affects peak cavity pressure and, since it changes how far the screw travels, it influences both cushion and screw-recovery time.

The estimated standard deviation for each variable can be determined from the average range of the runs for each variable. Figure 10.9 shows all the numbers for peak cavity pressure, the summary numbers for all the variables and the numbers used to calculate the monitoring limits for each variable.

CP1	CP2	CP3	CP4	CP5	CP6	CP7	CP8	CP9	CP10	Average CP	Range
1601	1585	1581	1593	1606	1573	1618	1615	1608	1598	1598	45
1656	1633	1628	1667	1607	1633	1622	1655	1638	1623	1636	60
3294	3310	3292	3324	3316	3320	3322	3321	3304	3308	3311	32
3314	3315	3296	3319	3323	3332	3316	3332	3324	3311	3318	36
2044	2014	2006	2006	2019	1998	2004	2020	2010	1997	2012	47
2041	2039	2023	2040	2009	2017	2033	2032	2030	2023	2029	32
2957	2920	2920	2940	2928	2966	2943	2924	2943	2937	2938	46
2906	2915	2915	2917	2937	2921	2901	2917	2904	2932	2917	36
1828	1822	1810	1822	1800	1821	1815	1798	1805	1819	1814	30
1827	1850	1838	1813	1817	1821	1797	1798	1830	1821	1821	53
3559	3529	3543	3539	3540	3552	3541	3529	3547	3543	3542	30
3486	3517	3485	3496	3506	3537	3494	3511	3485	3500	3502	52
2190	2199	2209	2213	2184	2192	2216	2212	2234	2226	2208	50
2213	2216	2220	2215	2238	2238	2240	2193	2197	2186	2216	54
3098	3144	3138	3136	3133	3117	3123	3128	3114	3135	3127	46
3112	3080	3098	3115	3101	3116	3090	3138	3110	3101	3106	58
										Average Range =	44.2

	r	r^2	$3/r^2$	min($3/r^2$, 6)	Avg Range	d_2	$\tilde{\sigma}_x$	Half-Limit Width
Weight Avg, g	0.71	0.50	6.0	6.0	0.0232	2.326	0.00997	0.0598 g
Cushion, in	-0.61	0.38	7.9	6.0	0.0773	3.078	0.0251	0.1506 in
Transfer Press, psi	-0.85	0.72	4.2	4.2	98.8	3.078	32.1	134 psi
SRT, s	0.72	0.51	5.9	5.9	0.257	3.078	0.0835	0.489 s
Peak CP, psi	0.60	0.36	8.3	6.0	44.2	3.078	14.36	86 psi

Figure 10.9 Calculation of average range for peak cavity pressure and determining the width of the starting monitoring limits for each variable

The half-limit widths would be added and subtracted from the baseline average value once the process has been started up and stabilized with the recommended settings from the DOE. With some monitoring systems, such as RJG, it would only be necessary to input the number of estimated standard deviations on each side of the average, namely min($3/r^2$, 6), and the software will then calculate the limits. Again, these limits are starting values, and they are only as good as they are effective – it is recommended not to "set and forget" when monitoring the injection molding process.

11 Injection Molding Process Set-Up Strategies

Before doing a DOE or performing a capability study to qualify injection molded parts, a baseline process must be developed. This author has discovered that there are multiple ways that good injection molded parts can be made. Since entering the injection molding industry in 1986, the author has used, at various points in time, the techniques discussed in this chapter to make good parts. There were microprocessor-controlled machines in the facility in 1986 but, as the youngest engineer, this author was assigned the thermoset bay, where it was hot and dirty. The thermoset bay had the oldest machines that were controlled mostly by turning knobs, moving toggle switches, and adjusting timers. The phenolic thermoset materials were black or dark green and came as a powder that was very dusty. All the thermoset molds were run at over 300 °F. There were about twelve injection molding machines in the bay, and they weren't all thermoset: about half ran products made with engineering grade thermoplastics.

11.1 Open-Loop, Pressure-Limited Injection Molding

The molding technique historically used on those machines is what will be referred to in this book as open-loop, pressure-limited injection molding. The main problem with open-loop control for injection molding is that the machine did not change anything to maintain injection velocity or pressure when the viscosity of the plastic material or the hydraulic fluid varied. Valves were manually opened to a given position. A schematic of 1970s vintage machine and what some typical injection pressure and injection velocity curves might have looked like (no process curves or monitoring on those machines) is shown in Figure 11.1.

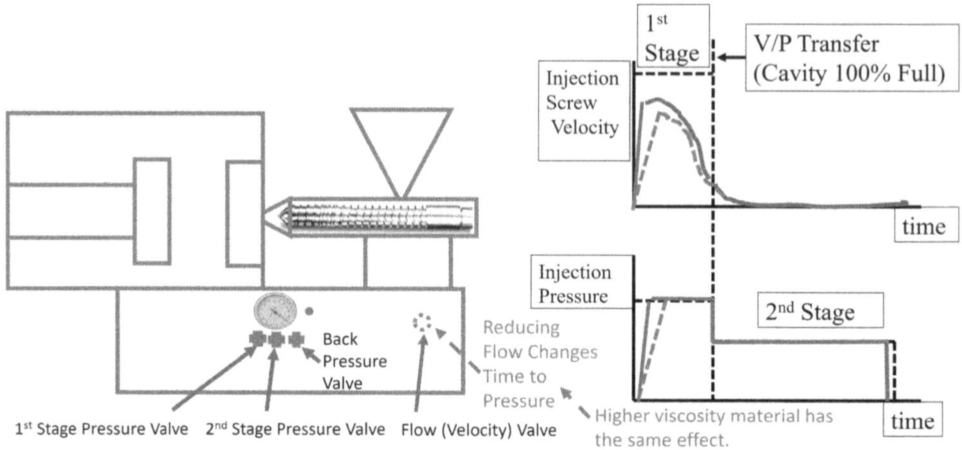

Figure 11.1 Schematic of 1970s vintage machine and representative pressure and velocity graphs

The dial shown in Figure 11.1 is an analog pressure gauge whose needle would show the pressure if the button next to the dial was pushed. Adjustment to a set pressure was done through trial and error by pushing the button and opening or closing the appropriate valve (1st stage, 2nd stage, or back pressure) to increase or decrease the pressure. Back pressure was typically set to 50 psi and was rarely changed from that. When determining the process for a new mold, this author learned to use the following steps for thermoplastics – the main difference from thermosets was a much longer cure time (cure instead of cooling).

1. Starting conditions: Set the shot-size toggle switch as needed or to about half the maximum if it hadn't been calculated beforehand, close the 2nd stage pressure valve, open the 1st stage valve about a quarter of the way, open the flow control valve all the way (it was around the back of the machine and was only adjusted if absolutely necessary), set the 1st stage timer to 1 second, the hold timer to 10 seconds, and the cooling timer to 20 seconds. Screw RPMs were adjusted to make sure shot size was reached a few seconds before the mold opened.

2. Start running and adjust the 1st stage pressure until the part is just full. Then keep increasing the pressure until the part barely flashes or is overpacked. Set the pressure to halfway between just full and just flashed. Take care to ensure that a cushion is maintained at all pressures, and the shot-size toggle switch is adjusted to give about a half inch of cushion for the final pressure. V/P transfer was time-based. At this point, it is important to inspect the part to see if the first-stage timer needs to be increased. Jetting at the gate would be a reason to slow down at the beginning by partially closing the flow control valve. The flow control valve had colored rings and numbers so that the valve position could be documented (blue 7 for example). The dotted lines in Figure 11.1 show what could happen to the veloc-

ity and pressure if this was done – not shown is that the 1st stage time would have to be increased to obtain the same packed part. Very rarely was a faster time needed for the 1st stage timer, but that could be obtained by shortening the time and increasing the pressure.

3. Add 500 psi of hold pressure and adjust if needed. The cushion usually ended up being about a quarter of an inch.
4. Do a gate freeze study as discussed in the next section and set the hold pressure to about one second above gate freeze.
5. Minimize the cooling time, adjusting screw RPMs if needed to reach shot size a few seconds before the mold opened.

Retrofitted toggle switches for V/P transfer significantly improved these machines. Before that improvement, both 2nd stage pressure and 1st stage pressure would have to be adjusted often to compensate for material viscosity variations (see Equations 10.1 and 10.2) and their effect on injection speed. After the retrofit, adjustments were usually limited to 2nd stage pressure. Even without the retrofit, once these machines produced many very good parts once oil and machine temperatures stabilized and material viscosity remained consistent. The problem was that the machines needed frequent adjustment to compensate for fluctuations and the resulting scrap rates were higher on those machines than on the newer, computer-controlled machines.

Sometime in 1987, this author expanded his horizons past the open-loop machines in the thermoset bay. There were still some open-loop machines in the other areas of the facility, but the majority were closed-loop controlled and there were at least five different brands. In those days, it was not uncommon for processing strategies to vary by machine or even by shift. To standardize operations, the facility implemented the following process strategy for all new molding jobs starting in the late 1980s.

11.2 D2 Molding Process

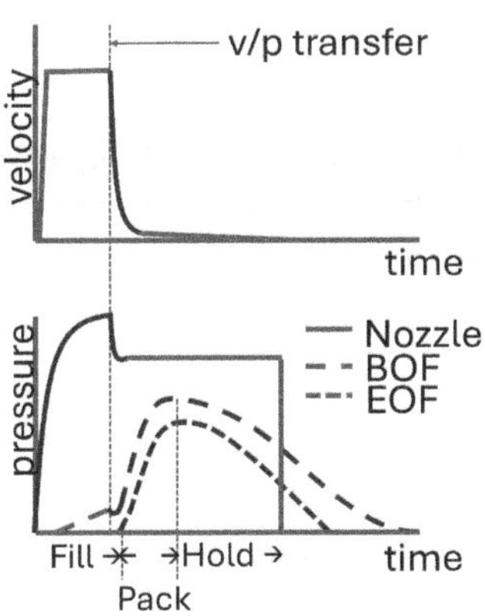

Figure 11.2
D2 velocity and pressure curves

The remainder of this chapter will discuss methods used on today's closed-loop control molding machines. Anyone who has taken an injection molding class at Penn State Behrend in the Plastics Engineering Technology program or in a non-credit workshop since 1994 will recognize this as a 2-stage set-up. In RJG training, this is known as DECOUPLED MOLDING II. This method decouples the fill and pack stages, 1st stage velocity controls 95–98% of the fill while the hold pressure completes the remaining fill and manages the pack and hold portions of injection, as shown in Figure 11.2. For this book, this author has decided to use D2, because it is shorter and because many in industry are familiar with the D2, D3 and D1 terminology.

This is the method that this author starts with for any new mold; consequently, the following steps will be reviewed more thoroughly. While using a single injection velocity and a single hold pressure is the simplest process to set up and repeat, there are occasions where profiled velocity and/or pressure or a different processing strategy is justified.

1. Set the clamp
2. Get parts 95% full at maximum injection velocity
3. Set the injection velocity
4. Set the hold pressure
5. Set the hold time
6. Minimize the cycle
7. Set other parameters

Note: For this initial set-up, the barrel temperatures and mold or mold coolant temperature should be set mid-range of the material supplier's recommendation. Check that the purge melt temperature is within 5 °F of the setting for the front zone barrel temperature. If it is not, adjust the rear and middle zones until the purge temperature falls within this five-degree range.

11.2.1 Set the Clamp

This is the step that technical people are most likely to get right without knowing anything about plastics. However, it is also an area that is not always optimized with respect to minimizing cycle time. The primary objectives here are:

- Open and close the mold fast, but smoothly
- Open only as much as needed for the parts to drop
- Eject the parts fast without damaging them
- Minimize ejector strokes (normally shouldn't need more than one)

- Set mold protection to optimize protection and cycle time
- Set the clamp force as needed based on the projected area of the part and the material being molded (typically around 3 ton/in.^{2} of projected area)

The machine manuals typically provide good instructions on how to determine machine settings to accomplish these objectives. Most have either controls that will ramp speeds when accelerating or decelerating or that allow profiled speed settings during mold and ejector forward and backward movements. Sudden starts or stops of the clamp can cause the molding machine to jump around or cause damage if the plates slam together. Sudden starts when ejecting can damage the plastic parts. An underutilized feature of many molding machines is "ejection-on-the-fly". This feature allows for ejection to start before the mold is completely open and will shorten cycle time in most instances.

When running a mold for the first time, it is a good idea to dry-cycle it first. This is done by disabling the injection unit and cycling without injecting plastic. Water, or other cooling medium, should be circulated through the mold and, if there is a hot runner manifold, the heaters should be turned on. The flow rate of the water through each cooling circuit should be checked to guarantee that it has sufficient turbulence (see Figure 11.3). Dry-cycling should initially be done at slow speeds, with low pressure and clamp force to minimize the chance of damaging anything that is not installed or fitted correctly. It is especially important to start with a low mold-protection pressure. As the mold cycles, listen for unusual sounds that may indicate binding or galling. Do a visual check for this when the mold is open. Gradually increase speeds and pressures so as to minimize the dry-cycle time without compromising mold protection. Record this dry-cycle time to serve as a baseline for comparison once plastic injection begins.

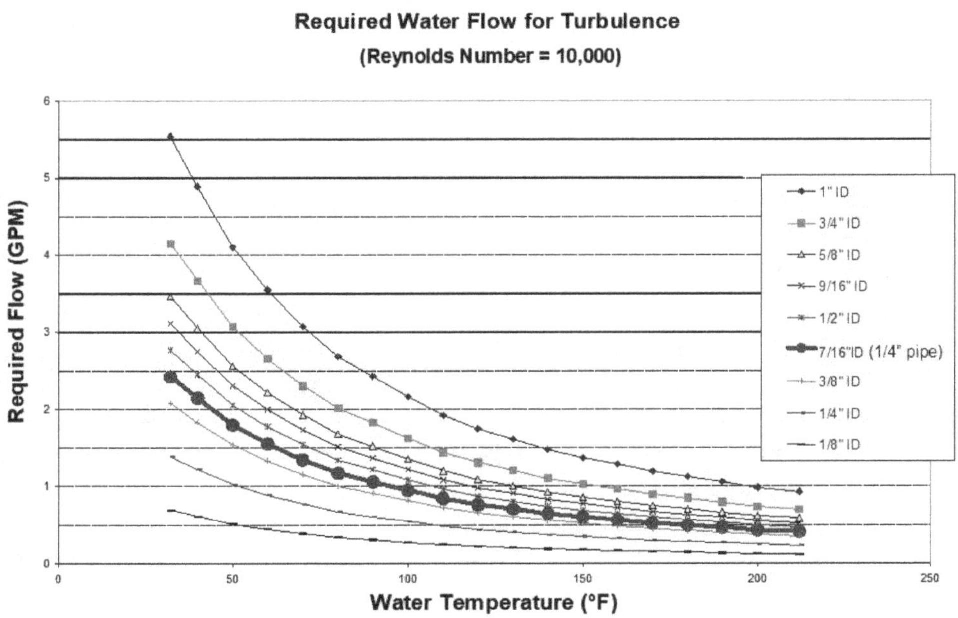

Figure 11.3 Required water flow for various cooling line diameters

11.2.2 Get Parts 95% Full at Maximum Injection Velocity

Table 11.1 Initial Settings

Transfer position	0.5 to 1.0 in.
Shot size	To fill about 75% of the part
Injection velocity	Mid-range
Injection pressure limit	Mid-range
Hold (2nd stage) pressure	0
Hold time	At least a couple seconds longer than anticipated gate seal time (typically 10–20 s)
Cooling timer	Set so that total cycle time is about 25% higher than quoted

Note: Some machines will have a minimum hold pressure even though they allow a setting of zero. If this hold pressure is enough to cause the screw to continue its forward motion during the hold time, set hold time to zero and set screw rotation delay to the original hold time. You must remember to reverse these two times when you get to Step 4 (setting hold pressure).

The goal here is to get the plastic flow front (in the first cavity to fill) just short (generally 95–98% full) of the last place in the mold to fill while injecting the plastic as fast as possible. It is usually not advisable to set the injection velocity to the maximum setting for the first shot into the mold, especially with molds sensitive to overpacking. A mold is overpacked when too much plastic is forced into the mold under excessive pressure. This can break fragile mold inserts or make the part difficult to eject from the mold. It can also end your day with that mold, as it may need to be disassembled and reassembled to clean out flash between inserts. Gradually adjust the shot size, injection velocity, and injection pressure limit until the part is just short. When at the maximum velocity, make sure that the injection pressure limit is at least 2,000 psi (at the nozzle) above the pressure it takes to reach the 95–98% full point. This should guarantee that the pressure does not limit the velocity, that is, that the velocity and pressure are not coupled.

Another sign of overpacking is flash at the mold parting line or between inserts. A mold, especially when new, should not flash unless it is overpacked. That is why we fill the mold to only approx. 95%, instead of all the way. To use an analogy: when Humpty Dumpty fell off the wall, it wasn't the speed of the fall that broke him, it was that sudden stop at the bottom. Similarly, the pressure inside a mold typically does not build up rapidly until the plastic hits that last place to fill in the mold. If the melt hits that last place too fast, the result is a sudden pressure spike in the mold and, quite often, flash. The practice of transferring to hold pressure when the part is slightly short (95–98% full) allows the plastic melt to slow down and, in effect, acts as a safety net. The lower velocity results in slower pressure build-up in the mold, preventing flash. If Humpty Dumpty had landed on a safety net, or at least a bush, he might still be together today.

Once you get the part(s) 95% full, there are a few things to look for before proceeding to the next step. It is possible that the mold may need some fixes before it is ready to run in production. In most cases, it is better to correct mold design and construction issues now, rather than during part qualification or after releasing the mold to production. Some things to look for include:

a) Flash – As previously discussed, short shots should typically not have flash. If flash occurs at the parting line, it is likely that either the clamp pressure is not set high enough or that the molding machine lacks the necessary clamp force for the mold. If the flash occurs because the slides are being pushed back, the root causes are likely either insufficient clamp force or a poor fit between the wedge block and the slide. Flash appearing between inserts or around ejector pins indicates too much clearance or deflection of the mold walls. Any of these problems should be addressed before proceeding. If the problems are minor, you may be able to finish the process set-up and fix the flash problem afterwards.

b) Burning – If the burn is on the surface and is caused by trapped air, venting in the mold needs to be addressed. These types of burns can eat away the steel and cause major problems. If the burn is just underneath the surface and is caused by high shear rates from injecting at high velocity, this should be noted as a possible limit to the maximum injection velocity. Burns caused by high shear rates are most common with PVC.

c) Jetting, gate blush or flow lines – These defects can be caused by an injection velocity that is either too fast or too slow in certain areas of the mold. They can generally be "processed out" by profiling the injection velocity settings, but it is usually better to modify the mold or gate design. At this stage, these defects should just be noted.

d) Cavity balance – If running a multi-cavity mold, the parts will ideally all fill at the same time. Since this is not practical, the goal should be to get the weights of the fill-only shots all within 5% of each other when the fullest part is 95–98% full. If the flow converges to an area of the part that is thinner than the nominal wall at end-of-fill, then the balance should be checked before the plastic reaches the thin sections of the first cavity to fill.

11.2.3 Set the Injection Velocity

The part is now 95–98% full at the maximum velocity. If none of the problems mentioned in the previous section is present, is this the optimum speed to run in production? Doubtful. There are other quality problems that may be more difficult to detect.

This step will detail a method to determine an initial injection velocity. Faster and/or slower speeds will then usually be tried in a DOE to see how injection velocity affects the CTQ characteristics of the part.

There are other methods to determine an initial injection velocity. As previously noted, this author's first experience with setting up molding processes used a "magic" fill time of one second. Some software flow analysis programs will automatically determine an "optimum." In simulation, the primary goal when selecting a velocity for a given mold configuration is for the software to calculate the most uniform melt temperature at the instant of fill that leads to the lowest overall stress in the part. Too fast will shear-heat the plastic as it fills and the areas close to the gate will freeze off before the end-of-fill areas. Too slow and too much heat is lost, which can lead to high stresses toward the end of fill. The theory is sound, but the results are based on imperfect material data and imperfect mathematical models of the flow and heat transfer that is occurring. As mentioned previously, George Box famously said, "all models are wrong, but some are useful". Today's commercial flow analysis programs are useful, but the simulation results are not always the same distance from reality. The velocity chosen is just a starting point, so this strategy saves material and machine time over the procedure that will be reviewed here. In 2013, an article by Hoffman & Beaumont was published in *Plastic Technology* that contained some good points about the method that will be reviewed here (use of a relative viscosity versus a relative shear rate curve). Their studies with their Therma-flo apparatus examined pressure drop across a rectangular flow channel at varying thicknesses and mold and melt temperatures so as to find stable flow rates where the pressure gradient across the part remains constant. From this author's experience, previous molding setups with parts of similar geometry, materials, and runner and gate systems trumps these methods. However, absent that previous knowledge, the method reviewed here has proven to be very effective in most cases. This author also likes it because it encourages processors to look at and refer to a polymer melt's most important flow property: viscosity. Other considerations include cases of an imbalanced multi-gated or multi-cavity mold where injection velocity can sometimes shift the balance. If the mold cannot be altered to fix the imbalance, the velocity that yields the best balance usually also provides the most robust process.

The viscosity of flowing plastic decreases with increasing velocity. Viscosity is defined as resistance to flow and is a very important property when filling the mold. By running a series of injection velocities, a plot of viscosity versus velocity can be obtained. A flow parameter that is directly proportional to injection velocity, called the shear rate, will be used instead of injection velocity in the graph. The equations for shear rate ($\dot{\gamma}$) of non-Newtonian fluids through rectangular and cylindrical flow channels are shown in Equation 11.1 and Equation 11.2.

$$\dot{\gamma} = \frac{6Q}{Wh^2}\left(\frac{2n+1}{3n}\right) \tag{11.1}$$

Where:

Q = volumetric flow rate

W = channel width

h = channel thickness

n = power law index for the material

$$\dot{\gamma} = \frac{32Q}{\pi d^3}\left(\frac{3n+1}{4n}\right) \tag{11.2}$$

Where:

d = channel diameter

Equations 10.1 and 10.2 are shown again here:
$\Delta P = \frac{12\eta LQ}{Wh^3}$ for rectangular and $\Delta P = \frac{128\eta LQ}{\pi d^4}$ for cylindrical channels.
Volumetric flow rate is equal to the volume of a flow section divided by the time (t) that it takes to flow the length (L) of the section. So $Q = \frac{LWh}{t}$ for a rectangular volume and $Q = \frac{L\pi d^2/4}{t}$ for a cylindrical volume. Substituting these formulas for Q into Equations 10.1, 10.2, Equation 11.1, and Equation 11.2 and solving for viscosity and shear rate gives:

Table 11.2 Viscosity and Shear Rate Equations for Rectangular and Cylindrical Channels

	Rectangular	Cylindrical
Viscosity, η =	$\Delta Ph^2t/(12L^2)$	$\Delta Pd^2t/(32L^2)$
Shear rate, $\dot{\gamma}$ =	$\frac{6L}{ht}\left(\frac{2n+1}{3n}\right)$	$\frac{8L}{dt}\left(\frac{3n+1}{4n}\right)$

From the preceding table, the only non-constant terms for a given mold and material are ΔP and t. It can also be seen that, as plastic flows through various cross-sections at a constant injection speed (constant volumetric flow rate), each section will have a different viscosity and a different shear rate.

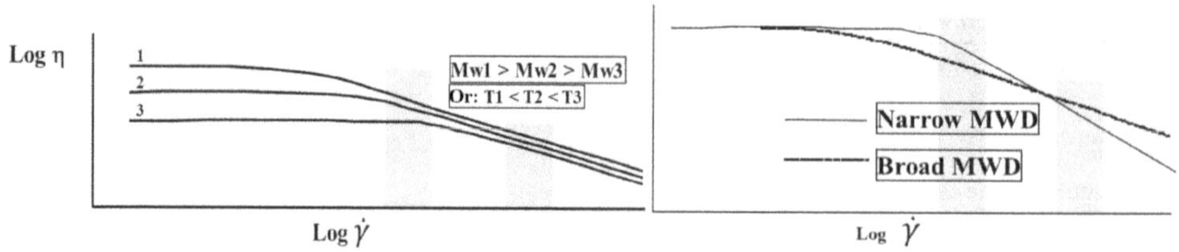

Figure 11.4 Graphs showing how viscosity varies with temperature (T), molecular weight (Mw), and molecular weight distribution (MWD)

The graph on the left in Figure 11.4 shows how variations in temperature or in molecular weight of the incoming material can affect viscosity. The shaded region on the left tends to be more typical of shear rates through the part, while the right shaded region is more typical of shear rates through gates or, possibly, the primary runner in a multi-cavity mold. The graph on the right shows what can happen if the molecular weight distribution of the incoming material changes – the viscosity through the runner might be higher while the viscosity through the part is lower. Regardless, material variation typically leads to part variation and process adjustments.

Removing all the constants from the equations in Table 11.2 allows the following two terms to be defined:

Relative viscosity

$$\eta_r = \Delta P \times t \tag{11.3}$$

and

Relative shear rate

$$\dot{\gamma}_r = 1/t \tag{11.4}$$

Where:

t = fill time

ΔP = V/P transfer pressure

A paper by Wareham & Ratzlaff and presented at the 1999 SPE ANTEC discussed an equation by which the injection velocity to use can be selected from a plot of relative viscosity versus relative shear rate. It can be rewritten as:

$$\text{Selected } \eta_r = (\max.\eta_r - \min.\eta_r) \times 0.025 + \min.\eta_r \tag{11.5}$$

The straight line in the graph on the left in Figure 11.5 is the selected relative viscosity. Its intersection with the curve gives the selected relative shear rate and injection velocity. This equation was developed to take the subjectivity out of where to pick a point on the graph.

Determination of Initial Injection Velocity

Note: Yellow highlighted areas are user-defined.

Transfer Pressure		Fill Time		Injection Velocity		Relative Shear Rate	Relative Viscosity
13600	psi	0.21	s	11.80	in/s	4.76	2856
12320	psi	0.30	s	7.87	in/s	3.33	3696
10650	psi	0.44	s	5.25	in/s	2.27	4686
9500	psi	0.65	s	3.50	in/s	1.54	6175
8450	psi	0.98	s	2.34	in/s	1.02	8281
7850	psi	1.41	s	1.56	in/s	0.71	11069
7120	psi	2.25	s	1.04	in/s	0.44	16020
6610	psi	3.22	s	0.69	in/s	0.31	21284
6350	psi	4.50	s	0.46	in/s	0.22	28575
6550	psi	7.54	s	0.31	in/s	0.13	49387
7220	psi	11.24	s	0.21	in/s	0.09	81153
	psi		s	0.14	in/s	#N/A	FALSE
	psi		s	0.09	in/s	#N/A	FALSE

SELECTED VELOCITY = **5.1** in/s

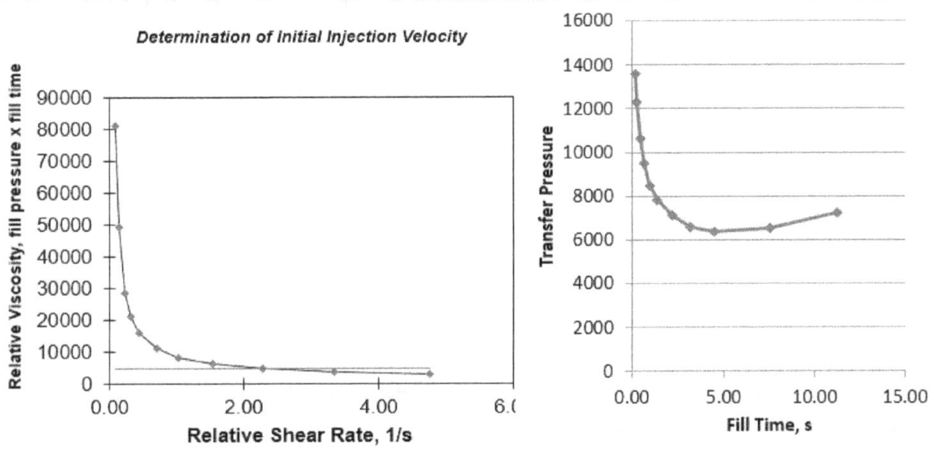

Figure 11.5 Determining the injection velocity using the relative viscosity versus relative shear rate graph

The Excel worksheet shown in Figure 11.5 that does the calculations is included with this book. The user inputs the maximum injection velocity that is not pressure-limited and the spreadsheet is then populated with the velocity to run for the study. Each velocity is 2/3 of the one before it. For each velocity, the fill pressure and fill time are recorded. The spreadsheet will not select a velocity until the slowest fill time is greater than 10 seconds or 40 times the fastest fill time.

As can be seen in Figure 11.5, the selected viscosity is on the flatter portion of the curve. The reasoning behind selecting on the flat region is that it provides a process less sensitive to process perturbation. An example is shown in Figure 11.6. The velocity study was done with an optical disk grade of polycarbonate at three different mold temperatures. A variotherm mold was used so that a mold temperature well above the glass transition temperature (300 °F) of the amorphous polycarbonate could be run. The main reason that the pressure to fill would be higher is that the frozen layer is thicker in the case of a colder mold, which leads to the effective flow thickness being smaller. Equation 10.2 shows that the pressure drop is inversely proportional to the thickness cubed; consequently, a small change in thickness results in a large change in pressure drop. At the high velocities, the change in frozen layer thickness is very small and, hence, the change in pressure on the right side of Figure 11.6 and the change in relative viscosity on the left side are very small. The changes are much larger at the lower fill speeds (low shear rates) and, if operated at those lower speeds, the process would be more sensitive to temperature change. Molecular weight variations, which affect the viscosity term in term in Equation 10.1, typically have a similar effect on the relative viscosity curve, with larger differences at lower shear rates. A paper by Mertes et al. that was presented at the 2001 SPE ANTEC showed that the flat portion of the curve yields a more robust process.

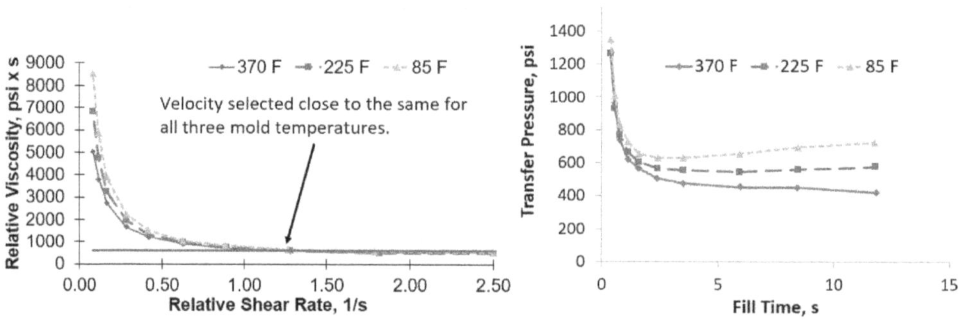

Figure 11.6 Relative viscosity versus relative shear rate for three mold temperatures of an optical disk grade of polycarbonate

It is important to look at the parts and the final graph when doing these studies. Jetting is a common defect that may require a lower velocity or a profiled injection speed – if the gate can't be altered to fix it. Flow lines and other surface defects may occur at some velocities and not others. Unless short shots are difficult to remove from the mold, this study is normally run without hold pressure; therefore, the presence of sink marks and voids, especially at the high velocities, should not be a concern at this point. Once hold pressure is applied, sinks and voids toward the end-of-fill could indicate shear heating during fill and that a lower velocity is needed.

Something to watch for on the relative viscosity versus relative shear rate graph is a truncated graph. This is shown in Figure 11.7 and is distinguishable by the straight viscosity selection line crossing the curve very close to the highest shear rate. This usually occurs when the machine's maximum injection rate is not high enough for the mold or the machine's maximum injection pressure limits the highest-possible velocity. If a machine is intended to run on several different machines in production, the selected velocity must be achievable on all of them.

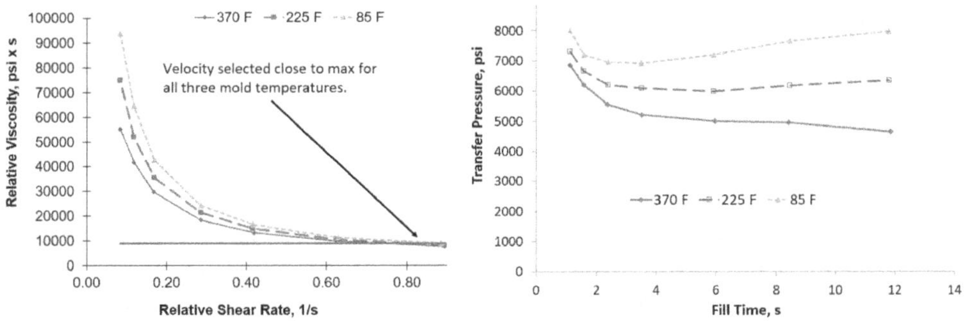

Figure 11.7 Truncated relative viscosity versus relative shear rate graph

One final point on the relative viscosity curve spreadsheet is that the highest shear rate should have the lowest relative viscosity. If the relative viscosity starts to rise at the highest velocities, the machine controller is likely anticipating a need to slow down the screw quickly and so does not maintain the set velocity all the way to the V/P transfer. Alternatively, it could be a problem with the method used by the controller to capture the V/P transfer pressure. In that event, the spreadsheet will select the point to the right if the viscosity selection line crosses the curve twice; however, the point to the left or the minimum may make more sense.

After the selected velocity is set on the machine, the transfer position should be adjusted to regain the original 95–98% full part(s). The parts should be as full as possible without hitting the last place to fill repeatedly. If defects on the short shots such as jetting, gate blush, or flow lines persist and cannot be fixed by modifying the mold or part design, a profiled injection velocity may have to be considered to eliminate them. While sometimes necessary, this is not desirable, because these defects often resurface when the mold is run in production. Profiled velocities are especially difficult to repeat if the same mold is run in several different molding machines.

11.2.4 Set the Hold Pressure

The goal here is to identify a hold pressure window and to set the pressure in the middle of the window. As a rule, this transfer pressure usually amounts to between 50 and 70% of the transfer pressure, but it is not rare for this rule to be broken. The bottom and top of the window are found as follows:

Minimum hold pressure:	Part(s) should be filled, should not have "unacceptable" sink marks or any other signs of underpacking
Maximum hold pressure:	Part(s) should not have "unacceptable" flash, push marks or any other signs of overpacking

By setting the hold pressure in the middle of this window, the chance of experiencing these visual defects during production is minimized. It is especially important not to run with flash on your parts, because continued running of parts will cause the mold to wear and make the problem worse. It is acceptable to operate not exactly in the middle of this window if other criteria, such as dimensional requirements, dictate this. If the difference between the min. and max. hold pressures is more than 4,000 psi (at the nozzle), consideration could be given to operating at a higher pressure to minimize shrinkage. However, tooling adjustments should be considered before making deviations that would require the hold pressure to be set near either the top or the bottom of the window.

If it is discovered that there is a very small hold pressure window or – worse – no window, some reasons could be:

- Insufficient clamp tonnage on the machine
- Mold not polished correctly
- Insufficient draft angle to release the part
- Thick sections in the part
- Poor gate location (flowing from thin-to-thick)
- Poor fit of mold inserts
- Cavity imbalance

These problems need to be addressed before releasing the mold to production.

For simplicity, the process is set up with one constant hold pressure. A decreasing hold pressure profile can sometimes be beneficial to pack the part out more evenly across the part and is generally only needed in special cases. An increasing hold pressure profile is usually a "Band-Aid" for a mold that flashes or burns easily and corrections to the mold are preferable.

11.2.5 Set the Hold Time

The hold pressure is typically applied until the gate freezes. If the gate and runners have been sized correctly, the gate will not freeze until enough plastic has been pushed into the cavity to compensate for the shrinkage of the plastic. Once the gate is frozen, the continued application of hold pressure only wastes energy and, possibly, increases cycle time. The parts will typically have less variation if the gate freeze is accomplished on every shot. Therefore, a time slightly longer than the gate freeze time (typically ½ to 2 s) is usually set on the hold timer. The most widely used method of determining the gate freeze time is plotting hold time against part weight, as shown in Figure 11.8 for the example from Chapter 9.

HOLD TIME (GATE FREEZE) DETERMINATION

Hold Time	Part Weight
20	28.32
18	28.33
16	28.32
14	28.33
12	28.3
10	28.21
8	28.14
6	28.06
4	27.93
2	27.68

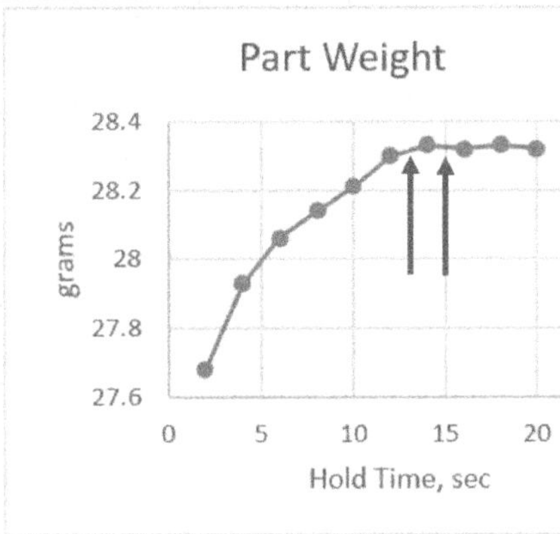

Figure 11.8 Gate-freeze study. The gate was frozen at about 13 seconds and hold time was set at 15 seconds

When collecting the data for this graph, it is recommended to start with a hold time estimated to exceed the gate freeze and work down to 1 s, collecting parts for each hold time setting. As time is taken off the hold timer, add an equal amount of time to the cooling timer. This will keep the overall cycle time and residence time in the barrel constant, both of which can affect part weight. It is also important to weigh only the parts, not the runner. For multi-cavity molds, it is preferable to measure each cavity individually, as variation in gate seal time between cavities can indicate tooling problems.

If a cavity pressure transducer is located near the gate, gate freeze can be determined more quickly. Figure 11.9 shows a cavity pressure curve for a hold time of 5 s. When hold time is set at 5 s, the pressure drops abruptly, because the gate has not yet frozen and material can flow back out of the part into the runner. If the hold time were set at 6 s or longer, the pressure would continue to decrease smoothly, as shown in the dotted line.

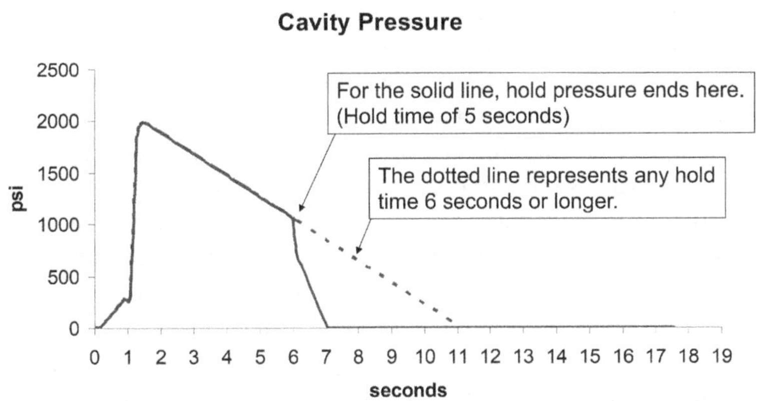

Figure 11.9 Cavity pressure method for determining hold time

For hot runner molds, the gate sometimes does not freeze completely. However, the slope of the part weight versus hold time still changes, as shown in Figure 11.10. Variation can be minimized by choosing a hold time on the flatter portion of the curve where the weight is not changing as fast.

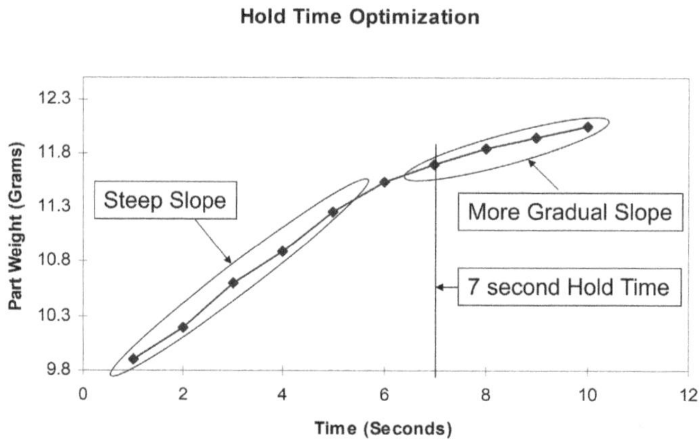

Figure 11.10 Part weight versus hold time for hot runner

11.2.6 Minimize the Cycle

As anyone in the manufacturing business knows, time is money. The goal of this step is to minimize cooling time on the press. Two constraints typically determine how low this time can be set:

1. The time needed so that parts consistently maintain the desired shape when ejected from the mold, or
2. The time it takes for the screw to recover.

Screw recovery, the second reason, is usually only a constraint for thin-walled, fast-cycling parts. In high-volume applications, such cases usually justify new or upgraded molding machines. A more powerful screw motor would allow faster screw recovery; also, the implementation of a shut-off nozzle would allow screw recovery to continue while the mold is opening and closing. For example, if a new screw motor can reduce a 4 s cycle to 3 s, production rate is increased by 25% (four million parts can be made in the time it previously took to make three million parts). The money saved by making more parts in less time will pay for new equipment. Also, most fast-cycling molding machines will be equipped with shut-off nozzles and the sequence will allow the screw to still be recovering while the mold is opening and closing.

However, the first reason stated above, the time needed so that parts consistently maintain the desired shape when ejected from the mold, is the most common cycle limiter. The cooling timer is reduced until the part(s) show(s) signs of being too warm and then it is set slightly higher. The time required for the part to be cool enough to eject will vary with the geometry of the part, material, mold design and construction, and mold temperature. A rule of thumb is that the part surface should be at least 30 °C below the heat deflection temperature (HDT) at 66 psi when ejected from the mold. HDT values are provided in Table 12.17 of the Appendix. However, it is always better to have data specific to the grade being molded.

11.2.7 Set Other Parameters

Other parameters that should be checked and possibly adjusted include the following:

11.2.7.1 Cushion

The minimum cushion should be the amount of plastic left in front of the screw when the hold time ends. If the screw ever goes all the way forward (zero position) during the hold phase, the cavity pressure response is very similar to that described for the 5 s hold time illustrated in Figure 11.9. In other words, if the screw position goes to zero, the effective hold pressure is zero at that point, because at that point the applied pressure on the screw is pushing on metal, rather than on plastic. The cushion should be run as small as possible, as the material left in the cushion is being kept hot longer and is, therefore, more susceptible to degradation. The cushion will vary shot-to-shot, primarily due to variations in how the non-return valve shuts off after every shot. Generally, about 10% of the shot size is an acceptable cushion.

11.2.7.2 Injection Pressure Limit and Fill Time Limit

These two are grouped together because they serve a single purpose: protecting the mold from being overpacked, especially in multi-cavity applications. Under normal processing conditions, neither of these two parameters should ever influence the process. The injection pressure limit should be set at about 2000 psi (200 psi hydraulic pressure) above the pressure needed to reach the V/P transfer position. The pressure needed to reach the transfer position will vary shot-to-shot, primarily due to variations in the non-return-valve shut-off and plastic viscosity. If monitored closely, this pressure can sometimes be set lower on molds that are very sensitive to overpacking. Care must be taken that the injection pressure limit never affects the fill time readout under normal conditions.

The fill time limit can be set very close to the actual fill time. On a high-quality, closed-loop control machine, the fill time readout does typically not vary by more than 0.02 to 0.03 s, and so the fill time limit can be set as close as 0.04 s above the actual fill time on molds sensitive to overpacking. If the mold is less sensitive, 10 to 20% above the actual fill time is sufficient.

Figure 11.11 shows the cavity pressure, injection pressure, and injection velocity curves for three scenarios in a four-cavity mold: 1) standard operation with the two limits set correctly; 2) a shot where one gate gets blocked, but the two limits are set correctly; 3) a shot where one gate gets blocked, but both limits are set incorrectly. The red lines represent the two limits. It should be noted that the cavity pressure was measured in one of the three cavities that did not have the blocked gate. It is the pressure spike inside the part cavity that can potentially do damage to the mold.

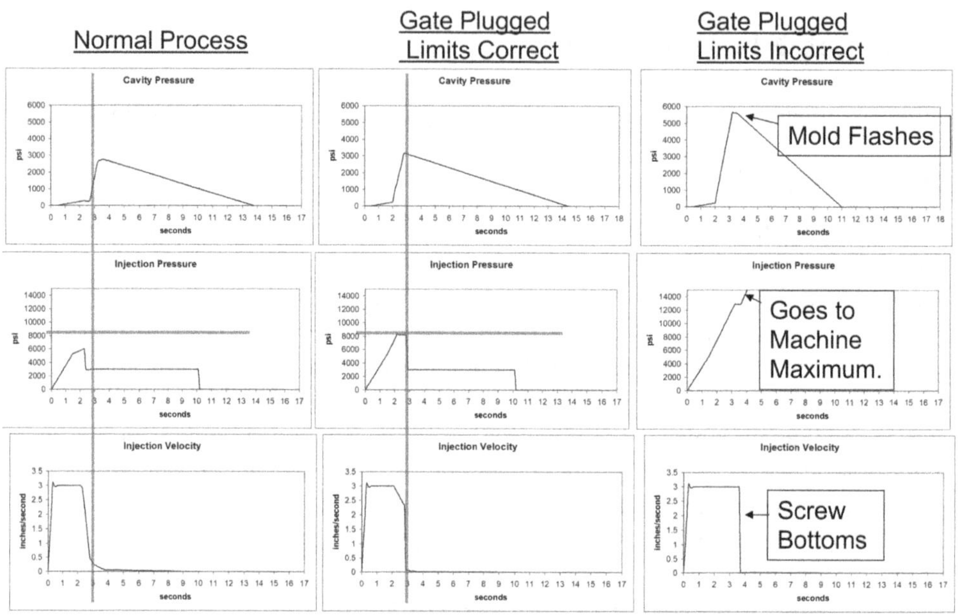

Figure 11.11 Setting the injection pressure and fill-time limits

11.2.8 Alternative V/P Transfer Methods

Most molding machines offer several options for the V/P transfer. The standards on almost every molding machine are screw position, time, and pressure. The problems with using time for an open-loop controlled machine were discussed earlier in this chapter. However, for closed-loop velocity control, screw position and time should be the same, as a constant speed over a set distance should always take the same amount of time. Despite this, it is standard practice to use time as a safety setting to prevent overpacking. The pressure limit is also used as part of that safety, but Equations 10.1 and 10.2 reveal why it should never serve as a V/P transfer method for a D2 process.

A common option and sometimes standard feature on molding machines is the use of an external signal for V/P transfer. This usually involves providing either a contact closure or a voltage signal. The most common signals are a specific level of cavity pressure or the arrival of the melt detected by a surface thermocouple. When using the melt-arrival method, typically an additional screw travel or time is needed to reach the desired transfer position. Papers presented at the 2013 (Johnson) and 2019 (Johnson & Johnson) SPE ANTEC conferences evaluated which methods provided the most robust process when variations in material viscosity and check ring shut-off occurred. A new method was also introduced, a contact switch (MeltSwitch) located on the surface of the cavity or behind a pin, which proved to be equivalent to the surface thermocouple as an effective melt-arrival technique.

For each transfer method, the full factorial DOE shown in Table 11.3 was run, with part weight of the fill-only parts serving as the response variable. The initial processes were set up with the low-viscosity material, and included decompression to ensure that all the fill-only parts would be short shots. The variation in decompression was designed to simulate check ring shut-off variation; without decompression, there is more leakage past the check ring before it shuts off. At the injection velocity used for this study, there was close to a 20% difference in the relative viscosities of the two materials.

Table 11.3 DOE for V/P Transfer Studies

Run #	Material Viscosity	Decompression (Variation in Check-Ring Shut-Off)
1	Low	0
2	Low	0.25 inch
3	High	0
4	High	0.25 inch

The following points summarize the studies and recommendations:

- Screw-position transfer – This was the only method significantly affected by decompression. This demonstrates why a process using this method should be set up closer to 95% rather than 98% full. The low-viscosity parts were slightly heavier, most likely because the check ring did not shut off quite as fast with the higher-viscosity material.
- Cavity-pressure transfer – The decompression had no effect; as expected, the screw would simply travel further to compensate for check ring leakage until the required pressure is reached. However, the viscosity had a large effect, as predicted by Equations 10.1 and 10.2. As was the case for injection pressure, this author recommends never using this method for a D2 process.
- Melt-arrival transfer – The change in viscosity and decompression had no effect on the weight when using a surface thermocouple or contact switch (MeltSwitch) for V/P transfer. This is therefore the recommended method when installing sensors or switches into a mold.

11.3 D3 Molding Process

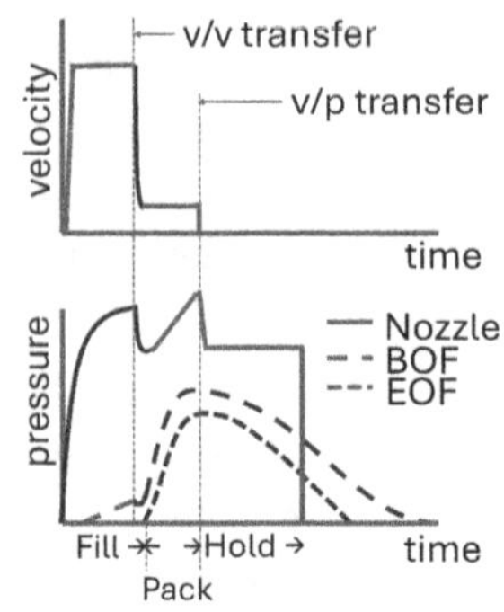

Figure 11.12
D3 velocity and pressure curves

The following steps are used when setting up a D3 (also called 3-stage or RJG DECOUPLED III) process. Steps 4 and 5 are the only ones different from the D2 process and will be discussed here. Typical process curves are shown in Figure 11.12.

1. Set clamp
2. Get parts 95% full at maximum injection velocity
3. Set the injection velocity
4. **Set the pack velocity**
5. **Set the hold pressure to eliminate bounce**
6. Set the hold time
7. Minimize the cycle
8. Set other parameters

11.3.1 Set the Pack Velocity

After the initial filling velocity, a second velocity profile is set. On most newer machines, the velocity change can be slowed down when the part is 95% full. However, the built-in ramping as the speed changes from a relatively fast fill to the slow pack may require that the machine start the transfer a little sooner; if the velocity is too high at the beginning of pack, the part will flash. A pack velocity that is 10% of the initial fill velocity is typically the starting point. The pack velocity may need to be reduced if the starts flash when just full or may need to be increased if the pack time is "excessively" long or record-grooving or other surface defects appear at the end of fill. Pack velocity would usually be included as a DOE factor for a D3 process, as would V/P transfer. The initial V/P transfer is set in a similar way as hold pressure for the D2 process. For D3, the window is between the V/P transfer that gives a just full part and the V/P transfer that shows signs of overpacking or flash. Note that sink marks or voids should be ignored here, as the hold phase is intended to eliminate those defects.

11.3.2 Set the Hold Pressure to Eliminate Bounce

The term "hold pressure" is highly descriptive within the context of a D3 process. In the absence of hold pressure, at V/P transfer the injection screw typically bounces back very noticeably. This occurs because the melt is compressed it front of the screw during pack and then acts like a spring when the pressure is relieved. Consequently, enough hold pressure should be applied to eliminate most of this bounce and the cushion at the end of the hold phase should typically be just slightly smaller than where it was at the moment of V/P transfer.

The "small frame" example from Chapter 9 illustrates where a D3 process might be beneficial, as the velocity at the end of fill changed the flow pattern and the quality of the part. The D3 process gives added control over the pack phase during viscosity fluctuations and, as will be seen in the V/P transfer discussion, usually yields more repeatable parts when the viscosity of the material varies. However, the trade-off is that it is sometimes harder to duplicate the same process on another machine. It is also usually possible to pack the parts out to a higher cavity pressure with D3 than with D2. This is because when hold pressure is increased in a D2 process, it simultaneously raises the final fill and initial pack speeds, which often leads to part flashing.

11.3.3 Alternative V/P Transfer Methods

The same studies conducted for the D2 process were also performed for the D3 process with the exception that the response was of "fill and pack-only" part weight instead of "fill-only" part weight. It was found that the viscosity change did not significantly impact any of these V/P transfer methods. The following summarizes the effects of decompression or check-ring variation and recommendations:

- Screw-position transfer – As expected, the packed parts with decompression weighed about 0.8% more because they were fuller at the point of velocity change.
- Cavity-pressure transfer – The parts without decompression weighed about 0.8% more than the parts with decompression. This difference can be seen in Figure 11.13. Since the part was less full at transfer, the "dip" is more than twice as long before the part starts to pack. There are several theories as to why that altered the part weight despite identical peak pressures. It is likely related to a thicker frozen layer that formed during the extended time required to fill the final portion of the part. The important thing is that the parts are statistically different.

- Melt-arrival transfer – In this case, since screw position was the method used for the V/V transfer, the parts with decompression were fuller when the velocity change occurred, and the results were about the same as for screw-position transfer.

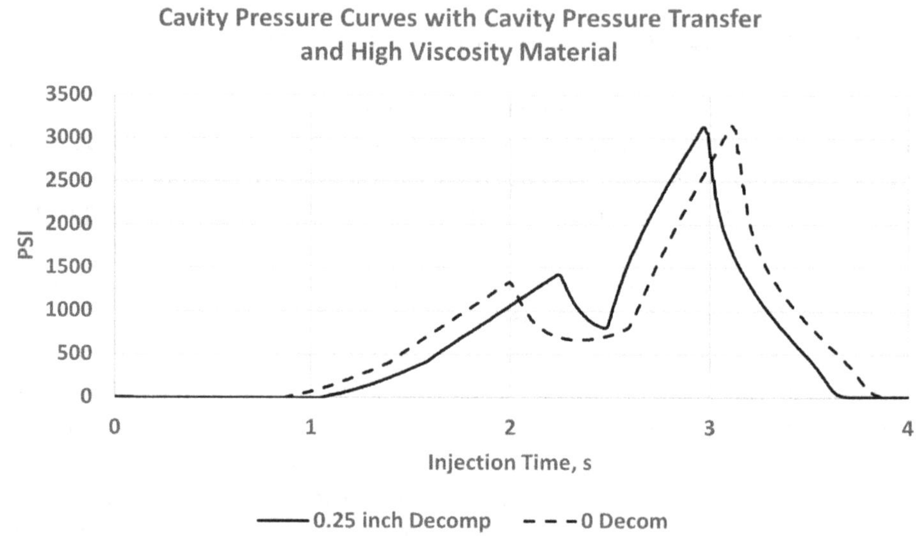

Figure 11.13 D3 Process with cavity-pressure transfer at two levels of decompression

None of the three transfer methods distinguished themselves as being better for the standard D3 process. However, the following modified-D3 process was found to be insensitive to both viscosity variation and decompression or check-ring variation. The modified approach performed similarly to the standard D3 for screw-position transfer, but was much less effective for cavity-pressure transfer, for the same reasons cited for the D2 process.

11.3.4 Modified-D3 Process

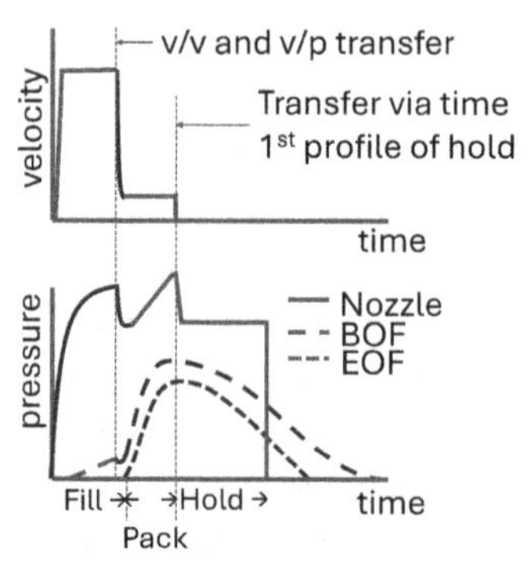

Figure 11.14 Modified-D3 process

Comparison of Figure 11.14 with Figure 11.12 reveals the differences in setting up the modified-D3 process versus the standard D3 approach. The modified-D3 process can only be used on machines equipped with a velocity limit setting during hold pressure. If this feature is available, the pack velocity can be set after V/P transfer as the first profile of what is a typically pressure-controlled phase; consequently, what the machine identifies as V/P transfer really is not the transfer point when this method is used. The pressure during the first profile is set high enough to ensure it is never reached. The pack window is set by adjusting the duration of the first profile, ideally positioned midway between the time required to just fill the part and the time that gives rise to overpack or flash. Then, hold pressure is applied via a second profile. The only time that this process offers an advantage over the standard D3 process is when a melt-arrival method (thermocouple or MeltSwitch) is employed for transfer.

11.4 D1 Molding Process

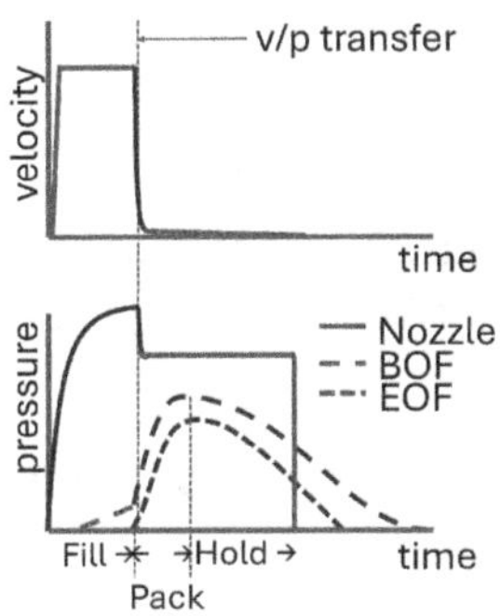

Figure 11.15
D1 Velocity and Pressure Curves

The process curves for the D1 molding process look very similar to those of the D2 process when comparing Figure 11.15 with Figure 11.2. The key differences are the missing "dip" at the end of the 1st stage in the beginning-of-fill cavity pressure curve and the fact that V/P transfer occurs after pack has begun. The primary difference in the process set-up lies in how V/P transfer is identified and how the hold pressure is set.

11.4.1 Set V/P Transfer

The V/P transfer is set in the window half-way between the point where the part is just full and the point where overpacking or flash begins. It is critical that this transfer occurs after pack has started. The momentum of the screw finishes the pack and determines the peak cavity pressure. To ensure that the transfer occurs after the start of pack, it is often necessary to reduce the injection velocity from its normal setting to decrease the momentum of the screw. Lower velocities also tend to broaden the V/P transfer window. This factor should be considered if a mold is to be run on machines with different screw sizes, as the momentum could be different.

11.4.2 Set the Hold Pressure

The hold pressure should be set just below the point where it starts to increase the peak cavity pressure.

11.4.3 Alternative V/P Transfer Methods

The same studies conducted for the D3 process were also performed for the D1 process. The results and recommendations are presented below.

- Screw-position transfer – The viscosity and decompression both influenced the "fill and pack" part weights. The weights were higher as viscosity decreased and when decompression was used. This method is not recommended for D1.
- Cavity-pressure transfer – As long as transfer occurred after V/P transfer, neither viscosity nor decompression (check-ring shut-off variation) affected the weight of the "fill and pack" part weights.
- Melt-arrival transfer – Neither viscosity nor decompression (check-ring shut-off variation) affected the weight of the "fill and pack" part weights.

11.5 iMFLUX Molding Process

For completeness, a "new" pressure-controlled process will be reviewed – an example of "what goes around comes around" in pressure-controlled molding. There are other variants of this method, but iMFLUX is the method that this author has used and is most familiar with. The company iMFLUX was formed in 2013 to develop and sell this "low-pressure molding" system. In this author's experience, the cavity pressure isn't always lower than that of a D2 process, but the injection pressure during fill is always lower than the V/P transfer pressure required in a D2 set-up.

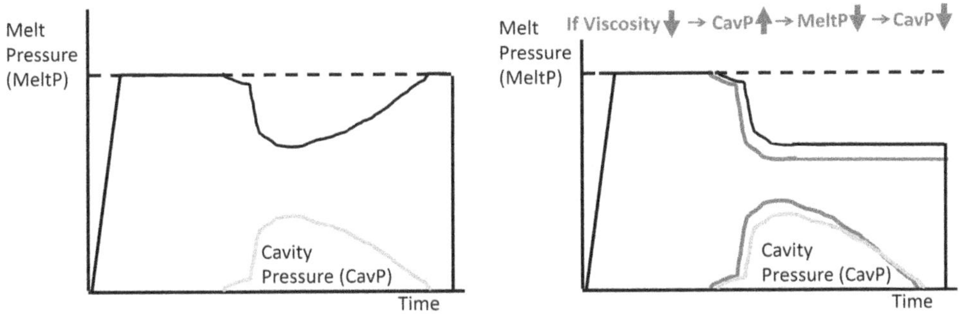

Figure 11.16 Typical iMFLUX injection velocity, pressure and cavity pressure curves (left) and how how the process adjusts for a decrease in viscosity (right)

Although it is an option on some molding machines, a distinctive feature of the iMFLUX process is that it uses the melt pressure at the nozzle, rather than force or pressure behind the injection screw, to control the process. From a control standpoint, the nozzle should be a better option for all processes, but placing a pressure transducer in this area generally increases maintenance costs, as the nozzle is subject to significant wear in both production and university lab environments. Figure 11.16 shows what this author thinks are the best features of iMFLUX. When the mold is equipped with a cavity pressure transducer, a portion or all the cavity pressure can be subtracted from or to the applied melt pressure using what is called Process Factor A (PFA), as shown in Equation 11.6.

$$\text{Melt pressure applied} = \text{Melt pressure set} + \text{PFA} \times \text{Cavity pressure} \quad (11.6)$$

In Figure 11.16, PFA = −1.0 and the entire cavity pressure curve is subtracted from the set melt pressure. Unless molding very thick parts, the PFA is always a negative value. The melt pressure curve on the right side of Figure 11.16 shows how this helps keep parts more consistent during fluctuations in material viscosity. If the viscosity of the material decreases slightly, this real-time control will start decreasing the melt pressure sooner because, with a lower viscosity and constant pressure, the flow rate will increase (as seen in the velocity graph). Further, since the cavity pressure will go up, the system will decrease the melt pressure, which will subsequently keep the cavity pressure lower that what it would have been otherwise. Also shown on the right side of Figure 11.16 is the PFA limiter, which is usually used with a negative PFA. After the cavity pressure starts to decrease, the limiter prevents it from rising again, causing the melt pressure to "straight-line" at the peak cavity pressure. PFA is typically an important factor for a DOE. It functions similarly to hold pressure in a D2 process.

Figure 11.16 also shows a typical injection velocity curve for the pressure-controlled process. Velocity is very fast at the beginning of fill, frequently approaching or reaching the machine's maximum capacity. Then, velocity declines as the plastic is pushed further. When the PFA brings the melt pressure down, the velocity decreases faster. The curve on the right shows how the velocity increases with a decrease in viscosity.

The following procedure is a basic set-up for establishing a baseline iMFLUX injection process in a mold equipped with a cavity pressure transducer close to the last place to fill. There are numerous tuning parameters that currently need to be adjusted to optimize the response of the pressure curve; these will not be discussed here (and will ideally be hopefully be automated in the future). This basic set-up also assumes that a baseline D2 process has already been established.

1. Initial settings:
   a. Set the melt pressure at the higher of the hold-pressure setting or 67% of the V/P transfer for the D2 process
   b. Set the step time 30% higher than the inject time + hold time for the D2 process
   c. Set PFA = −3.0 (this should create a "fill only" part)
   d. Turn on the PFA limiter
2. Start molding
3. If it takes more than 3 times the D2 process time to reach the cavity pressure transducer, increase melt pressure until the arrival time is about 3 times longer
4. Find the PFA window. Increase PFA (toward zero) until the parts achieve an acceptable appearance. Then continue increasing the PFA until the parts start to flash or show signs of overpacking. Set PFA to the midpoint of the window.
5. Perform a gate freeze by weighing parts at different step times. Increase the cooling time as the step time is decreased; this will maintain a constant cycle.

A study was performed to compare an iMFLUX process against a D2 process when a large viscosity change was introduced. The D2 process utilized two grades of polycarbonate, with the viscosity of the low-viscosity material being about half that of the high-viscosity material. The mold was an eight-cavity disk that was not very well balanced. A pressure transducer located at the end-of-fill of one of the first cavities to fill was used to control the process.

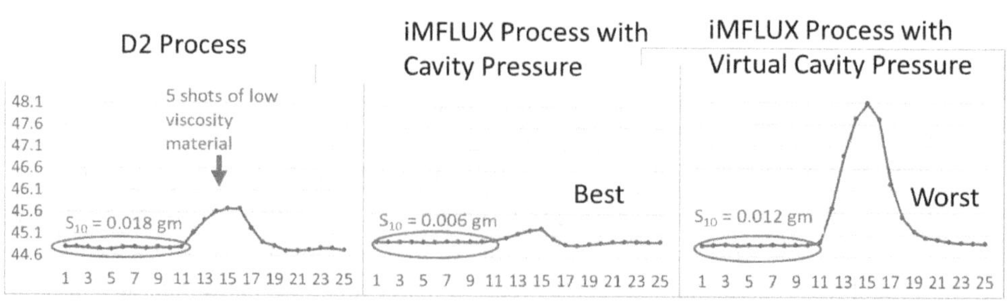

Figure 11.17 Results of a study comparing a D2 process with iMFLUX with a large viscosity change

The study was run by starting the process with the high-viscosity material and keeping it continuously running to stabilize for 15 minutes before parts were collected for measurement. A "slug" of low-viscosity material sufficient for 5 shots was introduced in the middle of the run by draining the material hopper, feeding in the low-viscosity material in and then refilling the hopper with the high-viscosity grade once the low-viscosity material reached the feed throat.

The most obvious result seen in Figure 11.17 was that the iMFLUX process using a cavity pressure transducer was the most robust process, since the part weight changed the least. The worst-performing process was the iMFLUX process without the cavity-pressure transducer. The "virtual cavity pressure" is a set amount of pressure that is subtracted (for a negative PFA) from the melt pressure after the screw travels a set distance. It is basically the same as a hold pressure setting in a D2 process. Another result that is noteworthy is that the sample standard deviation of the first ten shots with the D2 process was three times higher than the best iMFLUX process (0.018 g/0.006 g) and so, when molding with a constant viscosity material, the iMFLUX process produced more consistent parts. The F statistic = $0.018^2/0.006^2 = 9$ which is greater than the $F_{critical}$ of 3.18. It should also be pointed out that this study only measured weight which sometimes correlates with CTQ characteristics.

12 Appendix

Table 12.1 Bias Correction Factors for Estimating σ_x

Subgroup Size	R Factor d_2	s Factor c_4
2	1.128	0.7979
3	1.693	0.8862
4	2.059	0.9213
5	2.326	0.9400
6	2.534	0.9515
7	2.704	0.9594
8	2.847	0.9650
9	2.970	0.9693
10	3.078	0.9727
11	3.173	0.9754
12	3.258	0.9776
13	3.336	0.9794
14	3.407	0.9810
15	3.472	0.9823
16	3.532	0.9835
17	3.588	0.9845

Table 12.1 Bias Correction Factors for Estimating σ_x (*continued*)

Subgroup Size	R Factor d_2	s Factor c_4
18	3.640	0.9854
19	3.689	0.9862
20	3.735	0.9869
21	3.778	0.9876
22	3.819	0.9882
23	3.858	0.9887
24	3.895	0.9892
25	3.931	0.9896
30	4.086	0.9915
35	4.213	0.9927
40	4.322	0.9936
45	4.415	0.9943
50	4.498	0.9949
60	4.639	0.9957
70	4.755	0.9963
80	4.854	0.9968
90	4.939	0.9972
100	5.015	0.9975

Using Range: $\hat{\sigma}_X = {}^{R}/_{d_2}$ or ${}^{\overline{R}}/_{d_2}$
Using Sample Standard Deviation: $\hat{\sigma}_X = {}^{s}/_{c_4}$ or ${}^{\overline{s}}/_{c_4}$

Table 12.2 Control Chart Constants

N	A_2	D_3	D_4
2	1.880	–	3.268
3	1.023	–	2.574
4	0.729	–	2.282
5	0.577	–	2.114
6	0.483	–	2.004
7	0.419	0.076	1.924
8	0.373	0.136	1.864
9	0.337	0.184	1.816
10	0.308	0.223	1.777
11	0.285	0.256	1.744
12	0.266	0.283	1.717
13	0.249	0.307	1.693
14	0.235	0.328	1.672
15	0.223	0.347	1.653

Table 12.3 The Standard Normal Distribution

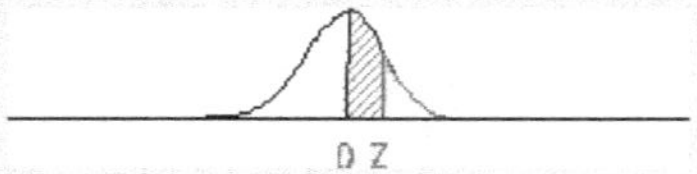

Entry represents the area under the standard normal distribution from the mean to Z.

Z	0.00	0.01	0.02	0.03	0.04	0.05	0.06	0.07	0.08	0.09
0.0	0.0000	0.0040	0.0080	0.0120	0.0160	0.0199	0.0239	0.0279	0.0319	0.0359
0.1	0.0398	0.0438	0.0478	09517	0.0557	0.0596	0.0636	0.0675	0.0714	0.0753
0.2	0.0793	0.0832	0.0871	0.0910	0.0948	0.0987	0.1026	0.1064	0.1103	0.1141
0.3	0.1179	0.1217	0.1255	0.1293	0.1331	0.1368	0.1406	0.1443	0.1480	0.1517
0.4	0.1554	0.1591	0.1628	0.1664	0.1700	0.1736	0.1772	0.1808	0.1844	0.1879
0.5	0.1915	0.1950	0.1985	0.2019	0.2054	0.2088	0.2123	0.2157	0.2190	0.2224
0.6	0.2257	0.2291	0.2324	0.2357	0.2389	0.2422	0.2454	0.2486	0.2517	0.2549
0.7	0.2580	0.2611	0.2642	0.2673	0.2704	0.2734	0.2764	0.2794	0.2823	0.2852
0.8	0.2881	0.2910	0.2939	0.2967	0.2995	0.3023	0.3051	0.3078	0.3106	0.3133
0.9	0.3159	0.3186	0.3212	0.3238	0.3264	0.3289	0.3315	0.3340	0.3365	0.3389
1.0	0.3413	0.3438	0.3461	0.3485	0.3508	0.3531	0.3554	0.3577	0.3599	0.3621
1.1	0.3643	0.3665	0.3686	0.3708	0.3729	0.3749	0.3770	0.3790	0.3810	0.3830
1.2	0.3849	0.3869	0.3888	0.3907	0.3925	0.3944	0.3962	0.3980	0.3997	0.4015
1.3	0.4032	0.4049	0.4066	0.4082	0.4099	0.4115	0.4131	0.4147	0.4162	0.4177
1.4	0.4192	0.4207	0.4222	0.4236	0.4251	0.4265	0.4279	0.4292	0.4306	0.4319
1.5	0.4332	0.4345	0.4357	0.4370	0.4382	0.4394	0.4406	0.4418	0.4429	0.4441
1.6	0.4452	0.4463	0.4474	0.4484	0.4495	0.4505	0.4515	0.4525	0.4535	0.4545
1.7	0.4554	0.4564	0.4573	0.4582	0.4591	0.4599	0.4608	0.4616	0.4625	0.4633

Z	0.00	0.01	0.02	0.03	0.04	0.05	0.06	0.07	0.08	0.09
1.8	0.4641	0.4649	0.4656	0.4664	0.4671	0.4678	0.4686	0.4693	0.4699	0.4706
1.9	0.4713	0.4719	0.4726	0.4732	0.4738	0.4744	0.4750	0.4756	0.4761	0.4767
2.0	0.4772	0.4778	0.4783	0.4788	0.4793	0.4798	0.4803	0.4808	0.4812	0.4817
2.1	0.4821	0.4826	0.4830	0.4834	0.4838	0.4842	0.4846	0.4850	0.4854	0.4857
2.2	0.4861	0.4864	0.4868	0.4871	0.4875	0.4878	0.4881	0.4884	0.4887	0.4890
2.3	0.4893	0.4896	0.4898	0.4901	0.4904	0.4906	0.4909	0.4911	0.4913	0.4916
2.4	0.4918	0.4920	0.4922	0.4925	0.4927	0.4929	0.4931	0.4932	0.4934	0.4936
2.5	0.4938	0.4940	0.4941	0.4943	0.4945	0.4946	0.4948	0.4949	0.4951	0.4952
2.6	0.4953	0.4955	0.4956	0.4957	0.4959	0.4960	0.4961	0.4962	0.4963	0.4964
2.7	0.4965	0.4966	0.4967	0.4968	0.4969	0.4970	0.4971	0.4972	0.4973	0.4974
2.8	0.4974	0.4975	0.4976	0.4977	0.4977	0.4978	0.4979	0.4979	0.4980	0.4981
2.9	0.4981	0.4982	0.4982	0.4983	0.4984	0.4984	0.4985	0.4985	0.4986	0.4986
3.0	0.49865	0.49869	0.49874	0.49878	0.49882	0.49886	049889	0.49893	0.49896	0.49900
3.1	0.49903	0.49906	0.49910	0.49913	0.49916	0.49918	0.49921	0.49924	0.49926	0.49929
3.2	0.49931	0.49934	0.49936	0.49938	0.49940	0.49942	0.49944	0.49946	0.49948	0.49950
3.3	0.49952	0.49953	0.49955	0.49957	0.49958	0.49960	0.49961	0.49962	0.49964	0.49965
3.4	0.49966	0.49968	0.49969	0.49970	0.49971	0.49972	0.49973	0.49974	0.49975	0.49976
3.5	0.49977	0.49978	0.49978	0.49979	0.49980	0.49981	0.49981	0.49982	0.49983	0.49983
3.6	0.49984	0.49985	0.49985	0.49986	0.49986	0.49987	0.49987	0.49988	0.49988	0.49989
3.7	0.49989	0.49990	0.49990	0.49990	0.49991	0.49991	0.49992	0.49992	0.49992	0.49992
3.8	0.49993	0.49993	0.49993	0.49994	0.49994	0.49994	0.49994	0.49995	0.49995	0.49995
3.9	0.49995	0.49995	0.49996	0.49996	0.49996	0.49996	0.49996	0.49996	0.49997	0.49997

Table 12.4 Student's t Distribution, $P(t > t_{\alpha,df})$

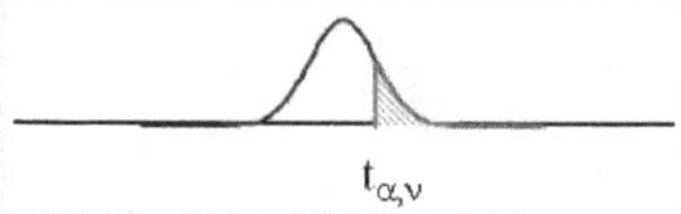

Degrees of Freedom	α				
	0.1	0.05	0.025	0.01	0.005
1	3.0777	6.3137	12.7062	31.8210	63.6559
2	1.8856	2.9200	4.3027	6.9645	9.9250
3	16377	2.3534	3.1824	4.5407	5.8408
4	1.5332	2.1318	2.7765	3.7469	4 6041
5	1.4759	2.0150	2.5706	3.3649	4.0321
6	1.4398	1.9432	2.4469	3.1427	3.7074
7	1.4149	1.8946	2.3646	2.9979	3.4995
8	1.3968	1.8595	2.3060	2.8965	3.3554
9	1.3830	1.8331	2.2622	2.8214	3.2498
10	1.3722	1.8125	2.2281	2.7638	3.1693
11	1.3634	1.7959	2.2010	2.7181	3.1058
12	1.3562	1.7823	2.1788	2.6810	3.0545
13	1.3502	1.7709	2.1604	2.6503	3.0123
14	1.3450	1.7613	2.1448	2.6245	2.9768
15	1.3406	1.7531	2.1315	2.6025	2.9467
16	1.3368	1.7459	2.1199	2.5835	2.9208
17	1.3334	1.7396	2.1098	2.5669	2.8982
18	1.3304	1.7341	2.1009	2.5524	2.8784
19	1.3277	1.7291	2.0930	2.5395	2.8609
20	1.3253	1.7247	2.0860	2.5280	2.8453
21	1.3232	1.7207	2.0796	2.5176	2.8314
22	1.3212	1.7171	2.0739	2.5083	2.8188
23	1.3195	1.7139	2.0687	2.4999	2.8073
24	1.3178	1.7109	2.0639	2.4922	2.7970
25	1.3163	1.7081	2.0595	24851	2.7874
26	1.3150	1.7056	2.0555	2.4786	2.7787

Degrees of Freedom	α				
	0.1	0.05	0.025	0.01	0.005
27	13137	1.7033	2.0518	2.4727	2.7707
28	1.3125	17011	2.0484	24671	2.7633
29	1.3114	1.6991	2.0452	2.4620	2.7564
30	1.3104	1.6973	2.0423	2.4573	2.7500
31	1.3095	1.6955	2.0395	2.4528	2.7440
32	1.3086	1.6939	2.0369	2.4487	2.7385
33	1.3077	1.6924	2.0345	2.4448	2.7333
34	1.3070	1.6909	2.0322	2.4411	2.7284
35	1.3062	1.6896	2.0301	2.4377	2.7238
36	1.3055	1.6883	2.0281	2.4345	2.7195
37	1.3049	1.6871	2.0262	2.4314	2.7154
38	1.3042	1.6860	2.0244	2.4286	2.7116
39	1.3036	1.6849	2.0227	2.4258	2.7079
40	1.3031	1.6839	2.0211	2.4233	2.7045
41	1.3025	1.6829	2.0195	2.4208	2.7012
42	1.3020	1.6820	2.0181	2.4185	2.6981
43	1.3016	1.6811	2.0167	2.4163	2.6951
44	1.3011	1.6802	2.0154	2.4141	2.6923
45	1.3007	1.6794	2.0141	2.4121	2.6896
46	1.3002	1.6787	2.0129	2.4102	2.6870
47	1.2998	1.6779	2.0117	2.4083	2.6846
48	1.2994	1.6772	2.0106	2.4066	2.6822
49	1.2991	1.6766	2.0096	2.4049	2.6800
59	1.2961	1.6711	2.0010	2.3912	2.6618
69	1.2939	1.6672	1.9949	2.3816	2.6490
79	1.2924	1.6644	1.9905	2.3745	2.6395
89	1.2911	1.6622	1.9870	2.3690	2.6322
99	1.2902	1.6604	1.9842	2.3646	2.6264
99999999	1.2816	1.6449	1.9600	2.3264	2.5758

Table 12.5 Chi-Square Distribution, $P(\chi^2 > \chi^2_{\alpha,df})$

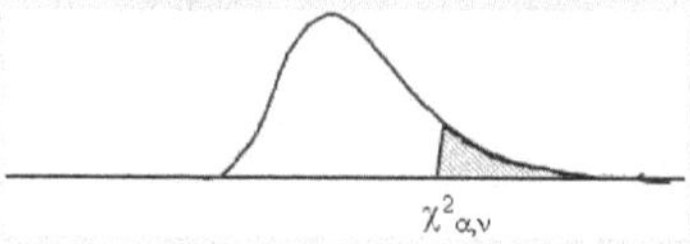

Degrees of Freedom	α							
	0.995	0.990	0.975	0 950	0.050	0.025	0.010	0.005
1	0.0000	0.0002	0.0010	0.0039	3.8415	5.0239	6.6349	7.8794
2	0.0100	0.0201	0.0506	0.1026	5.9915	7.3778	9.2104	10.5965
3	0.0717	0.1148	0.2158	0.3518	7.8147	9.3484	11.3449	12.8381
4	0.2070	0.2971	0.4844	0.7107	9.4877	11.1433	13.2767	14.8602
5	0.4118	0.5543	0.8312	1.1455	11.0705	12.8325	15.0863	16.7496
6	0.6757	0.8721	1.2373	1.6354	12.5916	14.4494	16.8119	18.5475
7	0.9893	1.2390	1.6899	2.1673	14.0671	16.0128	18.4753	20.2777
8	1.3444	1.6465	2.1797	2.7326	15.5073	17.5345	20.0902	21.9549
9	1.7349	2.0879	2.7004	3.3251	16.9190	19.0228	21.6660	23.5893
10	2.1558	2.5582	3.2470	3.9403	18.3070	20.4832	23.2093	25.1881
11	2.6032	3.0535	3.8157	4.5748	19.6752	21.9200	24.7250	26.7569
12	3.0738	3.5706	4.4038	5.2260	21.0261	23.3367	26.2170	28.2997
13	3.5650	4.1069	5.0087	5.8919	22.3620	24.7356	27.6882	29.8193
14	4.0747	4.6604	5.6287	6.5706	23.6848	26.1189	29.1412	31.3194
15	4.6009	5.2294	6.2621	7.2609	24.9958	27.4884	30.5780	32.8015

Degrees of Freedom	α							
	0.995	0.990	0.975	0 950	0.050	0.025	0.010	0.005
16	5.1422	5.8122	6.9077	7.9616	26.2962	28.8453	31.9999	34.2671
17	5.6973	6.4077	7.5642	8.6718	27.5871	30.1910	33.4087	35.7184
18	6.2648	7.0149	8.2307	9.3904	28.8693	31.5264	34.8052	37.1564
19	6.8439	7.6327	8.9065	10.1170	30.1435	32.8523	36.1908	38.5821
20	7.4338	8.2604	9.5908	10.8508	31.4104	34.1696	37.5663	39.9969
21	8.0336	8.8972	10.2829	11.5913	32.6706	35.4789	38.9322	41.4009
22	8.6427	9.5425	10.9823	12.3380	33.9245	36.7807	40.2894	42.7957
23	9.2604	10.1957	11.6885	13.0905	35.1725	38.0756	41.6383	44.1814
24	9.8862	10.8563	12.4011	13.8484	36.4150	39.3641	42.9798	45.5584
25	10.5196	11.5240	13.1197	14.6114	37.6525	40.6465	44.3140	46.9280
26	11.1602	12.1982	13.8439	15.3792	38.8851	41.9231	45.6416	48.2898
27	11.8077	12.8785	14.5734	16.1514	40.1133	43.1945	46.9628	49.6450
28	12.4613	13.5647	15.3079	16.9279	41.3372	44.4608	48.2782	50.9936

Table 12.6 The F Distribution: Values of $F_{0.05}$

Denominator Degrees of Freedom	Numerator Degrees of Freedom														
	1	2	3	4	5	9	24	29	39	49	59	69	79	99	999
1	161.4	199.5	215.7	224.6	230.2	240.5	249.1	250.0	251.1	251.7	252.2	252.5	252.7	253.0	254.2
2	18.51	19.00	19.16	19.25	19.30	19.38	19.45	19.46	19.47	19.48	19.48	19.48	19.48	19.49	19.49
3	10.13	9.55	9.28	9.12	9.01	8.81	8.64	8.62	8.60	8.58	8.57	8.57	8.56	8.55	8.53
4	7.71	6.94	6.59	6.39	6.26	6.00	5.77	5.75	5.72	5.70	5.69	5.68	5.67	5.66	5.63
5	6.61	5.79	5.41	5.19	5.05	4.77	4.53	4.50	4.47	4.45	4.43	4.42	4.42	4.41	4.37
6	5.99	5.14	4.76	4.53	4.39	4.10	3.84	3.81	3.78	3.76	3.74	3.73	3.72	3.71	3.67
7	5.59	4.74	4.35	4.12	3.97	3.68	3.41	3.38	3.34	3.32	3.31	3.29	3.29	3.28	3.23
8	5.32	4.46	4.07	3.84	3.69	3.39	3.12	3.08	3.05	3.02	3.01	3.00	2.99	2.98	2.93
9	5.12	4.26	3.86	3.63	3.48	3.18	2.90	2.87	2.83	2.80	2.79	2.78	2.77	2.76	2.71
10	4.96	4.10	3.71	3.48	3.33	3.02	2.74	2.70	2.66	2.64	2.62	2.61	2.60	2.59	2.54
11	4.84	3.98	3.59	3.36	3.20	2.90	2.61	2.58	2.53	2.51	2.49	2.48	2.47	2.46	2.41
12	4.75	3.89	3.49	3.26	3.11	2.80	2.51	2.47	2.43	2.40	2.39	2.37	2.36	2.35	2.30
13	4.67	3.81	3.41	3.18	3.03	2.71	2.42	2.39	2.34	2.32	2.30	2.29	2.28	2.26	2.21
14	4.60	3.74	3.34	3.11	2.96	2.65	2.35	2.31	2.27	2.24	2.22	2.21	2.20	2.19	2.14
15	4.54	3.68	3.29	3.06	2.90	2.59	2.29	2.25	2.21	2.18	2.16	2.15	2.14	2.12	2.07
16	4.49	3.63	3.24	3.01	2.85	2.54	2.24	2.20	2.15	2.13	2.11	2.09	2.08	2.07	2.02
17	4.45	3.59	3.20	2.96	2.81	2.49	2.19	2.15	2.11	2.08	2.06	2.05	2.04	2.02	1.97
18	4.41	3.55	3.16	2.93	2.77	2.46	2.15	2.11	2.07	2.04	2.02	2.00	1.99	1.98	1.92

Denominator Degrees of Freedom	Numerator Degrees of Freedom														
	1	2	3	4	5	9	24	29	39	49	59	69	79	99	999
19	4.38	3.52	3.13	2.90	2.74	2.42	2.11	2.08	2.03	2.00	1.98	1.97	1.96	1.94	1.88
20	4.35	3.49	3.10	2.87	2.71	2.39	2.08	2.05	2.00	1.97	1.95	1.93	1.92	1.91	1.85
21	4.32	3.47	3.07	2.84	2.68	2.37	2.05	2.02	1.97	1.94	1.92	1.90	1.89	1.88	1.82
22	4.30	3.44	3.05	2.82	2.66	2.34	2.03	1.99	1.94	1.91	1.89	1.88	1.87	1.85	1.79
23	4.28	3.42	3.03	2.80	2.64	2.32	2.01	1.97	1.92	1.89	1.87	1.85	1.84	1.82	1.76
24	4.26	3.40	3.01	2.78	2.62	2.30	1.98	1.95	1.90	1.86	1.84	1.83	1.82	1.80	1.74
29	4.18	3.33	2.93	2.70	2.55	2.22	1.90	1.86	1.81	1.78	1.76	1.74	1.73	1.71	1.65
39	4.09	3.24	2.85	2.61	2.46	2.13	1.80	1.76	1.70	1.67	1.65	1.63	1.62	1.60	1.53
49	4.04	3.19	2.79	2.56	2.40	2.08	1.74	1.70	1.64	1.61	1.58	1.56	1.55	1.53	1.45
59	4.00	3.15	2.76	2.53	2.37	2.04	1.70	1.66	1.60	1.57	1.54	1.52	1.51	1.49	1.40
69	3.98	3.13	2.74	2.50	2.35	2.02	1.68	1.63	1.57	1.54	1.51	1.49	1.48	1.45	1.37
79	3.96	3.11	2.72	2.49	2.33	2.00	1.66	1.61	1.55	1.51	1.49	1.47	1.45	1.43	1.34
99	3.94	3.09	2.70	2.46	2.31	1.98	1.63	1.58	1.52	1.48	1.45	1.43	1.42	1.39	1.30
999	3.85	3.00	2.61	2.38	2.22	1.89	1.53	1.48	1.41	1.37	1.33	1.31	1.29	1.26	1.11

Note: $F_{0.95} = 1/F_{0.05}$

Table 12.7 The F Distribution: Values of $F_{0.025}$

Denominator Degrees of Freedom	Numerator Degrees of Freedom														
	1	2	3	4	5	9	24	29	39	49	59	69	79	99	999
1	647.8	799.5	864.2	899.6	921.8	963.3	997.2	1000.8	1005.3	1007.9	1009.7	1010.9	1011.8	1013.1	1017.7
2	38.51	39.00	39.17	39.25	39.30	39.39	39.46	39.46	39.47	39.48	39.48	39.48	39.49	39.49	39.50
3	17.44	16.04	15.44	15.10	14.88	14.47	14.12	14.09	14.04	14.01	13.99	13.98	13.97	13.96	13.91
4	12.22	10.65	9.98	9.60	9.36	8.90	8.51	8.47	8.42	8.38	8.36	8.35	8.34	8.32	8.26
5	10.01	8.43	7.76	7.39	7.15	6.68	6.28	6.23	6.18	6.15	6.12	6.11	6.10	6.08	6.02
6	8.81	7.26	6.60	6.23	5.99	5.52	5.12	5.07	5.02	4.98	4.96	4.94	4.93	4.92	4.86
7	8.07	6.54	5.89	5.52	5.29	4.82	4.41	4.37	4.31	4.28	4.26	4.24	4.23	4.21	4.15
8	7.57	6.06	5.42	5.05	4.82	4.36	3.95	3.90	3.84	3.81	3.79	3.77	3.76	3.74	3.68
9	7.21	5.71	5.08	4.72	4.48	4.03	3.61	3.57	3.51	3.47	3.45	3.43	3.42	3.40	3.34
10	6.94	5.46	4.83	4.47	4.24	3.78	3.37	3.32	3.26	3.22	3.20	3.18	3.17	3.15	3.09
11	6.72	5.26	4.63	4.28	4.04	3.59	3.17	3.13	3.07	3.03	3.01	2.99	2.98	2.96	2.89
12	6.55	5.10	4.47	4.12	3.89	3.44	3.02	2.97	2.91	2.87	2.85	2.83	2.82	2.80	2.73
13	6.41	4.97	4.35	4.00	3.77	3.31	2.89	2.85	2.78	2.75	2.72	2.70	2.69	2.67	2.60
14	6.30	4.86	4.24	3.89	3.66	3.21	2.79	2.74	2.68	2.64	2.62	2.60	2.58	2.57	2.50
15	6.20	4.77	4.15	3.80	3.58	3.12	2.70	2.65	2.59	2.55	2.53	2.51	2.49	2.47	2.40
16	6.12	4.69	4.08	3.73	3.50	3.05	2.63	2.58	2.51	2.47	2.45	2.43	2.42	2.40	2.32
17	6.04	4.62	4.01	3.66	3.44	2.98	2.56	2.51	2.45	2.41	2.38	2.36	2.35	2.33	2.26

Denominator Degrees of Freedom	Numerator Degrees of Freedom														
	1	2	3	4	5	9	24	29	39	49	59	69	79	99	999
18	5.98	4.56	3.95	3.61	3.38	2.93	2.50	2.45	2.39	2.35	2.32	2.30	2.29	2.27	2.20
19	5.92	4.51	3.90	3.56	3.33	2.88	2.45	2.40	2.34	2.30	2.27	2.25	2.24	2.22	2.14
20	5.87	4.46	3.86	3.51	3.29	2.84	2.41	2.36	2.29	2.25	2.23	2.21	2.19	2.17	2.09
21	5.83	4.42	3.82	3.48	3.25	2.80	2.37	2.32	2.25	2.21	2.18	2.16	2.15	2.13	2.05
22	5.79	4.38	3.78	3.44	3.22	2.76	2.33	2.28	2.21	2.17	2.15	2.13	2.11	2.09	2.01
23	5.75	4.35	3.75	3.41	3.18	2.73	2.30	2.25	2.18	2.14	2.11	2.09	2.08	2.06	1.98
24	5.72	4.32	3.72	3.38	3.15	2.70	2.27	2.22	2.15	2.11	2.08	2.06	2.05	2.03	1.94
29	5.59	4.20	3.61	3.27	3.04	2.59	2.15	2.10	2.03	1.99	1.96	1.94	1.92	1.90	1.82
39	5.43	4.06	3.47	3.14	2.91	2.46	2.02	1.96	1.89	1.85	1.82	1.79	1.78	1.75	1.66
49	5.35	3.98	3.40	3.06	2.84	2.39	1.94	1.88	1.81	1.76	1.73	1.71	1.69	1.66	1.56
59	5.29	3.93	3.35	3.01	2.79	2.34	1.89	1.83	1.75	1.71	1.67	1.65	1.63	1.60	1.50
69	5.25	3.89	3.31	2.98	2.76	2.30	1.85	1.79	1.72	1.67	1.63	1.61	1.59	1.56	1.45
79	5.22	3.87	3.29	2.95	2.73	2.28	1.82	1.76	1.69	1.64	1.60	1.58	1.56	1.53	1.42
99	5.18	3.83	3.25	2.92	2.70	2.25	1.79	1.73	1.65	1.60	1.56	1.54	1.52	1.49	1.37
999	5.04	3.70	3.13	2.80	2.58	2.13	1.65	1.59	1.51	1.45	1.41	1.38	1.35	1.32	1.13

Note: $F_{0.975} = 1/F_{0.025}$

Table 12.8 d_2^*

k	n=2	n=3	n=4	n=5	n=6	n=7	n=8	n=9	n=10
1	1.414	1.906	2237	2.477	2.669	2.827	2.961	3.076	3.178
2	1.276	1.806	2.149	2.404					
3	1.227	1.767	2.120	2.378					
4	1.206	1.749	2.105	2.365					
5	1.189	1.738	2.096	2.358					
6	1.179	1.731	2.090	2.352					
7	1.172	1.726	2.086	2.349					
8	1.167	1.722	2.082	2.346					
9	1.163	1.718	2.080	2.344					
10	1.159	1.716	2.078	2.342					
11	1.157	1.714	2.076	2.341					
12	1.154	1.712	2.075	2.339					
13	1.152	1.711	2.073	2.338					
14	1.151	1.709	2.072	2.337					
15	1.149	1.708	2.071	2.337					
16	1.148	1.707	2.071	2.336					
17	1.147	1.707	2.070	2.335					
18	1.145	1.706	2.069	2.335					
19	1.145	1.705	2.069	2.334					
20	1.144	1.705	2.068	2.334					
21	1.143	1.704	2.068	2.333					
22	1.143	1.704	2.068	2.333					
23	1.142	1.703	2.067	2.333					
24	1.141	1.703	2.067	2.333					
25	1.141	1.702	2.066	2.332					
26	1.141	1.702	2.066	2.332					
27	1.140	1.701	2.065	2.332					
28	1.140	1.701	2.065	2.331					
29	1.139	1.701	2.065	2.331					
30	1.139	1.700	2.065	2.331					
>30	1.128	1.693	2.059	2.326					

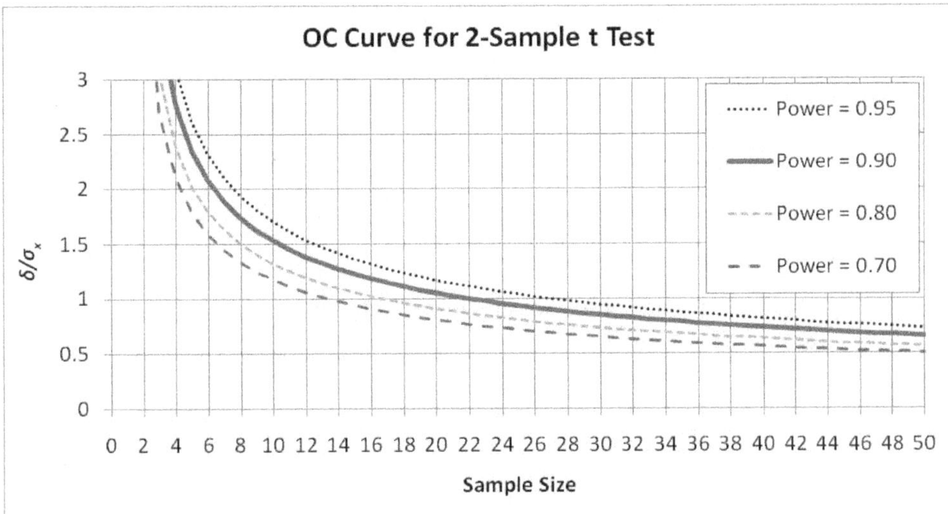

Figure 12.1 OC curve for a 2 sample t test (see also Table 12.9); δ is the minimum difference that can be detected, and σ_x is the standard deviation of the populations that the samples are taken from

For a 2-sample t-test, to see whether the means of two samples are different, the following Table 12.9 indicates how many samples of each are needed for α = 0.05 and different β levels:

Table 12.9 Table for 2-Sample t-Test with α = 0.05 (see OC curve in Figure 12.1)

Standard Deviations to Be Able to Detect	β = 0.10 Power = 0.90	β = 0.20 Power = 0.80	β = 0.30 Power = 0.70
0.25	338	253	199
0.50	86	64	51
0.75	39	29	23
1.00	23	17	14
1.50	11	9	7
2.00	7	6	5
2.50	5	4	4
3.00	4	4	3

All numbers from Minitab (Stat, Power & Sample Size). Powers are approximate.

Table 12.10 Confounding Pattern for 2^2, 4-Run DOE (see Table 12.11)

Number of Factors	Recommended Factors	$2^{(k-p)}$	Fraction	Design Resolution
2	A B	2^2	Full factorial	Full factorial
3	A B C	2^{3-1}	½	III

Table 12.11 4-Run DOE (see confounding pattern in Table 12.10)

Run	A BC	B AC	C AB
1	+1	+1	+1
2	+1	−1	−1
3	−1	+1	−1
4	−1	−1	+1

Table 12.12 Confounding Pattern for 2^3, 8-Run DOE (see Table 12.13)

Number of Factors	Recommended Factors	$2^{(k-p)}$	Fraction	Design Resolution
3	A B C	2^3	Full factorial	Full factorial
4	A B C G	2^{4-1}	½	IV
5	A B C D E	2^{5-2}	¼	III
6	A B C D E F	2^{6-3}	1/8	III
7	A B C D E F G	2^{7-4}	1/16	III

Table 12.13 8-Run DOE (see confounding pattern in Table 12.12)

Run	A BD CE FG	B AD CF EG	C AE BF DG	D AB CG EF	E AC BG DF	F AG BC DE	G AF BE CD
1	+1	+1	+1	+1	+1	+1	+1
2	+1	+1	−1	+1	−1	−1	−1
3	+1	−1	−1	−1	−1	+1	+1
4	+1	−1	+1	−1	+1	−1	−1
5	−1	−1	+1	+1	−1	−1	+1
6	−1	−1	−1	+1	+1	+1	−1
7	−1	+1	−1	−1	+1	−1	+1
8	−1	+1	+1	−1	−1	+1	−1

Table 12.14 Confounding Pattern for 2^4, 16-Run DOE (see Table 12.15)

Number of Factors	Recommended Factors	$2^{(k-p)}$	Fraction	Design Resolution
4	A B C D	2^4	Full factorial	Full factorial
5	A B C D O	2^{5-1}	½	V
6	A B C D K L	2^{6-2}	¼	IV
7	A B C D K L M	2^{7-3}	1/8	IV
8	A B C D K L M N	2^{8-4}	1/16	IV
9	A B C D E K L M N	2^{9-5}	1/32	III
10	A B C D E F K L M N	2^{10-6}	1/64	III
11	A B C D E F G H I K N	2^{11-7}	1/128	III
12	A B C D E F G H I J K N	2^{12-8}	1/256	III
13	A B C D E F G H I J K M N	2^{13-9}	1/512	III
14	A B C D E F G H I J K L M N	2^{14-10}	1/1028	III
15	A B C D E F G H I J K L M N O	2^{15-11}	1/2048	III

Table 12.15 16-Run DOE (see confounding pattern in Table 12.14)

Run	A BE CF DG HK IN JM LO	B AE CH DI FK GN JL MO	C AF BH DJ EK GM IL NO	D AG BI CJ EN FM HL KO	E AB CK DN FH GI JO LM	F AC BK DM EH GJ IO LN	G AD BN CM EI FJ HO KL	H AK BC DL EF GO IJ MN	I AN BD CL EG FO HJ KM	J AM BL CD EO FG HI KN	K AH BF CE DO GL IM JN	L AO BJ CI DH EM FN GK	M AJ BO CG DF EL HN IK	N AI BG CO DE FL HM JK	O AL BM CN DK EJ FI GH
1	+1	+1	+1	+1	+1	+1	+1	+1	+1	+1	+1	+1	+1	+1	+1
2	+1	+1	+1	−1	+1	+1	−1	+1	−1	−1	+1	−1	−1	−1	−1
3	+1	+1	−1	−1	+1	−1	−1	−1	−1	+1	−1	+1	+1	−1	+1
4	+1	+1	−1	+1	+1	−1	+1	−1	+1	−1	−1	−1	−1	+1	−1
5	+1	−1	−1	+1	−1	−1	+1	+1	−1	−1	+1	+1	−1	−1	+1
6	+1	−1	−1	−1	−1	−1	−1	+1	+1	+1	+1	−1	+1	+1	−1
7	+1	−1	+1	−1	−1	+1	−1	−1	+1	−1	−1	+1	−1	+1	+1
8	+1	−1	+1	+1	−1	+1	+1	−1	−1	+1	−1	−1	+1	−1	−1
9	−1	−1	+1	+1	+1	−1	−1	−1	−1	+1	+1	−1	−1	+1	+1
10	−1	−1	+1	−1	+1	−1	+1	−1	+1	−1	+1	+1	+1	−1	−1
11	−1	−1	−1	−1	+1	+1	+1	+1	+1	+1	−1	−1	−1	−1	+1
12	−1	−1	−1	+1	+1	+1	−1	+1	−1	−1	−1	+1	+1	+1	−1
13	−1	+1	−1	+1	−1	+1	−1	−1	+1	−1	+1	−1	+1	−1	+1
14	−1	+1	−1	−1	−1	+1	+1	−1	−1	+1	+1	+1	−1	+1	−1
15	−1	+1	+1	−1	−1	−1	+1	+1	−1	−1	−1	−1	+1	+1	+1
16	−1	+1	+1	+1	−1	−1	−1	+1	+1	+1	−1	+1	−1	−1	−1

Power is the probability of detecting a significant effect or difference when one truly exists. The Beta error is the probability of missing a significant effect. The Alpha error is the probability of "detecting" a significant effect or difference that does not truly exist.

For factorial experiments, the total replicates equal the number of runs times the replicates per run. For $\alpha = 0.05$ the following Table 12.16 indicates how many total replicates are needed for different β levels:

Table 12.16 Sample Size Determination for Factorial DOEs

Standard Deviations to Be Able to Detect	$\beta = 0.10$ Power = 0.90	$\beta = 0.20$ Power = 0.80	$\beta = 0.30$ Power = 0.70
0.25	676	508	400
0.50	172	128	104
0.75	80	60	48
1.00	48	36	28
1.50	24	20	16
2.00	16	12	12
2.50	12	12	8
3.00	12	8	4

For example, if historical data or a baseline study shows that a good estimate of standard deviation is 0.004 and it is desired to be able to detect a 0.003 difference, 0.003/0.004 = 0.75. For 90% Power, 80 total replicates would be needed. That would be 20 replicates/run of a 4-run experiment, 10 replicates/run of an 8-run experiment, or 5 replicates/run of a 16-run experiment. For a 32-run experiment, either Power would have to be sacrificed (2 replicates/run would give a total of 64, just over 80% Power) or extra replicates would have to be made (3 replicates/run would give a total of 96, Power > 90%).

Table 12.17 Material Properties

Material	Melt Density, g/cm^3	Solid Density, g/cm^3	Glass Transition Temp, °F	Melting Temp, °F	Deflection Temp at 66 psi, °F
Amorphous:					
ABS	0.97	1.05	230		203
Polycarbonate	1.02	1.20	300		280
Polystyrene	0.97	1.05	210		180
Polysulfone	1.16	1.25	370		345
Polyvinyl chloride – rigid	1.23	1.35	175		183

Material	Melt Density, g/cm³	Solid Density, g/cm³	Glass Transition Temp, °F	Melting Temp, °F	Deflection Temp at 66 psi, °F
Semi-crystalline:					
Nylon 6/6	0.97	1.14	135	510	464
Nylon 6/10	0.97	1.08	120	440	340
Polyacetal	1.17	1.41	10	350	336
PBT polyester	1.11	1.34	110	435	240
PET polyester	1.16	1.36	155	510	252
High-density polyethylene	0.72	0.95	−180	280	186
Low-density polyethylene	0.76	0.92	−180	240	113
Polypropylene	0.75	0.90	15	350	204

References

[1] Benbow, D. W., Kubiak, T. M., *The Certified Six Sigma Black Belt*, 3rd ed., Milwaukee: ASQ Quality Press, 2017.

[2] Hoffman, D., Beaumont, J., A New Look at Evaluating Fill Times for Injection Molding, Plastics Technology, 7/29/2013.

[3] Johnson, B., Determining Which In-Mold Sensors Should be Used for V/P Transfer During Injection Molding for Three Different Injection Strategies, SPE ANTEC Tech. Papers (2013).

[4] Johnson, B., Johnson, N., Novel V/P Transfer Actuation Method and Injection Molding Strategies and Their Comparison to Traditional Methods, SPE ANTEC Tech. Papers (2019).

[5] Johnson, S., Johnson, B., Using Infrared Temperature Sensors to Study Temperature Changes of PVC during Flow with the Incorporation of Melt Rotation Technology, SPE ANTEC Tech. Papers (2016).

[6] Johnston, S., Kazmer, D., Gao, R., Estimation of Melt Temperature from In-Mold Temperature Sensor Data, SPE ANTEC Tech. Papers, 52, 2393–2397 (2006).

[7] Launsby, R. G., Weese, D. L., *Straight Talk on Designing Experiments*, Colorado Springs: Launsby Consulting, 1995.

[8] Levine, D. M., Ramsey, P. P., Smidt, R. K., *Applied Statistics for Engineers and Scientists*, Upper Saddle River, NJ: Prentice Hall, 2001.

[9] Mertes, S., Carlson, C., Bozzelli, J., Groleau, M., What Position on the Viscosity Curve Is the Most Repeatable with Respect to End of Fill Cavity PSI, SPE ANTEC Tech. Papers (2001) 620.

[10] Montgomery, D. C., Runger, G. C., *Applied Statistics and Probability for Engineers*, 6th ed., New York: Wiley, 2013.

[11. Montgomery, D. C., *Introduction to Statistical Quality Control*, 7th ed., New York: Wiley, 2020.

[12] Neave, H. R., *The Deming Dimension*, Knoxville: SPC Press, 1990.

[13] Ross, P. J., *Taguchi Techniques for Quality Engineering*, 2nd ed., McGraw-Hill: New York, 1995.

[14] Wareham, J. R., Ratzlaff, J. D., Computing an Injection Fill Speed from a Relative Viscosity Curve, SPE ANTEC Tech. Papers (1999).

[15] Wheeler, D. J., *The Process Evaluation Handbook*, Knoxville: SPC Press, 2000.

[16] Wheeler, D. J., *The Six Sigma Practioner's Guide to Data Analysis*, 2nd ed., Knoxville: SPC Press, 2010.

[17] Wheeler, D. W., *Understanding Industrial Experimentation*, Knoxville: SPC Press, 1988.

[18] Wheeler, D. J., Chambers, D. S., *Understanding Statistical Process Control*, 2nd ed., Knoxville: SPC Press, 1992.

[19] Juran, J. M., Gryna, F. M., *Juran's Quality Control Handbook*, New York: McGraw-Hill, 1988.

Index

S

T

U

V

W

Z

www.ingramcontent.com/pod-product-compliance
Lightning Source LLC
LaVergne TN
LVHW060527081026
849713LV00006B/44
* 9 7 8 1 5 6 9 9 0 5 6 1 6 *